E. Bompiani (Ed.)

Geometria del calcolo delle variazioni

Lectures given at the
Centro Internazionale Matematico Estivo (C.I.M.E.),
held in Saltino (Firenza), Italy,
August 21-30, 1961

C.I.M.E. Foundation
c/o Dipartimento di Matematica "U. Dini"
Viale Morgagni n. 67/a
50134 Firenze
Italy
cime@math.unifi.it

ISBN 978-3-642-10957-7 e-ISBN: 978-3-642-10959-1
DOI:10.1007/978-3-642-10959-1
Springer Heidelberg Dordrecht London New York

Reprint of the 1st ed. C.I.M.E., Ed. Cremonese, Roma, 1961
With kind permission of C.I.M.E.

Printed on acid-free paper

Springer.com

CENTRO INTERNATIONALE MATEMATICO ESTIVO
(C.I.M.E)

Reprint of the 1st ed.- Saltino, Italy, August 21-30, 1961

GEOMETRIA DEL CALCOLO DELLE VARIAZIONI

CENTRO INTERNAZIONALE MATEMATICO ESTIVO
(C.I.M.E.)

H. BUSEMANN

THE SYNTHETIC APPROACH TO FINSLER SPACES IN THE LARGE

ROMA - Istituto Matematico dell'Università - 1961

THE SYNTHETIC APPROACH TO FINSLER SPACES IN THE LARGE

1. INTRODUCTION. CURVES AND SEGMENTS.

The geodesics of a Riemann space can be obtained as curves which are locally shortest connections or as the auto-parallel curves of the distinguished affine connexion of Levi-Civita. Parallel displacement along a curve maps the local geometry at one point of the curve isometrically on that at another.

Direct extension of the second method to Finsler spaces where the line element has the form

$$ds = F(x^1, \ldots, x^n, dx^1, \ldots, dx^n) = F(x, dx)$$

and $F(x; dx)$ satisfies certain standard conditions, is impossible because the local geometry of a Finsler space is Minkowskian and two n-dimensional Minkowski spaces are in general not isometric.

Nevertheless, generalizations of parallel displacement have played a major role in the theory of Finsler spaces in two different approaches. The space may be considered as a set of line elements rather than points to which local euclidean geometries are attached. Since this will be the topic of Professor Davies' lectures I will not dwell on it here.

The second approach is to consider the space as a point, hence locally Minkowskian space and to face the concomitant analytical difficulties as well as imperfection inherent to any concept of parallel displacement in Finsler spaces. This is the topic of Professor Wagner's lectures.[1)]

Thirdly, one may start from the definition of geodesic as a locally shortest join and avoid the analytical complications by not using analysis. Although this seems to contradict the very name "differential geometry" synthetic arguments partly topological and partly similar to those of euclidean geometry have proved very successful in particular when dealing with problems in the large. These lectures will give an introduction to the field. For simplicity we restrict ourselves to the case of symmetric distances ($F(x, dx) = F(x, -dx)$), but much of the material can and has been extended to the non-symmetric case, see Busemann [1] and Zaustinsky [1] .

Proofs are expected in these cycles. Some proofs in the present theory are long and the technicalities are uninteresting. In particular at the beginning, proofs become necessary only because the axioms are chosen as weak as possible, whereas all consequences of the axioms would have to be postulated, if they were not contained in the axioms. Doubts have been expressed regarding the power of synthetic methods in differential geometry. The only way of combatting the doubts is to exhibit the efficiency of these methods in many different areas. We therefore will outline the proofs only in those cases where they elucidate the reason for the superiority of geometric arguments.

1) Readers who should see these notes without those of Professors Davies and Wagner are referred to Cartan [1] and Rund [1] .

H.Busemann

We are interested in the intrinsic geometry in the large of complete Finsler spaces, and our method is axiomatic. The distance is for us not the infinitesimal distance given by a line element, but the finite intrinsic distance of two points in the manifold. Therefore our first axiom is :

I. <u>The space, R, is metric</u>.

The distance of two points x,y is denoted by xy and satisfies the standard conditions $xx = 0$, $xy = yx > 0$ for $x \neq y$ and $xy + yz \geqslant 0$.

A curve $x(t)$, $\alpha < t < \beta$ is a continuous map of the interval $[\alpha, \beta]$ in R . Its length is defined in the natural way: if D_t : $\alpha = t_0 < t_1 < \ldots < t_n = \beta$ is a partition of $[\alpha, \beta]$ we put

$$L(x, D_t) = \sum_{i=1}^{n} \left| x(t_{i-1}) - x(t_i) \right|$$

and

$$L(x) = L_{\alpha}^{\beta}(x) = \sup_{D_t} L(x, D_t). \quad \text{Notice } L(x) \geqslant x(\alpha)x(\beta).$$

If $\|D_t\| = \sup_i |t_i - t_{i-1}|$ then the usual theorem

$L(x, D_t) \rightarrow L(x)$ for $\|D_t\| \rightarrow 0$ holds, see G , p.19 (G refers here and elsewhere to Busemann [2]). An important implication of this fact is the <u>additivity of arclength</u>

$$\sum_{i=1}^{n} L_{t_{i-1}}^{t_i}(x) = L_{\alpha}^{\beta}(x) \quad \text{for any partition } D_t .$$

Moreover, length is <u>lower semicontinuous</u> (G p.20) : If $x_\nu(t)$, $\alpha \leq t \leq \beta$, $\nu = 0, 1, 2, \ldots$ are curves and $x_\nu(t) \rightarrow x_0(t)$ for each t, then

(1.1) $L(x_0) \leq \liminf L(x_\nu)$

The curve x(t) is rectifiable if L(x) is finite. We can then introduce the arclength as parameter : y(s), $0 < s < L(x)$, is the point x(t) for which $s = L_0^t(x)$. A class of rectifiable curves x(t) whose representations y(s) in terms of arclength are identical is a <u>geometric (rectifiable) curve</u> C. The elements of the class are the <u>parametrizations of C</u> and the properties common to all parametrizations of C are the properties of C (initial point, end point, length are examples).

There are many important theorems for which a freedom in the choice of the parameter and hence the concept of geometric curve rather than parametrized curve is essential. The most important is this : We call a subset M of a metric space <u>finitely compact</u> if every bounded infinite set in M has an accumulation point in M , that is, if M satisfies the Bolzano-Weierstrass Theorem.Then the following selection theorem (G p.24) holds which is fundamental for much of the present theory, although we will not always mention the theorem explicitly.

(1.2) <u>If M is a finitely compact subset of a metric space and</u> $C_1, C_2, \ldots$ <u>are geometric curves in R with</u> $L(C_\nu) < \beta$ <u>and whose initial points form a bounded sequence, then a suitable subsequence</u> $C_{\nu_1}, C_{\nu_2}, \ldots$ <u>of</u> C_ν <u>tends uniformly to a geometric curve C and</u>

$$L(C) \leq \liminf L(C_{\nu_n})$$

The theorem means that parametrizations $x_n(t)$ of C_{ν_n}

and x(t) of C, $\alpha \leq t \leq \beta$, exist such that $x_n(t)$ tends uniformly to x(t).

A curve x(t), $\alpha \leq t \leq \beta$, satisfying $L(x) = x(\alpha)x(\beta)$ is a shortest curve from $x(\alpha) = a$ to $x(\beta) = b$ and is called a <u>segment</u> T(a, b) from a to b. If we want to specify that the segment is oriented from a towards b we use the notation $T^+(a, b)$. If y(s) represents T(a, b) in terms of arclength, then for $0 \leq s_1 < s_2 \leq ab = \beta$

$$ab \leq y(0)y(s_1) + y(s_1) + y(s_2)y(\beta) \leq$$

$$\leq L_0^{s_1}(y) + L_{s_1}^{s_2}(y) + L_{s_1}^{\beta}(y) = L_0^{\beta}(y) = ab$$

hence $y(s_1)y(s_2) = s_2 - s_1$, so that every subarc of a segment is a segment and $y(s) \longrightarrow s$ maps the segment isometrically on a segment of the real axis, whence the name. Allowing a shift of the origin we call representation of a T(a, b) curve z(t) with

$$\alpha \leq t \leq \beta = \alpha + ab, \quad z(\alpha) = a, \quad z(\beta) = b \quad \text{and}$$

$$z(t_1)z(t_2) = |t_1 - t_2|, \quad \alpha \leq t_i \leq \beta. \tag{1.3}$$

Let (xyz) indicate that $x \neq y$, $y \neq z$ and $xy + yz = xz$. Then

(1.4) (wxy) and (wyx) imply (xyz) and (wxz).

For

$$wz = wy + yz = wx + xy + yz > wx + xz > wz.$$

(Notice that (wxy) and (xyz) do not imply (wyz) or (wxz)).
The following trivial remark is often useful :

(1.5) <u>If</u> (xyz) <u>and</u> $T_1 = T(x,y)$ <u>and</u> $T_2 = T(y,z)$ <u>exist, then</u> $T_1 \cup T_2$ <u>is a</u> $T(x,z)$.

A segment connecting two given points will in general not exist. To insure their existence we add two axioms :

II <u>The space is finitely compact.</u>

III <u>For any two distinct points x,z a point y with (xyz) exists.</u>

It is easily seen that a segment T(x,y) <u>exists for any two points x,y</u> (G p.29). Finite compactness is much stronger than would be necessary for the existence of segments. However, I will insure in conjunction with the remaining two axioms that the space has those properties of finite dimensional spaces which are necessary to obtain differential geometric results, whether the axioms actually imply finite dimensionality is not known.

2. GEODESICS.

A <u>geodesic</u> is a locally isometric map of the entire real axis into the space R . This means that it can be represented in the form $x(t)$, $-\infty < t < \infty$ and that for each real t_0 a positive $\varepsilon(t_0)$ exists such that

$$x(t_1)x(t_2) = |t_1 - t_2| \quad \text{for} \quad |t_i - t_0| \leq \varepsilon(t_0) .$$

This trivially (G p.32) implies that t is arclength

$$L_{t_1}^{t_2}(x) = t_2 - t_1 \quad \text{for any} \quad t_1 < t_2$$

x(t) and y(t) represent the same geodesic if $\delta = +1$ and a real β exist such that

$$y(t) = x(\delta t + \beta) \quad \text{for all } t .$$

Frequently we will consider <u>oriented geodesics</u>. Then it is understood that in a representation x(t) increasing t corresponds to traversal in the positive sense. A second representation y(t) of the oriented geodesic has the form $y(t) = x(t + \beta)$.

A geodesic is a <u>straight line</u> if it is an isometric image in the large of the real axis, i.e. $x(t_1)x(t_2) = |t_1 - t_2|$ holds for any t_1, t_2.

A compact convex subset of a euclidean space satisfies axioms I, II, III but geodesics do not exist. We must have an axiom of prolongability. To include the usual objects of differential geometry the axiom must be a local requirement. We denote the <u>open sphere</u> with radius $\rho > 0$ about a point, i.e. the set of points x with $px < \rho$, by $S(p, \rho)$. The triangle inequality implies that

$$(2.1) \qquad S(p, \rho) \supset S(q, \rho - pq) \text{ if } \rho > pq .$$

We can then formulate our axiom as follows :

IV. <u>For each point p there is a positive</u> ρ_p <u>such that for any two distinct points x,y in</u> $S(p, \rho_p)$ <u>a point z with (xyz) exists</u>.

The function ρ_p may be erratic, however then IV implies the existence of a well behaved function satisfying the axiom. For put

$$\rho(p) = \sup \rho_p, \text{ where } \rho_p \text{ satisfies IV for fixed p .}$$

Then either $\rho(p) = \infty$, which means that z with (xyz) exists for any distinct x,y and hence $\rho(q) = \infty$ for any q.

(2.2) $\begin{cases} \rho(p) \quad , \quad \text{or} \quad 0 < \rho(p) < \infty \quad \text{and} \\ |\rho(p) - \rho(q)| \leq pq \;. \end{cases}$

The latter follows at once from (2.1), G p.33.

The existence of geodesics means in ordinary differential geometry that every line element lies on a geodesic. This implies that a segment, or shortest geodesic join, can be extended to a geodesic, and for us this will mean existence of geodesics, (G pp.34, 35)

(2.3) <u>If Axioms I to IV hold and</u> $x(t)$, $\alpha \leq t \leq \beta$, $\alpha < \beta$ <u>represents a segment, then a geodesic</u> $y(t)$ $(-\infty < t < \infty)$ <u>exists such that</u> $y(t) = x(t)$ <u>for</u> $\alpha \leq t \leq \beta$.

<u>If</u> $\rho(p) \equiv \infty$ <u>then</u> $y(t)$ <u>may be chosen as a straight line</u>.

According to a recent observation of Szenthe [1] it may happen that $y(t)$ extending $x(t)$ to a geodesic exists which is not a straight line even when $\rho(p) \equiv \infty$.

Axioms I to IV do not contain any uniqueness properties for segments or geodesics. The simplest example showing this is the (x_1,x_2)-plane the metric

$$xy = |x_1 - y_1| + |x_2 - y_2| \;.$$

If $x_1 < y_1$ and $x_2 < y_2$ then any curve $z(t) = (z_1(t), z_2(t))$ from x to y for which both $z_1(t)$ and $z_2(t)$ are non-decreasing will be (but not necessarily represent) a segment and monotone continuation will provide straight lines. Our final axiom will therefore be a uniqueness postulate. Observing that, in differential geometry, the shortest geodesic join is not necessarily unique, but that its continuation to a geodesic is, we require.

Axiom V. *If* (xyz_1), (xyz_2) *and* $yz_1 = yz_2$ *then* $z_1 = z_2$.

The spaces satisfying all five axioms are called G-spaces, the G alluding to geodesic. Unfortunately the word G-space has lately also been used in a different sense, where the G alludes to group. In a G-space the extension of a proper segment to a geodesic is unique. Therefore, *if* $\rho(p) \equiv \infty$ *then all geodesics are straight lines and the space is called straight*. There are many important straight spaces besides the euclidean and hyperbolic spaces. In the terminology of the calculus of variation all simple connected spaces without conjugate points are straight.

We do not postulate the local uniqueness of $T(x,y)$ because this important property is contained in the axioms.

(2.4) *If* (xyz) *then* $T(x,y)$ *and* $T(y,z)$ *are unique*.

For if two segments T_1, T_2 from y to z existed then $z_i \in T_i$ with $z_1 \neq z_2$ and $yz_1 = yz_2$ would exist and satisfy (xyz_i) by (1.4) contradicting V. In particular

(2.5) $T(x,y)$ *is unique for* $x,y \in S(p, \rho(p))$.

Regarding the $\varepsilon(t_0)$ occurring in the definition of a geodesic it can easily be proved (G p.38) that :

(2.6) *If* $x(t)$ *represents a geodesic then it represents a segment for* $|t - t_0| \leq \rho(x(t_0))$.

In the absence of differentiability we define a <u>lineal element at p</u> as a segment with center p and length min($\rho(p)/2,1$). The <u>multiplicity</u> of a geodesic at a point (on the geodesic) is the cardinal number of distinct lineal elements at p lying on the geodesic. (A lineal element represented twice, hence infinitely often, for different t-intervals if a representation counts only once).

<u>(2.7) The multiplicity of a geodesic at a point is finite or countable.</u>

For in a representation x(t) of a geodesic different lineal elements at the same point correspond to disjoint intervals of the t-axis. A similar simple argument shows (G p.44)

(2.8) A geodesic has an at most countable number of multiple points.

Here we use the usual terminology to call a point multiple if the multiplicity is greater than 1, simple if it is one. The geodesic is simple if all its points are simple. A standard argument (G p.45) shows

(2.9) x(t) <u>represents a simple geodesic if and only if</u> $x(t_1) = x(t_2)$ <u>implies</u> $x(t_1 + t) = x(t_2 + t)$ <u>for all t</u>.

We eliminated the zero-dimensional G-spaces as trivial and want to do the same for one-dimensional spaces. A simple but most useful lemma is needed.

(2.10) <u>If</u> $x,y \in S(p,\rho)$ <u>then</u> $T(x,y) \subset S(p, 2\rho)$.

For let $w \in T(x,y)$. Then

$$\min(wx,wy) \leq xy/2 \leq (xp + py)/2 < \rho$$

and if $wx = \min(wx,wy)$ then

$$pw \leq px + xw < 2\rho \ .$$

Strangely enough this crude estimate is the best possible even on the sphere: If p and w are antipodal points on a sphere of radius 2, choose x and y on the same great circle through p and w with $ox = py = \pi + \varepsilon$, $0 < \varepsilon < \pi$. Then T(x,y) passes through w and $pw = \pi$. Whereas spheres are locally convex under the usual assumption for Finsler spaces, they are not necessarily so even in straight G-spaces.

A "great circle of length β " in a G-space is a geodesic isometric to a circle of length β . Its representation is distinguished by

$$x(t_1)x(t_2) = \min_{|\nu|=0,1,2} |t_1 - t_2 + \nu\beta| \ .$$

Straight lines and great circles contain with any two points x,y a segment T(x,y) and this property is characteristic:

(2.11) <u>If a geodesic G contains with any two points x,y a segment T(x,y) then it is a straight line or a great circle</u>.

First we show that G is simple. If G contained two lineal elements L_1, L_2 at the same point p , let a_i be an endpoint of L_i. Since $pa_i < \rho(p)/3$ the segment $T(a_1,a_2)$ is unique (see (2.5)) and lies therefore on G , moreover $T(a_1, a_2) \subset S(p, \rho(p))$, hence T(p,x) is unique for $x \in T(a_1,a_2)$ and lies on G , so that the multiplicity of G at p would not be countable.

If x(t) represents G then by (2.9) $x(t_1) = x(t_2)$ implies $x(t_1 + t) = x(t_2 + t)$ for all t. If $x(t_1) \neq x(t_2)$ for $t_1 < t_2$ then the arc $t_1 \leq t \leq t_2$ is the only arc in G from $x(t_1)$ to $x(t_2)$ and therefore by hypothesis a segment or $x(t_1)x(t_2) = |t_1 - t_2|$. Or there are two arcs on G from $x(t_1)$ to $x(t_2)$ and one of them must be a segment which leads to a great circle, for details see G p.46.

This proof shows that G-space containing two distinct line elements L_1, L_2 at p has at least dimension 2 because $\bigcup T(p,x)$, $x \in T(a_1,a_2)$ is homeomorphic to a triangle.

(2.12) A one-dimensional G-space is a straight line or a great circle.

For, a point of the space is center of at most one lineal element, therefore all geodesics are simple and no two geodesics intersect. If two different geodesics G_1, G_2 existed then for $p_i \in G_i$ a segment $T(p_1,p_2)$ would lie on a geodesic intersection G_1 and G_2.

Thus there is only one geodesic and this contains with any two points x,y a segment T(x,y) because the space has this property. The assertion now follows from (2.11).

Since zero- and one-dimensional G-spaces are trivial we will often tacitly assume that the space has dimension at least two. It can be proved that a two-dimensional G-space is a topological manifold (G pp.52-53), i.e. every point has a neighborhood homeomorphic to E^2. The corresponding problem for higher dimensions is open. As mentioned before, it is not known whether a G-space always has a finite dimension.

3. SPACES IN WHICH THE GEODESIC THROUGH TWO POINTS IS UNIQUE.

A further corollary of (2.11) is

(3.1) <u>If the geodesic through any two distinct points of a G-space is unique, then each geodesic is either a straight line or a great circle</u>.

For if the geodesic G contains x and y ($x \neq y$) it must contain every T(x,y) because a geodesic containing T(x,y) exists and the geodesic through x and y is unique.

The statement (3.1) can be considerably improved. For this and other purposes we need the concept of universal covering space. The mapping α of the G-space R' on the G-space R is <u>locally isometric</u> if a positive function $\beta_{p'}$ exists such that α maps $S(p', \beta_{p'})$ isometrically on $S(p'\alpha, \beta_{p'})$. Putting $p'\alpha = p$ it can be proved (G p.171) that α <u>maps $S(p', \rho(p)/2)$ isometrically on $S(p, \rho(p)/2)$</u> so that in contrast to the topological theory of covering spaces a $\beta_{p'}$ exists which is not only independent of the choice of p' in $p\alpha^{-1}$ but also of the space R' (and the mapping α). A consequence of this fact is that <u>a locally isometric map of a compact G-space on itself is an isometry in the large or a motion</u> (G p.172).

If a locally isometric map α of R' on the R exists then R' is a covering space of R. The cardinal number of points in $p\alpha^{-1}$ is independent of p and is usually called the number of sheets of R' (over R). This number is at most countable, because any two distinct points in $p\alpha^{-1}$ is at least $2\rho(p)$, (G pp.171,172).

If $x'(t)$ is a geodesic in R' then $x'(t)\alpha = x(t)$ is a geodesic in R. Conversely, a geodesic in R can be lifted : given a geodesic $x(t)$ in R and a point $p' \in x(t_0)\alpha^{-1}$ there is a unique geodesic $x'(t)$ in R' such that $x'(t_0) = p'$ and $x'(t)\alpha = x(t)$. In general it is not true that there is only one geodesic G' through p' with $G'\alpha = G$ where $x(t)$ represents G. For $x(t)$ may have a multiple point at $x(t_0)$ and different lineal elements of G at $x(t_0)$ may lead to different G'. However, if $x(t_0)$ is a simple point of G, then G' is unique (G p.169).

If $x(t)$, $\alpha \leq t \leq \beta$, represents a segment then the corresponding part of $x'(t)$ is a segment. The converse is obviously not true. If $T(x(\alpha),x(\beta))$ is unique then so is $T(x'(\alpha),x'(\beta))$ (G p.169).

Among the covering space of R' there is a simply connected one, which is unique up to isometries and is called the universal covering space of R (G. Section 28).

We now come to the improvement of (3.1). We observe first

(3.2) If two distinct geodesics G_1, G_2 each contain with two points x,y a segment $T(x,y)$ and have two common points a, a' then G_1 and G_2 are great circles of the same length and a, a' are antipodal on both, moreover $G_1 \cap G_2 = a \cup a'$.

For both G_i contain segments $T(a,a')$ and these are distinct, hence no point c with (abc) can exist, see (2.4) which proves that G_1, G_2 are great circles of the same length, (2.4) also shows that they cannot have any other common points.

A G-space is sphere like if the geodesics are great circles and the geodesics through a given point a all pass through a second point $a' \neq a$, which we call the antipode to a.

It follows from (3.2) that all geodesics through a given point a have the same length, moreover that a is the antipodal to a'. All geodesics have the same length, because for any two non-intersecting geodesics a third intersecting both exists.

A G-space is of the <u>elliptic type</u> if the geodesics are great circles of the same length and the geodesic through to distinct points is unique.

Identification of antipodal points in a spherelike space of dimension $\geqslant 2$ yields a space of the elliptic type. The argument is quite elementary and may be found in G p.129. We can now prove

(3.3) <u>If dim R ≥ 2 and R is not simply connected, if, moreover each geodesic in R contains with any two points x,y a segment T(x,y), then R is of the elliptic type and has a spherelike space as two-sheeted universal covering space.</u>

Proof. We know from (2.11) that all geodesics in R are great circles or straight lines. Since R is not simply connected its universal covering space R' has at least two sheets.

We show first that a geodesic G' in R' contains with two points a', b' at least one segment T(a',b'). If G is the image of G' under the local isometry of R' on R, assume first that $a = a'\alpha$ and $b = b'\alpha$ are neither identical nor antipodal in G. If G' did not contain a T(a',b'), then a geodesic H' containing a T(a',b') would exist and G and $H'\alpha$ would be two different geodesics through a and b which contradicts (3.2) (G is simple, hence only one geodesic G' with $G'\alpha = G$ through a exists, see above). If a = b or a and b are antipodes on G , choose c' close to b' so that a and c are neither identical nor antipodal. Then G' contains a T(a',b') and by continuity also a T(a',b'). Thus all geodesics in R' are straight lines or great circles.

Next we prove that there is only one geodesic G' in R' over a given geodesic in R. Let both G_1' and G_2' lie over G and choose $a_i' \in G_i$ so that a_1 and a_2 are neither identical nor antipodal on G. The image H' of a geodesic H' containing a_1' and a_2' intersects G in a_1 and a_2. We conclude from (3.2) that G = H and hence $G_1' = H' = G_2'$.

It now follows that if $G'\alpha = G$ and $p \in G$ then $p\alpha^{-1} \subset G'$. For through every point $p' \in p\alpha^{-1}$ there is a G' with $G'\alpha = G$ and there is only one G' mapped on G . Therefore two distinct geodesics through p' both contain all points in $p\alpha^{-1}$ and by (3.2) there cannot be more than two. Thus every geodesic passing through one of these points passes through the other and the space R' is spherelike. R is then of the elliptic type.

(3.4) <u>A space in which the geodesic through two points is unique and which contains a great circle is not simply connected</u>.

This can be proved for general G-spaces (G pp.201, 202), but is so simple under a minimum of differentiability hypotheses, that we prove it only for this case.

The geodesics are all straight lines or great circles and defining the length of a straight line as infinite it is obvious that the length L(G) of a geodesic G through a fixed point p depends continuously on the geodesic. Choose p such that at least one geodesic through p is a great circle.

Consider an elliptic space E of the same dimension as R in which the geodesics have length 1. Choose a point p' in E and map the tangent space of R at p linearly on the tangent space of E at p'. To a line element L of R at p now corresponds a line element L' of E at p' and thus to the geodesics G_L through the geodesic $G_{L'}$ through L'. We map G_L in $G_{L'}$ such that

$$p'x' = \frac{1}{2} \cdot \frac{px}{1 + px} \cdot \frac{2 + L(G)}{L(G)}$$

The point x' is not uniquely determined by this relation, but we can determine it in a neighborhood of p' such that the map is topological there. Then it becomes topological everywhere, and is meaningful also for the great circles, because the antipodal point to p on G_L becomes the antipodal point on $G_{L'}$. If G_L is a straight line, then its image is $G_{L'}$ with the antipodal point to p omitted.

Thus R is mapped topologically as a subset E_R of E . A projective line cannot be contracted to a point in E , hence still less in E_R. Therefore the great circles in R through p , which are mapped on projective lines, cannot be contracted. The improvement of (3.1) is a corollary of (3.3,4) :

(3.5) Theorem. *If the geodesic through two distinct points of a G-space of dimension greater than 1 is unique, then R is either straight or R is of the elliptic type and has a spherelike space as two sheeted universal covering space.*

4. INVERSE PROBLEMS.

A two-dimensional G-space is a topological manifold and if the geodesic through 2 points is unique it is either homeomorphic to E^2 and straight or homeomorphic to p^2 and of the elliptic type. In view of the fact that in the Riemannian case the geodesics determine the metric up to trivial transformations except in a few cases (Liouville's Theorem), it might seem reasonable to look for all metrizations of E^2 as a straight G-space or of P^2 as a space of the elliptic type. The following considerations will show that their problems are too general to be interesting, but determining the curve systems which occur as sets of geodesics proves interesting. One of the principal consequences of our investigation is an insight into the enormous variety of Finsler metrics.

For a system S of curves in E^2 to be the geodesics of a straight space the following is obviously necessary : If $e(x,y)$ is an auxiliary euclidean metrization of the plane then

a) <u>Each curve in S can be represented in the form</u> $p(t)$ $(-\infty < t < \infty)$, $p(t_1) \neq p(t_2)$ <u>for</u> $t_1 \neq t_2$, <u>and</u> $e(p(o),p(t)) \to \infty$ <u>for</u> $|t| \to \infty$.

b) <u>There is exactly one curve of S through two given distinct points of the plane</u>.

We will show that these trivially necessary conditions are also sufficient.

(4.1) <u>Theorem</u>. <u>Given a system S of curves in E^2 satisfying the conditions a) and b) then E^2 can be metrized</u> (in a great variety of ways) <u>as a G-space with the curves in S as geodesics</u>.

For $a \neq b$ let $G(a,b)$ be the curve in S through a and b, and $T(a,b)$ the subarc of $G(a,b)$ with endpoints a,b. If x is an interior point of $T(a,b)$ we write $[a\ x\ b]$. Put $T(a,a) = a$. First one establishes some simple topological properties of S, (G pp.57,58), in particular the Axiom of Pasch : If $c \notin G(a,b)$ and $[apb]$ then an S-curve H through p intersects $T(a,c) \cup T(c,b)$. If $a_n \to a$ and $b_n \to b$ then $T(a_n,b_n) \to T(a,b)$ in the sense of Hausdorff's closed limit, and also $G(a_\nu, b_\nu) \to G(a,b)$ provided $a \neq b$.

Let G^+ be an oriented S-curve and $p \notin G^+$. If x traverses G^+ in the positive sense then the oriented S-curve $G^+(p,x)$ (p precedes x in the orientation) tends to an oriented S-curve A^+ called <u>the asymptote to G^+ through p</u>. For any $q \in A^+$ the asymptote to G^+ through q is also A^+. For example, if q follows p on A^+, then the segment $T(q,x)$ tends to a ray R. If R did not lie on A^+, then $G(p,r)$ with $r \in R - q$ would not intersect G^+ and A^+ cannot be the asymptote to G^+ through p. The case where q precedes p is similar (G pp.59, 60). We will see later that the asymptote relation is in general neither symmetric nor transitive.

Here we need only the following consequence of this construction : <u>a given S-curve G can be imbedded in a simple family of S-curves</u>, i.e. one with the property that every point of the plane lies on exactly one curve in the family (the asymptotes to an orientation of G form a simple family).

To construct our metric consider a simple family F of S-curves. Fix a point z in the plane, let L_a be the curve in F through a. Denote one side of L_z by H^+ the other by H^{-1} and put for any $L \in F$

$$t(L) = \begin{cases} e(z,L) & \text{if} \quad L \subset H^+ \\ -e(z,L) & \text{if} \quad L \subset H^- \\ 0 & \text{if} \quad L = L_z \end{cases}$$

Then $t(L)$ depends monotonically on L . Put

$$d(a,b) = \left| t(L_a) - t(L_b) \right| .$$

Then

$$d(a,b) = d(b,a) \quad \begin{cases} = 0 & \text{if} \quad L_a = L_b \\ > 0 & \text{if} \quad L_a \neq L_b \end{cases}$$

$d(a,b) + d(b,c) \geqslant d(a,c)$ with equality if $L_a = L_b$ or $L_b = L_c$ or L_b lies between L_a and L_c.

$$d(a,z) \leq e(a,z)$$

because $d(a,z) = \left| t(L_a) \right| = e(z,L_a) \leq e(z,a)$.

Choose a denumerable number of simple families $F_1, F_2, \ldots$ such that $\bigcup F_i$ is dense among S-curves. For each F_i define the distance d_i (with the same point z) as d was defined for F and put

$$ab = \sum_{i=1} 2^{-i} d_i(a,b) .$$

This number is finite because

$$d_i(a,b) \leq d_i(a,z) + d_i(z,b) \leq e(a,z) + e(z,b) .$$

Clearly $ab = ba \geqslant 0$, and if $a \neq b$ then a S-curve separating a from b exists, hence also a curve in $\bigcup F_i$, say a curve in F_k. Then $d_k(a,b) > 0$ and $ab > 0$.

Trivially $ab + bc \geqslant ac$. We must show that the relations $[axb]$ and (axb) are equivalent. Let $[axb]$. For a given i either

$G(a,b) \in F_i$ and then $d_i(a,b) = d_i(a,x) = d_i(x,b) = 0$, or the curve in F_i through x in F_i lies between the curves through a and b, and then $d_i(a,x) + d_i(x,b) = d_i(a,b)$. Thus the latter relation holds for all i , which implies (axb).

We show next that $ax + xb > ab$ if a,x,b are distinct and $[axb]$ does not hold. If $x \in G(a,b)$ there is trivially a S-curve separating x from a and b . If $x \notin G(a,b)$ then $G(a',b')$ with $[aa'x]$ and $[bb'x]$ will separate x from a and b . Therefore a curve in some F_k will exist separating x from a and b . Then the curve in F_k through x is different from those through a and b and does not lie between them, hence

$$d_k(a,x) + d_k(x,b) > d_k(a,b)$$

and $ax + xb > ab$.

It remains to be shown that the Bolzano Weierstrass Theorem holds. I do not know whether this is always the case. But the parameter t(L) was largely arbitrary and by modifying it we can reach finite compactness (G p.62).

If the system S is well behaved we may replace the summation by an integration. We illustrate this by giving some interesting metrizations of the (x_1,x_2)-plane with the ordinary lines $ax_1 + bx_2 + c = 0$ as geodesics. Remember that the euclidean metric is by Beltrami's Theorem the only Riemannian metric with this property.

Let $g(t)$ $(t \geq 0)$ be continuous, non-negative, increasing with $g(t) \to \infty$ for $t \to \infty$. Put

$$f(x,\alpha) = \operatorname{sign}(x_1\cos\alpha + x_2\sin\alpha)g(x_1\cos\alpha + x_2\sin\alpha), \frac{-\pi}{2} < \alpha \leq \frac{\pi}{2}.$$

Observe that $|x_1\cos\alpha + x_2\sin\alpha|$ is the euclidean distance of the line through the point $x = (x_1,x_2)$ with normal α from the origin $z = (0,0)$. Thus $f(x,\alpha)$ corresponds to our $t(L)$ in the general case and $|f(x,\alpha) - f(y,\alpha)|$ to $d(x,y)$. Integrating instead of summing we form the distance

$$\rho_g(x,y) = \int_{-\pi/2}^{\pi/2} |f(x,\alpha) - f(y,\alpha)| \, d\alpha ,$$

for which the ordinary lines are the geodesics. This distance is invariant under the euclidean rotations about z, which are therefore also motions, more specifically rotations, for ρ_g. We consider some special choices of g.

1) $g(t) = \log(1 + t)$. Then the distance becomes relatively small as we move out and a simple estimate shows that for any two parallel lines L_1, L_2 the distance $x_1L_2 (x_1 \in L_1)$ tends to zero when x_1 traverses L_1 in either direction.

2) $g(t) = e^t$. The distance becomes large when moving out and $x_1L_2 \to \infty$ if x_1 traverses L_1 in either direction.

3) $g(t) = t^\beta$, $\beta > 0$. (For $\beta = 1$ this gives the euclidean metric). Then with $\delta x = (\delta x_1, \delta x_2) (\delta > 0)$

$$f(\delta x, \alpha) = \delta^\beta f(x,\alpha) .$$

If $k > 0$ is given we can determine $\delta > 0$ such that $\delta^\beta = k$ and we have for any two points x,y

$$\rho_g(\delta x, \delta y) = k \rho_g(x,y) .$$

Thus $x' = \delta x$ is a dilation for the metric ρ_g with the given dilation factor k. Thus we have a metric with the ordinary lines as geodesics, all dilations from z and all rotations about z which is not euclidean. Whereas the metrics in cases 1) and 2) are or can be made smooth everywhere, the metric 3) has in a well defined sense a singularity at z when $\beta \neq 1$, and this singularity cannot be eliminated without making the metric euclidean, see (10.3).

The superposition principle used to construct the metric in Theorem (4.1) can be used in many other ways. For example : define the function f(t) by

$$f(t) = \begin{cases} 0 & \text{for} \quad t < 0 \\ \sqrt{t} & \text{for} \quad 0 \leq t \leq 1 \\ 1 & \text{for} \quad t > 1 \end{cases}$$

then with $e(x,y) = \left[(x_1 - y_1)^2 + (x_2 - y_2)^2\right]^{1/2}$

$$xy = e(x,y) + \left| f(x_1) - f(y_1)\right| + \left| f(x_2) - f(y_2)\right|$$

is a metrization of the (x_1,x_2)-plane with the ordinary lines as geodesics. Let z = (o,o), a = (2t,o), b = (t,t), c = (o,2t). Then (abc) and for $0 < t \leqslant 1/2$

$$za = zc = 2t + \sqrt{2t} \qquad zb = \sqrt{2}t + 2\sqrt{t}$$
$$zb - za = \sqrt{t}\,(2 - \sqrt{2}\,)(1 - \sqrt{t}\,)$$

and zb > za = zc for $0 < t \leq 1/2$. Therefore no circle with radius ≤ 2 about z is convex. This hinges on the singularity of t at 0. e mentioned that in smooth Finsler spaces small spheres are convex (G p.162).

The result for P^2 corresponding to (4.1) is :

(4.2) Theorem. In the projective plane P^2 let a system S' of curves be given such that a') <u>Each curve in S' homeomorphic to a circle</u>

b') <u>There is exactly one curve of S' through given distinct points.</u>

<u>Then P^2 can be metrized as a G-space for which the curves in S' are the geodesics.</u>

This theorem was first proved by Skornyakov [1], a simpler proof is found in Busemann [4]. I will outline a proof of (4.2) and at the same time of

(4.3) <u>If P^n $(n \geq 2)$ is metrized on a G-space with the projective lines as geodesics, then this metric can be extended to P^{n+1} so that the projective lines in P^{n+1} are the geodesics.</u>

In both problems we pass from P^n to the spherical space S^n. In (4.2) we then obtain a system of curves S on S^2 such that each curve is homeomorphic to a circle and any curve on S through a point a also passes through the antipodal point a' to a on S^2. We may assume that one S-curve is the equator A of S^2 and the ordinary great circles through the north and south poles belonging to A are S-curves. We metrize A as a great circle such that antipodal points of S^2 are antipodal on A.

In (4.3) on S^{n+1} a great S^n denoted by A and metrization of A as a G-space with the ordinary circles as geodesics is given. In both cases we denote by H and H' the two open hemispheres bounded by A. If $p \in H$ then its antipode $p' \in H'$. For any point $x \neq p,p'$ there is exactly one semi-great circle (S-curve in (4.2)) with endpoints p,p' and containing x. It intersects A in a point x_p. If also $y \neq p,p'$ we put $f_p(x,y) = x_p y_p$. Notice that $f_p(x,y) = xy$ for

$x,y \in A$. Since $x \to x_p$ maps antipodal points onto antipodal points on A, each semi-great circle not passing through p is with $f_p(x,y)$ as metric a segment of the same length β. (Figure 1). Put $0 = f_p(p',x) = f_p(p,x) = f_p(x,p') = f_p(x,p)$.

Let $g(p)$ be any positive continuous function on H such that $\int_H g(p)dp = 1$ and put

$$xy = \int_H f_p(x,y)g(p)dp \, .$$

This integral exists as a Riemann integral, because $f_p(x,y)$ is bounded, continuous when x and y are different from p,p' and lower semicontinuous when x or y fall in p or p'. For any S-curve G: $\int_{G\cap H} f_p(x,y)g(p)dp = 0$, because G intersects each meridian in 2 points only. These facts yield very easily that each semi great circle is with the metric xy a segment of length β and the metric A is evidently the same as before.

The proof that $xy + yz > xz$ when x,y,z do not lie in this order on a semi great circle is very similar to the corresponding proof in (4.1).

The inverse problem has also been solved for the torus without conjugate points, i.e. a torus whose universal covering space (the plane) is straight. This problem differs from the preceding problems in that there are non-obvious necessary conditions. This will be discussed after the theory of parallels has been developed.

5. THE THEORY OF PARALLELS.

If R is a non-compact G-space and q any point then at least one ray with origin q exists, i.e. there is a geodesic $x(t)$ with $x(\alpha) = q$ for some α and

$$(5.R) \qquad x(t_1)x(t_2) = \left| t_1 - t_2 \right| \text{ for } t_i \geqslant \alpha .$$

For we need only take points q_n with $qq_n \to \infty$. On a segment $T(q,q_n)$ choose r_n with $qr_n = 1$ and select a subsequence $n_1, n_2, \ldots$ for which $T(q,r_{n_k})$ tends to a segment $T(q,r)$. Then $T(q,r_{n_k})$ tends to a ray.

If (5.R) represents a ray with origin q and p is any point, moreover $p_n \to p$ and $t_n \to \infty$, then by the same reasoning a suitable subsequence of $T(p_n, x(t_n))$ will tend to a ray C with origin p. Any ray obtainable in this way (i.e. varying the sequences p_n and t_n) is called a co-ray to R (from p). This co-ray is not necessarily unique, however, if $p_1 \in C - p$ then the co-ray from p_1 to R is unique and is a subray of C , (G section 22).

If R' is any ray contained in R or containing R then the co-rays to R' are obviously the same as those to R .

The oriented straight line A^+ is an asymptote to the oriented line L^+ is every positive subray of A^+ (i.e. a ray $a(t)$, $t > \alpha$ if $a(t)$ represents A^+) is a co-ray to some, and hence to every, positive subray of L^+.

According to our previous statement there is at most one asymptote A^+ through a given point to a given oriented line L^+. If the space is straight then A^+ always exists (G Section 23), and this definition of asymptote coincides with the one used previously in the special case of the plane.

If A and L are straight lines in a G-space and if with suitable orientations A^+, A^- of A and L^+, L^- of L the oriented lines A^+ and A^- are asymptotes to L^+ and L^- respectively then A is called <u>parallel</u> to L .

Symmetry and transitivity of the asymptote relation are defined in the obvious way

(5.1) <u>Transitivity of the asymptote relation implies symmetry. Even in a straight plane the asymptote relation or parallelism is not necessarily symmetric</u>.

To prove the first statement, let A^+ be an asymptote to L^+, and if $q \in L^+$, let B^+ be the asymptote to A^+ through q . If the asymptote relation is transitive, then B^+ is an asymptote to L^+, hence coincides with L^+, so that L^+ is an asymptote to A^+ and the relation is symmetric.

(Fig.2) To construct an example confirming the second statement, let H be the branch in $x < 0$ of the hyperbola $x_1x_2 = -1$ in the (x_1,x_2)-plane. Let S^1 be the set of all curves originating from H by translations. There is exactly one curve in S^1 through two distinct points x,y for which $(x_2 - y_2)/(x_1 - y_1) > 0$. Let S^0 consist of all lines $x_2 = mx_1 + b$ with $m \leq 0$ and of the lines $x_1 = a$. Then $S = S^0 S^1$ satisfies the hypotheses a) b) of (4.1) and hence occurs as the set of geodesics of a straight space.

Let L^+ and A^+ be the lines $x_1 = -1$ and $x_1 = 0$ oriented upwards. Then A^+ is the asymptote to L^+ through p = (0,0), but B^+

is not the asymptote to A^+ through $q = (-1,1)$, because H passes through q . Actually H^+ oriented upward is an asymptote to A^+.

Moreover, the curve in S^1 through p and $(-1,x_2)$ tends to A when $x_2 \to -\infty$, therefore A^- is an asymptote to L^-, so that A is parallel to L, but L is not parallel to A .

There is one important case in which parallels occur. A <u>motion</u> of a G-space R is a distance preserving (and hence topological) mapping M of R on itself : $xMyM = xy$ for all x,y . If x(t) is a straight line in R and M acts on this line as a proper translation, i.e.

$$x(t)M = x(t+\alpha) \quad \text{for all } t, \quad \alpha \neq 0$$

then L is called an <u>axis</u> of M.

(5.2) <u>If L is an axis of M , then it is an axis of M^n for $n \neq 0$ and</u>

$$ppM^n = \min_{x \in R} xxM^n \quad \underline{\text{for}} \quad p \in L \;,\; n \neq 0 .$$

It suffices to show this for n = 1 . Choose any point x put $x_k = xM^k$ and $p_k = pM^k$ for $k > 1$. Then

$$kpp_1 = pp_k \leq px + xx_1 + .. + x_{k-1}x_k + x_kp_k = 2px + kxx_1 \quad \text{or}$$

$$ppM \leq 2k^{-1}px + xxM$$

whence the assertion follows for $k \to \infty$.

Using this fact and some simple auxiliary facts from the general theory of co-rays which were not discussed here one finds (G p.267 for straight spaces and Busemann and Pedersen [1] in the general case) :

(5.3) Two axes of the same motion are parallel to each other.

The existence of axes can often be deduced from two simple but very important principles.

(5.4) If for a motion M of a G-space R a point p with $0 < ppM = \inf_{x \in R} xxM$ exists and T is any segment from p to pM then $\bigcup_{i=-\infty}^{\infty} TM^i$ is a geodesic. Moreover $yyM = ppM$ for any point of this geodesic.

Let y be an interior point of T. Then

$$ppM = py + ypM + pMyM \geqslant yyM \geqslant ppM$$

hence $(ypMyM)$ and $yyM = ppM$. By (2.4) the segments $T(y,pM)$ and $T(p,y)$ hence also $T(pM,yM)$ are unique, and $(ypMyM)$ shows that TM is a continuation of T. A corollary of (5.4) is

(5.4') If for a motion M of R a point p with $0 < ppM = \inf_{x \in R} xxM < \rho(p)/$ exists then $(ppMpM^2)$.

For $T(p,pM)$ is unique and $T(p,pM)\, T(pM,pM^2)$ has length less than $\rho(p)$ and is part of a geodesic, compare (2.6).

(5.2) and (5.4') yield

(5.5) If R is straight then p lies on an axis of the motion M, or equivalently $(ppMpM^2)$, if and only if $ppM = \inf xxR$.

The second principle is similar to (5.4')

(5.6) If for a motion M of the G-space R a point p with $0 < ppM = \sup_{x \in R} xxM < \rho(p)/2$ exists, then $(ppMpM^2)$.

For let pM be the midpoint of p and u (u exists because $ppM < \rho(p)/2$ and pM is the only midpoint of p and u). Then

$$ppM \geqslant uuM \geqslant pMuM - pMu = pu - pMu = ppM$$

Therefore

$$pMu + uuM = pMuM = 2pMu$$

so that u is the midpoint of pM and uM and must coincide with pM^2.

As a corollary of these results we mention

(5.7) <u>If R is straight and if for the motion M of R a point p with ppM = sup xxM exists, then every point lies on an axis of M and all the axes are parallel.</u>

For (5.6) shows that $(ppMpM^2)$ hence the line through p and pM is an axis. By (5.2) we also have ppM = inf xxM, so that xxM is constant. Therefore xxM is everywhere minimal so that by (5.5) each point lies on an axis, and these axes are parallel by (5.3). The result (5.7) will be applied after some other applications of (5.6) have been discussed.

If R is compact and M is not the identity then a point p with $0 < ppM = \sup xxM$ always exists, but may not satisfy $ppM < \rho(p)/2$. If $ppM < \rho(p)/2$ we know $(ppMpM^2)$, $(pMpM^2pM^3), \dots$. There will be a first exponent k for which $kppM > \rho(p)/2$. Then the points $p, pM, \dots, pM^k$ lie by (2.6) on a segment. Then in either case $ppM^k > \rho(p)/2$ for a suitable positive k .

If we use the usual distance

$$d(M,N) = \sup_{x \in M} xMxN$$

for motions of a compact space we can conclude :

(5.8) <u>If G is a group of motions of a compact G-space R which does not consist of the identity alone, then the diameter of G is at least</u> $\rho(R)/2 = \min \rho(x)/2$.

Therefore G has no small subgroups. On the other hand

the group of all motions of R is compact, hence it is a Lie group.

(5.9) <u>The group of motions of a compact G-space is a compact Lie group</u>.

The corresponding problem for non-compact G-spaces is open.

6. CLOSED GEODESICS.

If the G-space R is not simply connected let R' be its universal covering space and α a locally isometric map of R' on R. The fundamental group F of R can be realized as the group of motions M of R' which lie over the identity 1_R of R, i.e. $M\alpha = 1_R$.

An oriented closed curve G^+ in R which is not homotopic to 0 determines the class of closed curves in R which are freely homotopic to G^+ and in F a class $K(G) \neq 1_R$ of conjugate elements. Let G^+ be a geodesic, $p \in G$ and $p' \in p\alpha^{-1}$. If x(t) represents G^+ and $x(\delta) = p$ we construct the geodesic G'^+ over G^+ with $x'(t)\alpha = x(t)$ and $p' = x'(\delta)$. Traversing G^+ once, which may correspond to t traversing the interval $\delta \leq t \leq \delta + \beta$, $x(\delta) = x(\delta + \beta)$. Then $x'(\delta + \beta) = p'M$, $M \in K(G)$. Traversing G^+ once more we arrive in R' at $x'(\delta + 2\beta) = p'M^2$. Thus M <u>translates G' into itself</u>, $x'(t)M = x'(t + \beta)$ for all t.

If G' is a straight line then G' is an axis of M, and we know from (5.2) that $x'(t)x'(t)M^n = \min y'y'M^n$, $n \neq 0$. Therefore nG is a shortest curve in its free homotopy class. Other shortest curves in R freely homotopic to G^+ appear in R' as other axes of M.

If G is a shortest curve freely homotopic to $nC \not\sim 0$,

$n \neq 0$, then it is in general not true that $G^+ = nG_1^+$, where G_1^+ is freely homotopic to C . The Moebius strip with a euclidean metric provides a counter example.

The statement is true in the special cases as

(6.1) <u>If R is a orientable two-dimensional G-space which is not simply connected, and if the geodesic G^+ is freely homotopic to $nK \not\sim 0$ and a shortest curve in the class of curves freely homotopic to nC , then</u> $G^+ = nG_1^+$, <u>where</u> G_1^+ <u>is a shortest curve freely homotopic to C</u> .

For Riemannian metrics this theorem was proved by Morse [1] for compact surfaces of genus greater than 1 and for the torus by Hedlund [1] . The proof for (6.1) is found in Busemann and Pedersen [1] and is much shorter, in spite of the generality of (6.1), through application of the theory of parallels. I will indicate how parallels enter.

Assume the plane P is metrized as a G-space and that M is an orientation preserving motion of P for which a point p with $0 < ppM = \inf_{x \in R} xxM$ exists. If T is a segment from p to pM we know from (5.4) that $H = \bigcup_{i=-\infty}^{\infty} TM^i$ is a geodesic. There cannot be a second segment S from p to pM because the oriented Jordan curves obtained by traversing T from p to pM and then S from pM to p would by M be mapped either on a curve with the opposite orientation or on a curve containing the first and there would be a fixed point by Brouwer's Theorem. A second application of Brouwer's Theorem yields that the geodesic H is simple, and then it is not hard to see that it is a straight line. Thus

(6.2) <u>If M is an orientation preserving motion of the G-plane P for which a point p with</u> $0 < ppM = \inf xxM$ <u>exists then the segment</u> $T(p,pM)$ <u>is unique and</u> $H = \bigcup_{i=-\infty}^{\infty} T(pM^i, pM^{i+1})$ <u>is an axis of M</u> .

It follows from the hypotheses of (6.1) that the universal covering space of R is a G-plane P . We may assume $n > 1$. To G^+ there corresponds a point $p \in P$ such that $0 < ppM^n = \inf xxM^n$ when $M \in K(C)$. Hence M^n has an axis H . All we have to show is that M also has H as axis. If this were not the case then $H' = HM \neq H$ because M has no fixed points.

$$H'M^n = HM^{n+1} = HM^nM = HM = H' .$$

Since M^n has no fixed points H' is an axis of M^n, hence parallel to H . If W is the halfplane bounded by H containing H', then, because M preserves orientation, $H'M \subset WM$. WM is properly contained in W . It would follow that WM^n is properly contained in W and hence that $HM^n \neq H$.

A simple consequence of the results of the last section is

(6.3) <u>If R is compact and the universal covering space R' of R is straight and</u> $M \neq 1_R$ <u>lies in the center of F , then the closed geodesics belonging to M have not multiple points, have the same length and cover R simply</u>.

For because R is compact it has a bounded fundamental domain C' in R'. If $x' \in R'$ is given there is an $N \in F$ such that $x'N \in C'$ and because M lies in the center of F

$$x'x'M = x'Nx'MN = x'Nx'NM ,$$

so that x'x'M attains its maximum in the closure of C'. Then (5.7) shows that the axes of M cover R' simply and that x'x'M is constant. Each of these axes provides a closed geodesic of the same length in the free homotopy class K determined by M . Since the conjugate class to M consists of M alone, two intersecting geodesics in K or a geodesic on K with multiple points would correspond to intersecting axes of M .

We mention the corollary

(6.3') <u>If the fundamental group of R is abelian and R is compact, R' straight, then the assertion of (6.3) holds for every non-trivial homotopy class.</u>

We apply this to the torus T with a straight universal covering space P . We can realize P as an (x_1, x_2)-plane in which the elements of F are given by f(m,n) : $x_1' = x_1 + m$; $x_2' = x_2 + n$, m,n integers.

Since an axis of M is also an axis of M^n we see that a geodesic G in P which passes through x and xf(m,n) must also pass through the points xf(νm, νn) for all integers ν .

There is an axis through x and xf(m,n) for given (m,n) $\neq 0$ and each axis lies in a family of parallel lines. This makes it plausible that the lines in P satisfy the parallel axiom, and this is correct, but the proof (G pp.216-218) is not trivial because we do not know in advance, that the asymptote relation is transitive. The inverse problem for the torus without conjugate points can now be solved in this form :

(6.4) <u>Theorem.</u> <u>In the</u> (x_1,x_2)-<u>plane let a system S of curves be given satisfying the conditions a) b) of Theorem (4.1). In addition assume that S goes into itself under the translations f(m,n), that</u>

a curve in S passing through x and xf(m,n) also passes through xf(νm, νn) for all integers ν and that S satisfies the parallel axiom. Then the plane can be metrized as a G-space such that the curves in S become the geodesics and the f(m,n) are motions.

The example constructed for the asymmetry of the asymptote relation shows that there are systems S invariant under all translations but there is no metrization of the plane with the f(m,n) as motions and S as set of geodesics. It is possible to find metrizations of the plane as a straight space for which

$$x_1' = x_1 + \nu \ , \ x_2' = x_2 + n \ , \quad \nu \text{ real}, \quad n \quad \text{an integer}$$

are motions, and such that the geodesics do not satisfy Desargues' Theorem (G p.222). Such a metric yields a torus with a one-parameter group of motions. It is most interesting, and exhibits again the enormous generality of Finsler spaces, that according to E.Hopf [1], the only Riemannian metrizations of the torus without conjugate are euclidean.

The arguments used in the proof of (6.3) can be applied in many other cases, for example :

(6.5) Let R be compact, with a straight universal covering space R' and possess a one parameter group of motions M_t (written additively: $M_{t_1+t_2} = M_{t_1}M_{t_2}$). If M_t' is a group of motions of R' lying over M_t then the orbits of M_t' are parallel geodesics covering R' simply and $x'x'M_t'$ is independent of x'. Therefore the orbits of M_t are simple geodesics covering R' simply and for small t the distance xxM_t is independent of x .

For it is wellknown, (see G p.180) that the elements of

M'_t commute with the elements of the fundamental group F of R . As in the proof of (6.3) it is seen that the distance $x'x'M'_t$ attains for $M'_t \neq 1$ a positive maximum. By (5.5) $x'x'M'_t$ is independent of t and all points of R' lie on axes of M'_t. Therefore $x'M'_{nt}$ lies on an axis. The same holds for the points $xM'_{nt/2}$, therefore the axes are the orbits of M'_t.

It follows that the orbits xM_t are geodesics, and these are simple because if one point were multiple all would be.

7. PERPENDICULARITY.

For a non-empty set Q and any point p of a metric space we put

$$pQ = \inf_{x \in Q} px. \quad \text{Then} \quad |pQ - rQ| \leq pr .$$

For if $x \in Q$ then $pQ \leq px \leq pr + rx$, hence $pQ \leq pr + rQ$. If $f \in Q$ exists such that $pf = pQ$ then f is called a foot of p on Q. Returning to G-spaces we have the important lemma :

(7.1) <u>If f is a foot of p on Q and (pxf) then f is the unique foot of x on Q</u> .

For if $q \in Q$ then

$$xq \geq pq - px \geq pf - fx = fx ,$$

hence f is a foot of x . Equality holds only when (pxq) and $xq = xf$, but then $q = f$ by Axiom V . We need the trivial observation

(7.2) <u>If the foot f_p of p on Q exists and is unique for $p \in R$, then f_p depends continuously on p</u>.

A straight line L is perpendicular to the set Q if $f \in L$ Q exists such that every point $x \in L$ has f as foot on Q . Then trivially $f = L \quad Q$, and (7.1) shows that two perpendiculars to the same set Q cannot intersect except at a point of Q .

We prove

(7.3) *If R is straight and every point p of R has a unique foot on the line G , then every point p not on G lies on exactly one perpendicular to G .*

The uniqueness was observed before. It suffices to show : if f is the foot of p on G , and (pfy), then f is the foot of y on G .

(Fig.3) We determine a and b on G with (afb) and $af = fb = fy$. For (axf) or (fxb) we determine z with (axz) and $xz = xa$. Thus the points z form together with a, y, and b an arc A from a to b . Let f_z be the foot of z on G for $z \in A$. Since $f_a = a$, $f_b = b$ the arc A contains by (7.2) a point z with $f_z = f$. If $z \neq a, b, y$ then

$$zf > zx + xp - pf > zx \geqslant zf_z .$$

Therefore $f_z \neq f$ and f must be the foot of y .

The uniqueness of a foot on a line is equivalent to the convexity of spheres. This is true locally, (G pp. 119, 120) but for simplicity we discuss this here only in the large for straight spaces.

The locus $px = \rho$ is denoted by $K(p, \rho)$. We say that $K(p, \rho)$ or $S(p, \rho)$ is convex if $xp \leq \rho$ and $zp \leq \rho$ and (xyz) implies $yp \leq \rho$, strictly convex if $yp < \rho$. The sphere $K(p, \rho)$ has order 2 if a straight line intersects $K(p, \rho)$ in at most two points.

(7.4) <u>In a straight space the following statements are equivalent</u>

a) <u>the spheres $K(p, \rho)$ are convex</u>

b) <u>the spheres $K(p, \rho)$ are strictly convex</u>

c) <u>the foot of a point on a line is unique</u>

d) <u>the spheres $K(p, \rho)$ have order 2</u> .

We prove a) $\Rightarrow$ b) $\Rightarrow$ c) $\Rightarrow$ d) $\Rightarrow$ a) . If a) did not imply b) then points p, x, y, z with (xyz), $px \leq \rho$, $py \leq \rho$ and $pz = \rho$ would exist. Let (xpq). Then

$$qx < qp + px \leq qp + py = qy$$

and similarly $qz < qy$, so that K(q, max(qx, qz)) would not be convex.

If the point p has two feet f_1, f_2 as on a line, then x with $(f_1 x f_2)$ satisfies $px \geq pf_1 = pf_2$ and the sphere $K(p, pf_1)$ is not strictly convex, which proves b) $\Rightarrow$ c).

Let K(p, q) intersect a line G in three points a, b, c with (abc). We show that this contradicts c). If c) did hold then the foot f_p of p on G cannot be a, b, or c . Assume $(f_p bc)$, let x traverse T(p,c). The foot f_x of p changes continuously and since $(f_p bc)$ and $c = f_c$ there is a u with (puc) such that $f_u = b$. But this is impossible because

$$ub > pb - pu = pc - pu = uc \ .$$

Finally we show d) $\Rightarrow$ a). If $K(p, \rho)$ is not convex then points p, a, b, c with (abc), $pa < \rho$, $pb < \rho$ and $pc > \rho$ exist. If (acx) then $px \geq xa - pa$, hence $px \to \infty$ for $ax \to \infty$, so that a point x_0 with $px_0 = pb$ and (acx_0) exists. Similarly a point y_0 with (cay_0) and $py_0 = pb$ exists, so that the sphere K(p, pb) intersects G at least thrice. (Figure 4).

If A^n is the n-dimensional affine space with affine coordinates $x_1, \dots, x_n$, then a <u>Minkowski metric</u> in A^n is given by

$$m(x,y) = F(y - x) = F(y_1 - x_1, \dots, y_n - x_n) ,$$

where $F(x)$ satisfies the standard condition :

$$F(x) > 0 \quad \text{for} \quad x \neq 0 \text{ , } F(kx) = kF(x) \quad \text{for} \quad k > 0$$

and $F(x) + F(y) \geqslant F(x + y)$ which is equivalent to requiring that $F(x) = 1$ is a convex surface. Here we are only interested in the symmetric case $m(x,y) = m(y,x)$ which corresponds to the condition $F(x) = F(-x)$. The space then satisfies our axioms I to IV, but in general not V . It will satisfy V only if <u>the surface $F(x) = 1$ is strictly convex</u>.

<u>The space is then straight with the affine lines as geodesics</u>. The sphere $K(p, \rho)$ is given by $F(x - p) = \rho$ and is homothetic to $F(x) = 1$. Thus all spheres are strictly convex, so that according to our theory <u>the perpendicular to a given line through a given point $p \in G$ exists</u>.

Perpendicularity of lines is in general not a symmetric concept. In fact, if the dimension of the space exceeds 2, then according to Blaschke [1] <u>perpendicularity is symmetric only if the metric is euclidean</u>. The proof of this fact rests on the following theorem on closed convex hypersurfaces in A^n.

<u>If for every line G the supporting lines of the closed convex surface C parallel to G touch C in a set lying in a hyperplane, then C is an ellipsoid</u>.

Consider the sphere $K(p,1)$ of a Minkowski space and a supporting hyperplane H of $K(p,1)$ at f ; then the line $L(p,f)$ is perpendicular to every line in H through f . Let the hyperplane parallel to H through p intersect K in Z , then $L(p,f)$ is also perpendicular to any line $L(p,x)$ with $x \in Z$, therefore the parallel L_x through x to $L(p,f)$ is perpendicular to $L(p,x)$. By hypothesis $L(p,x)$ is perpendicular to L_x, hence L_x is a supporting line of $K(p,1)$. Thus the supporting lines of $K(p,1)$ parallel to $L(p,f)$ touch $K(p,1)$ in the points of Z which lies in a hyperplane, and Blaschke's Theorem gives the assertion.

<u>There are plane Minkowski metrics with symmetric perpendicularity which are not euclidean</u>; all these metrics have been determined by Radon [1] .

It is quite easy to see that (G pp.144, 155) that a metrization of A^n, $n \geqslant 2$, as G-space with the affine lines as geodesics and with convex spheres is Minkowskian. In view of the great variety of metrizations of A^n with the affine lines as geodesics this seems to show the strength of the condition that the spheres be convex. Actually, convexity of spheres is strong only in combination with the parallel axiom, there are many straight spaces satisfying stronger convexity conditions which are not Minkowskian, for example, all simply connected Riemann spaces or G-spaces (see next section) with non-positive curvature.

The desarguesian character of the affine space is not essential. Let us formulate the parallel axiom as follows :

<u>Parallel Axiom</u>. <u>If A^+ is an asymptote to G^+ then A^- is an asymptote</u>

to $\bar{G}$, i.e. A is parallel to G. Moreover G is parallel to A.

In two dimensions this is equivalent to the usual parallel axiom. Defining differentiability of spheres in a geometric way it can be shown (G pp.146-150)

(7.5) Theorem. If a straight G-space of dimension at least three has differentiable convex spheres and satisfies the Parallel Axiom, then it is a Minkowski space with differentiable spheres.

A corollary of this theorem is (Gp. 151)

(7.6) If a G-space of dimension at least three satisfies the parallel axiom and has convex spheres and if perpendicularity of lines is symmetric, then the space is euclidean.

The differentiability of the spheres follows in this case from the symmetry of perpendicularity.

The question whether (7.5) holds in a plane is open. An analogous result can be proved under a stronger convexity hypothesis (satisfied by Minkowski spaces) without the differentiability of circles (G pp.157-160)

(7.7) Theorem. If a straight plane satisfies the parallel axiom and if for any line x(t) and any point p not on the line the function $(px(t))^{\alpha}$ is convex, where α is a fixed number ≥ 1, then the metric is Minkowskian.

The proofs of all these theorems are too long to be given here. There is a theorem analogous to (7.5) for spaces of the elliptic type which is easy to prove. Convexity of spheres must be rephrased. Observe that in an elliptic space in which the great circles have length 2β the spheres $K(p, \beta)$ are hyperplanes and that the spheres $K(p, \rho)$ with $\rho < \beta$ are quadrics.

This and (7.5) suggest the following theorem :

(7.7) Theorem. Let R be a space of the elliptic type with geodesics of length 2β and dimension at least 3. If the spheres $K(p, \rho)$ with $\rho < \beta$ have order 2 then the space is elliptic.

Order 2 means, of course, that a geodesic does not intersect the sphere more than twice. It is very easily seen (G pp.124, 125) that this hypothesis implies that the loci $K(p) = K(p, \beta)$ are flat, i.e. are G-spaces with the metric obtained from that of R by restricting it to $K(p)$. Thus hyperplanes exist and it also follows readily that the space can be identified with P^n, for a suitable (finite) $n \geqslant 3$ with the projective lines as geodesics. (G pp.125, 126).

The principal step consists in showing that with $K(p)$ considered as plane at infinity the spheres $K(p, \rho)$, $\rho < \beta$ are ellipsoids with center p.

We call the point q conjugate to p if $pq = \beta$. This is equivalent to $q \in K(p)$ or $p \in K(q)$. Let $x \in K(p,q)$ and x' the conjugate point to x on $L(p,x)$. Then x is a foot of x' on $K(x')$ because xx' is constant for $x \in K(x')$. From (7.1) and (xpx') we conclude that x is the only foot of p on $K(x')$, which means that $K(x')$ is a supporting plane of $K(p, \rho)$ at x which touches $K(p, \rho)$ only at x. Therefore $K(p, \rho)$ is a strictly convex hypersurface.

Let z be any point at infinity, i.e. $z \in K(p)$. Then $K(z)$ contains p and intersects $K(p, \rho)$ in a set Z. If $x \in Z$ and x' is defined as before then $G(z,x) \subset K(x')$, because $x \in K(x')$ and

$p \in K(z)$, $x \in Z \subset K(z)$ imply $x' \in G(p,x) \subset K(z)$, hence also $z \in K(x')$ (Figure 5). Thus $G(z,x)$ is a supporting line of $K(p, \rho)$. The point x was arbitrary on Z and the lines $G(z,x)$ are all parallel ($z \in K(p)$). This means that the supporting lines of $K(p, \rho)$ through z or parallel to $G(z,x)$ touch $K(p, \rho)$ along their hyperplane set Z, Blaschke's Theorem yields now that $K(p, \rho)$ is an ellipsoid.

If we make differentiability hypothesis we can conclude that the "infinitesimal" spheres are ellipsoids, and that the metric is Riemannian. Beltrami's Theorem shows then that the metric has constant curvature and is therefore elliptic. The same can be shown without differentiability hypotheses by using projective geometry.

8. CURVATURE.

Theorems into which the concept of curvature enters seem inaccessible to our methods because it looks as though second derivatives were necessary. This is not the case.

A Riemann space has non-positive curvature (G p.269) if and only if every point has neighborhood $S(p, \beta_p)$, $0 < \beta_p < \rho(p)/2$ such that any three points a,b,c in $S(p, \beta_p)$ and the midpoints b' of ab and c' of ab satisfy

(8.1) $\qquad 2b'c' \leq bc \qquad$ (<u>curvature non positive</u>)

If the curvature is negative and a,b,c do not lie on a segment then

(8.14) $\qquad 2b'c' < bc \qquad$ (<u>curvature negative</u>)

The curvature vanishes if and only if

(8.1") $\qquad 2b'c' = bc \qquad$ (<u>curvature zero</u>)

Nothing prevents us from <u>using (8.1),(8.1') or (8.1") as definition for G-spaces with non-positive, negative or vanishing curvature</u>. In the Riemannian case (8.1') is slightly weaker than negative curvature. It turns out that practically all Riemannian results on spaces with non-positive curvature extend with these definitions to G-spaces.

Most of them even extend under <u>weaker conditions</u>. If T is a segment, denote by C_T^{α} the set of points satisfying $xT < \alpha$. In the Riemannian case (8.1) is equivalent to requiring that each point has a neighborhood $S(p, \beta_p)$, $0 < \beta_p < \rho(p)/2$ such that each set $C_T^{\alpha} \subset S(p, \beta_p)$ be convex, and (8.1') is equivalent to the strict convexity $C_T^{\alpha} \subset S(p, \beta_p)$. There is no obvious analogue to vanishing curvature. In Finsler spaces there are important geometries where the C_T^{α} are convex (or strictly convex) and (8.1) (or (8.1')) does not hold, in particular the <u>Hilbert geometries</u>. They are the analogue to Klein's model of hyperbolic geometry obtained by replacing the ellipsoid as absolute locus by a closed convex hypersurface C. (These geometries are discussed in G Section 18). If D is the interior of C and the affine line through two distinct points x,y in D intersects C in x',y' then the Hilbert distance is

$$h(x,y) = \frac{k}{2} \left| \log R(x,y,x',y') \right| ,$$

where R means cross ratio. D becomes a straight G-space with the intersections of the affine lines with D as geodesics when C is strictly convex (actually a weaker condition suffices). The spheres are convex, hence perpendiculars exist, but perpendicularity is symmetric only in the hyperbolic geometry (Kelly and Page [1]), which is also the only Hilbert geometry in which (8.1) holds, (Kelly and Strauss [1]). However the sets C_T^{α} are strictly convex for any segment T . That much of the theory holds, when the C_T^{α} are locally convex, was discovered by Petersen [1] , see also G Chapter V .

Because of lack of time I restrict myself to the implications of (8.1) or (8.1'). Just as we succeeded to pass from ρ_p in Axiom IV to the continuous function $\rho(p) = \sup \rho_p$ we can pass here from β_p to a continuous function $\beta(p) = \sup(\beta_p, \rho(p))$. The principal consequence of (8.1) is that it leads to a property in the large.

(8.2) Let x(t) and y(t) represent geodesics. If for suitable constants α, δ, $t' < t''$ the segment

$$T_t = T(x(t), y(\alpha t + \delta))$$

is unique for $t' < t < t''$ then $f(t) = x(t)y(\alpha t + \delta)$ is a convex function. If x(t), y(t) represent different geodesics then f(t) is strictly convex if either $\alpha = 0$ or (8.1') holds.

For a proof we put $y(\alpha t + \delta) = x'(t)$. Because of their uniqueness the segments T_t depend continuously on t . Therefore, if p(t, u) $(0 \leq u \leq 1)$ is the point of T_t for which $p(t,u)x(t) = ux(t)x'(t)$, then p(t,u) is continuous in t and u . It suffices to show that an $\varepsilon > 0$ exists such that

$$f((t_1 + t_2)/2) \leq (f(t_1) + f(t_2))/2 \quad \text{for} \quad |t_1 - t_2| < \varepsilon \quad .$$

Divide the segments T_{t_1} and T_{t_2} into n equal parts by points p_i, q_i

with $p_0 = x(t_1)$, $p_n = x'(t_1)$ $(p_{i-1}p_ip_{i+1})$ and similarly for $q_i \in T_{t_2}$. If n is sufficiently large and ε is sufficiently small then a sphere $S(p_i, \beta(p_i))$ will contain p_{i+1}, q_i, q_{i+1}. Let r_i be the midpoint of p_i and q_i and v_i the midpoint of p_{i+1} and q_i. Then by (8.1)(Figure 6)

$$(8.3) \qquad \begin{aligned} 2r_iv_i &\leq p_ip_{i+1} \\ 2v_ir_{i+1} &\leq q_iq_{i+1} . \end{aligned}$$

Adding these inequalities gives with $w = (t_1 + t_2)/2$

$$2x(w)x'(w) = 2r_0r_n \leq 2\sum_{i=0}^{n-1}(r_iv_i + v_iv_{i+1}) \leq$$

$$\leq \sum p_ip_{i+1} + \sum q_iq_{i+1} = x(t_1)x'(t_1) + x(t_2)x'(t_2) .$$

For equality of the left and right sides it is necessary that equality holds in all the relations (8.3). If (8.1') holds this means that p_i, q_i, p_{i+1} lie on a segment and also that $q_ip_{i+1}q_{i+1}$ lie on a segment for each i . Then x(t) and y(t) represent the same geodesics.

If $\alpha = 0$ then $p_n = q_n = x(w) = x'(w)$ we add all equations (8.3) except the last two giving as before

$$2r_0r_{n-1} \leq p_0p_{n-1} + q_0q_{n-1} .$$

If m is the midpoint of p_{n-1} and p_n then unless one of the segments T_{t_1} and T_{t_2} falls on the other we have

$$2p_nr_{n-1} < 2(p_nm + mr_{n-1}) \leq p_{n-1}p_n + p_nq_{n-1} .$$

Adding this relation to the preceding one gives the derived result

$2r_0r_n < p_0p_n + q_0q_n$.

We conclude from (8.2) that the spheres $K(p, \rho)$ with $\rho < \rho(p)/2$ are strictly convex. For if $xp < \rho$, $yp < \rho$ then $T(x,y)$ lies in $S(p, \rho(p))$. The segments $T(pu)$, $u \in T(x,y)$ are therefore unique and pu is a strictly convex function of xu . Of course it now follows that <u>$K(p, \rho)$ is strictly convex for $\rho < \rho(p)$</u>.

If a convex function $f(t)$, $\alpha \leq t \leq \beta$ has the property for some u, $0 < u < 1$ the equality

$$f((1 - u)t_1 + ut_2) = (1 - u)f(t_1) + uf(t_2)$$

holds, then $f(t)$ is linear. This leads to the following result (G pp.240-242) :

(8.4) <u>Under the hypotheses of (8.2) let $x(t)$ and $y(t)$ be segments. If $f(t)$ is linear then</u> $\bigcup T_t$ <u>is isometric to a trapezoid in a Minkowski plane</u>.

We apply these results to straight spaces. If $x(t)$ and $y(t)$ represent different lines then for any α , δ the function $f(t) = x(t)y(\alpha t + \delta)$ is convex and is strictly convex if $\alpha = 0$ or if the curvature is negative.

Since the spheres are convex the foot f_t of $x(t)$ on the line G represented by $y(t)$ is unique. If again $w = (t_1 + t_2)/2$ and m is the midpoint of f_{t_1} and f_{t_2} then by (8.2)

$$2x(w)f_w \leq 2x(w)m < x(t_1)f_{t_1} + x(t_2)f_{t_2}$$

hence $x(t)G$ is convex and is strictly convex if (8.1') holds.

In spaces with non-positive curvature the parallels are well behaved.

The asymptote A^+ to a line G^+ through $p \notin G^+$ is the limit of the lines $G^+(p,x(s))$ for $s \to \infty$ when $x(t)$ represents G^+. If $z_s(t)$ represents $G^+(p,x(s))$ with $z_s(0) = p$, then $z_s(t)G$ is convex and vanishes for $t = px(s)$, hence decreases for $t \leq px(s)$. If $a(t)$ represents A^+ with $a(0) = p$, then

$$a(t) = \lim_{s \to \infty} z_s(t)$$

hence $a(t)G$ <u>is convex and non-increasing</u>, therefore bounded for $t \geqslant 0$. Also, $z_s(t)x(ts/px(s))$ is by (8.2) a convex function of t for fixed s and vanishes for $t = px(s)$, moreover

$$s - px(0) \leq px(s) \leq ps + px(0)$$

therefore $s/pxs \to 1$ for $s \to \infty$, We conclude that <u>$a(t)x(t)$ is convex and non-increasing</u>.

If A is parallel to G then $a(-t)x(-t)$ is also non-increasing hence $a(t)x(t)$ is constant and we conclude from (8.4) that $\bigcup_{t=-\infty}^{\infty} T(a(t),x(t))$ is isometric to a strip in a Minkowski plane bounded by parallel lines. Therefore the curvature cannot be negative:

(8.5) <u>In a straight space with negative curvature (non-identical) parallel lines do not exist</u>.

(8.6) <u>The asymptote relation is transitive (and hence symmetric) in a straight space with non-positive curvature</u>.

This assertion is contained in the following :

(8.7) <u>$a(t)$ represents an asymptote A^+ to G^+ given by $x(t)$ if and only if $a(t)x(t)$ is bounded for $t \geqslant 0$</u> .

We know that boundedness of $a(t)x(t)$ is necessary. Conversely let $a(t)x(t)$ be bounded for $t \geqslant 0$ and let $b(t)$ represent any line different from $a(t)$ with $b(o) = a(o)$. Then $b(t)a(t)$ is convex and $b(o)a(o) = 0$ therefore $b(t)a(t) \to \infty$ for $t \to \infty$ and

$$b(t)x(t) \geqslant b(t)a(t) - a(t)x(t) \to \infty ,$$

therefore $b(t)$ is not an asymptote to G^+.

We say that a metric space has the property of <u>domain invariance</u>, if for any two homeomorphic sets in the space of which one is open the other is also open. Whether all G-spaces have this property is not known. For finite dimensional G-spaces this can be proved, for two dimensional spaces it follows from the fact that they are topological manifolds.

For straight spaces of non-positive curvature a very similar argument to the classical (compare Cartan 2 , Note III , for G-spaces see G , Section 38) yields :

(8.8) <u>If the G-space R has non-positive curvature and the property of domain invariance, then its universal covering space is straight.</u>

In any G-space R whose universal covering space R' is straight a geodesic curve, i.e. a part $\alpha \leq t \leq \beta$ of a representation $x(t)$ of a geodesic, is unique within the homotopy class of geodesic curves connecting $x(\alpha)$ and $x(\beta)$. From our discussion of closed geodesics we know that two closed geodesics in R within the same free homotopy class lead to parallel axes in R'. Therefore we have the following theorem which goes back to Hadamard :

(8.9) <u>Let R be a G-space with non-positive curvature and domain in-</u>

variance. Then there is exactly one geodesic curve connecting two given points a, b within a given homotopy class. If the curvature is negative then there is at most one closed geodesic in a given (non-trivial) free homotopy class, exactly one when R is compact.

We also proved in (6.3) and (6.5) that parallels occur in R' when R is compact and its fundamental group has a center, or if R passes a one-parameter group of motions. Therefore

(8.10) The fundamental group of a compact space with negative curvature does not have a center and the space possesses only a finite number of motions.

For the latter fact we use that the motions of a compact G-space form a compact Lie group, (5.9), which contains a one-parameter subgroup unless finite.

Another theorem of this type which can be proved as in the classical case (see G p.258) is :

(8.11) The fundamental group of a compact G-space with non positive curvature does not have a (non-trivial) finite subgroup.

These theorems show that many manifolds cannot be metrized as G-spaces with non-positive or negative curvature.

A simple consequence of (8.4) without the assumption of domain invariance is :

(8.12) The universal covering space of a G-space with vanishing curvature is Minkowskian.

The concept of a G-space with curvature greater than K has recently been introduced by Kann (not yet published) in a way which coincides with the usual concept for Riemann spaces. He proves in particular an analogue to Bonnet's Theorem, namely that a two-dimensional G-space whose curvature exceeds $K > 0$ is compact. With a different definition of curvature based on Cartan's connection in Finsler spaces, see Cartan [1], and hence on osculating Riemann metrics, Auslander [1] proved this theorem (as well as other Riemannian theorems) for Finsler spaces of arbitrary dimension. Kann's proof can be extended to higher dimensions if some of the consequences of the standard differentiability hypotheses are added as assumptions to the axioms of a G-space.

9. THE BISECTOR THEOREM.

The euclidean and hyperbolic spaces and the spherical spaces of dimension greater than 1 are simply connected spaces with constant curvature. We will call them briefly the elementary spaces. They have an extremely simple joint characterization. For two distinct points a, a' denote as <u>bisector</u> B(a,a') the locus of the points x satisfying xa = xa'.

(9.1) <u>Theorem</u>. <u>The elementary spaces are those G-spaces for which every bisector contains with any two points x,y at least one segment T(x,y)</u>.

As previously we call a subset M of a G-space R flat if M with the metric of R restricted to M is a G-space. Then an equivalent formulation of (9.1) is that the elementary G-spaces are those in which the bisectors are flat.

Theorem (9.1) has an important local version :

(9.2) <u>Bisector Theorem</u>. <u>If for any five distinct points a,a',x,y,z</u>

__of a sphere__ $S(p,\delta_p)$, $o < \delta_p < \rho(p)$ __in a G-space the relations__ __$ax = a'x$, $az = a'z$ and (xyz) imply $ay = a'y$__ (or, equivalently, $x,z \in B(a,a')$ implies $T(x,z) \subset B(a,a')$) __then__ $S(p,\delta_p)$ __is isometric to a sphere in an elementary space.__

The proof of this theorem is quite long, G pp.310-330, but the basic ideas are simple and we will indicate them here. In contrast to our previous theorems, differentiability would in this case reduce the proof very materially. It is interesting that just in investigations concerned with topics related to the present, differentiability has always been avoided, because (9.1) belongs in character to the foundations of geometry, where differentiability is traditionally felt as alien.

The ideas of the proof become clearest when we assume that the space is straight and $B(a,a')$ contains with any two points x,y the segment $T(x,y)$. It follows rather easily that $B(a,a')$ is flat, i.e. contains for $x \neq y$ the whole line $G(x,y)$, (G p.311).

The midpoint c of a and a' lies on $B(a,a')$ and is the foot of a and a' on $B(a,a')$ because for $x \in B(a,a') - c$,

$2a'x = 2ax = ax + xa' > aa' = 2ac = 2a'c.$

For reasons of continuity there is a point x' with (xcx') and $ax = ax'$. Since $x' \in B(a,a')$ we conclude

$$a'x = ax = ax' = a'x' ,$$

therefore $a,a' \in B(x,x')$, hence c must be the midpoint of x and x'. Thus c is the foot of any $x \in B(a,a')$ on $G(a,a')$, in particular $G(x,x')$ is perpendicular to $G(a,a')$, and $B(a,a')$ is the union of all lines perpendicular to $G(a,a')$ at c . It follows that

$B(a,a) = B(b,b')$ if b and b' lie on $G(a,a')$ and have c as midpoint. It also follows that $G(a,a')$ is perpendicular to $G(x,x')$, so that <u>perpendicularity of lines is symmetric</u>.

I mentioned that it is not known whether G-spaces have always a finite dimension, but in several of our theorems the assertion implied finite dimensionality. Therefore we prove it here in the way of an example. We begin with two points a_1, a_1' with $a_1 a_1' = 2$ and midpoint c. In $B_1 = B(a_1, a_1')$ we take a_2, a_2' with $a_2 a_2' = 2$ and midpoint c. Then $B_2 = B(a_2, a_2') \; B_1$ is the locus of the points x with $xa_2 = xa_2'$ in B_1 and B_2 is flat as the intersection of two flat sets. If $B_2 = c$ we stop. Otherwise we continue choosing $a_3, a_3' \in B_2$ with $a_3 a_3' = 2$ and midpoint c. We form $B_3 = B(a_3, a_3') \quad B_2$ and continue if $B_3 \neq c$.

Since c is the foot of a_i on $B(a_i, a_i')$ hence on B_i we have $a_i a_j > 1$ for $i < j$. On the other hand $a_i c = 1$. Therefore the sequence $a_1 a_2, \ldots$ is bounded and has no accumulation point, hence is finite. Assume $B_m = c$. Then B_{m-1} is the line $G(a_{m-1}, a'_{m-1})$ and B_{m-2} is, topologically, a plane. If B_{m-3} is regarded as the union of the perpendiculars to $L_3 = G(a_{m-3}, a'_{m-3})$ in B_{m-3} then for $b \in L_3$ and x, x' on L_3 with midpoint b the perpendiculars to L_3 at b form $B(x,x') \quad B_{m-3}$ hence a plane, therefore B_{m-3} is homeomorphic to E^3. We conclude consecutively that B_1 <u>is homeomorphic to E^{m-1} and the space to E^m</u>.

It is easy to see that a bisector through m given points exists, hence a plane, or a two-dimensional flat set passing through three given points exists too, so that R will satisfy the usual incidence axioms if $m \geqslant 3$. For $\dim R = 2$ we can, however, not conclude that Desargues' Theorem holds. In any case we <u>reduce the theorem to showing that straight planes in which the bisectors are straight lines are euclidean or hyperbolic</u>.

For a given line G we define an involutory map R_G: $a \to a'$ of the plane on itself by

$$a' = a \quad \text{for} \quad a \in G \quad \text{and} \quad G = B(a,a') \quad \text{for} \quad a \notin G .$$

The point a' exists in the latter case. For let f be the foot of a on G and (afa'), af = fa'. Then our previous discussion shows that G = B(a,a'). It suffices to see that R_G is a motion. For then R_G is the reflection in G and the full group of motions in euclidean or hyperbolic geometry is generated by the reflections in lines.

We show first that <u>a line L which intersects G is mapped by</u> R_G <u>isometrically on a line L' intersecting G</u> (in the same point). If $L \cap G = s$ and (asb), $a,b \in L$, then a' and b' lie on different sides of G, hence T(a',b') intersects G in a point s'. Then G = B(a,a') = B(b,b') yields

$$ab \leq as' + s'b = a's' + s'b' = a'b' \leq a's + sb' = as + sb = ab .$$

Therefore (a'sb') and ab = a'b'. Thus G(a',b') passes through s. If c is any other point of L then, unless c = s, let c lie on the same side of G as b and say (acb). Then ab = a'b', ac = a'c' and ac = ab - ac give b'c' = a'b' - a'c' = bc. Therefore also $c' \in G(a',b')$ so that $LR_G = G(a',b')$ and the map of L on L' isometric.

If $H \cap G = s$ and H = B(p,q) then $H' = HR_G = B(p',q')$ (Figure 7). To prove this we choose t so close to s on H that G(q,t) and G(p,t) intersect G. Then

$$q't' = qt = pt = p't' , \quad q's = qs = ps = p's$$

therefore $H' = B(p',q')$, moreover

$$pR_G = p' = q'R_{H'} = qR_GR_{H'} = pR_HR_GR_{H'} \text{ , hence}$$

$$(9.3) \qquad R_G = R_HR_GR_{H'} \text{ .}$$

From here the proof proceeds as follows : we want to show that R_G also maps a line H which does not intersect G isometrically on a line H'. We choose H_1 so that it intersects H at a point so close to H such that with $H_0 = H$ and $H_{i+1} = R_{H_i}H_{i-1}$ the lines H_{n-1}, H_n, H_{n+1} intersect G and n is even, put $H'_{n+1} = R_GH_{n+1}$, $H'_n = R_GH_n$ and define successively (for descending i) $H'_{i-1} = H'_{i+1}R_{H'_i}$ until we reach H'_0. For simplicity we put $R_i = R_{H_i}$ and $R'_i = R_{H'_i}$. We observe that

$$H_0R_1 \ldots R_nR_GR'_n \ldots R'_1 = H'_0 \text{ .}$$

R_1 maps H_0 isometrically on H_3 and R_2 leaves H_2 pointwise fixed, R_3 maps H_2 isometrically on H_4, so that $R_1 \ldots R_n$ maps H_0 isometrically on H_n, then R_G maps H_n isometrically on H'_n, we see that the above combined mapping maps H_0 isometrically on H'_0. We want to see that this map is R_G.

The relation (9.3) implies

$$R_{i-1}R_iR_{i+1} = R_i \quad \text{or} \quad R_{i-1} = R_iR_{i+1}R_i \quad \text{, similarly for } R'_1$$

and $R_iR_GR'_i = R_G$ for $i = n, n+1$. Therefore

$$R_{n-1}R_nR_GR'_nR'_{n-1} = R_{n-1}R_GR'_{n-1} = R_nR_{n+1}R_nR_GR'_nR'_{n+1}R'_n = R_G$$

and it is clear that we obtain in this way successively that

$R_1 \ldots R_nR_GR'_n \ldots R'_1 = R_G$.

The Bisector Theorem has in addition to its own interest, many important <u>applications</u>. In this section we mention only one which seems hard to prove directly :

In the two-dimensional elementary geometries there are formulae expressing the area $\alpha(abc)$ of a triangle in terms of the length of its sides A,B,C .

$A = bc$, $B = ca$, $C = ab$, Hero's Formula

$\alpha^2(abc) = s(s - A)(s - B)(s - C)$ with $2s = A + B + C$ in the euclidean geometry, and in the hyperbolic geometry with curvature -1

$$\tan^2 \frac{\alpha(abc)}{4} = \tanh \frac{s}{2} \tanh \frac{s - A}{2} \tanh \frac{s - B}{2} \tanh \frac{s - C}{2}$$

with the analogous formula of L'Huilier in the spherical geometry. The question arises whether there are other two-dimensional geometries in which $\alpha(abc)$ is a function of the sides or, equivalently, where $ab = a'b'$, $bc = b'c'$, $ca = c'a'$ implies $\alpha(abc) = \alpha(a'b'c')$.

First we must define area. $\alpha(abc)$ is a <u>triangle area</u> in the subset M of a G-surface if

1) $\alpha(abc)$ is defined for a,b,c in M and symmetric in a,b,c and non negative.

2) $\alpha(abc) = 0$ if and only if abc lie on a segment.

3) If a,b,c,x lie in M and (bxc) then

$$\alpha(abx) + \alpha(axc) = \alpha(abc) .$$

We then have

(9.4) <u>If every point p of a G-surface R has a neighborhood $S(p, \delta_p)$, $\delta_p > 0$, in which a triangle area can be defined such that isometric triples have the same area then the universal covering space of R is elementary.</u>

It suffices to prove that $S(p, \varepsilon)$ with $\varepsilon = \min(\delta_p, \rho(p)/4)$

is isometric to a disc in an elementary surface. Consider two distinct points a, a' and two points x, y which lie on different sides of the extension of T(a, a') with center a and length $3\rho(p)/2$ which exists because $\rho(a) \geqslant \rho(p) - pa > 3\rho(p)/4$. The triangle inequality then shows at once that T(x,y) must intersect T in a point q with (aqa'). We want to show that $B(a,a') \cap S(p, \varepsilon)$ is a segment and it evidently suffices to prove that q coincides with the midpoint m of T(a,a'), then the Bisector Theorem yields (9.4). We have

$$\alpha(axy) = \alpha(a'xy) \quad \text{because} \quad ax = a'x\ ,\ ay = a'y\ ,$$

hence by 3)

$$\alpha(axq) + \alpha(ayq) = \alpha(a'xq) + \alpha(a'yq)\ .$$

Assume $m \in T(q,a')$. Then by 1) and 3)

$$\alpha(axq) \leqslant \alpha(axq) + \alpha(xqm) = \alpha(axm) = \alpha(a'xm) \leqslant \alpha(a'xq).$$

Similarly $\alpha(ayq) \leqslant \alpha(a'yq)$. Therefore $\alpha(xqm) = 0$ and by 2) $q = m$.

Now the question arises whether in a straight plane say, the metric is euclidean if $\alpha(abc)$ is defined for all a,b,c and is a function f(bc,aG(b,c)). Strangely enough the answer is negative. This property characterizes the Minkowski geometries with symmetric perpendicularity. There f(u,v) also has the form kuv. The value of k distinguishes the euclidean metric. If area is so normalized that the unit disc has measure π, then $k > 1/2$ for the Minkowski metrics which are not euclidean (Petty [1]).

10. MOBILITY.

The most important applications of the Bisector Theorem

are to questions of mobility.

(10.1) <u>Let every point p of the G-space have a neighborhood</u> $S(p, \delta_p)$ <u>such that for 4 points x, y, a, a' on</u> $S(p, \delta_p)$ <u>with</u> $xa = xa'$ <u>and</u> $ya = ya'$ <u>a motion of</u> $S(x, \rho)$ <u>(i.e. an isometric map of</u> $S(x, \rho)$ <u>on itself) exists which leaves x and y fixed and takes a into a', then the universal covering space of R is elementary.</u>

If suffices to prove this for $\delta_p \leq \rho(p)$. If the points x, y, a, a' are distinct and (xzy) then, since T(x,y) is unique a motion M of $S(x, \rho)$ leaving x and y fixed also leaves z fixed. If $aM = a'$ then $az = aMzM = a'z$. Therefore the hypotheses of the Bisector Theorem are satisfied and $S(p, \delta_p)$ is isometric to a sphere in an elementary space. The connectedness properties of a G-space show that this is the same elementary space for different p and this establishes the theorem.

The corresponding theorem in the large can be given different forms

(10.2) <u>If for any four points x,y,a,a' with</u> $xa = xa'$ <u>and</u> $ya = ya'$ <u>of a G-space R a motion of R exists leaving x and y fixed and taking a into a' then R is elementary or a great circle.</u>

The elliptic space does not have this mobility property (G p.338), but it does have the property for triples of small diameter. Thus for example

(10.3) <u>A G-space R is elementary or elliptic if and only if a</u> $\delta > 0$ <u>exists such that for any two triples</u> a_i <u>and</u> a_i', $i = 1,2,3$ <u>with</u> $a_ia_j = a_i'a_j'$ <u>and</u> $a_ia_j < \delta$ <u>a motion of R exists which takes</u> a_i <u>into</u> a_i', $i = 1,2,3$.

The (not attained) upper bound for δ is two thirds of the diameter of R , see Rhodes [1] .

The next step of the theorems is to determine those G-spaces for which for every two pairs of points a,b and a'b' with ab = a'b' a motion exists taking a into a' and b into b'. It is trivial that the only two-dimensional spaces with this property are provided by the elementary and elliptic geometries.

In the general case it is easily seen that the spaces are finite dimensional, and according to a beautiful result of Wang [1,2] the elementary and the elliptic spaces provide the only solutions for all odd dimensions. In even dimensions there are other solutions all of which have been determined by Wang [1] and Tits [1] . They are the hermitian hyperbolic and elliptic spaces defined for all even $n \geq 4$, the quaternian elliptic and hyperbolic spaces defined for all $n \equiv 0 \pmod 4$, $n \geq 8$, and the Cayley hyperbolic and elliptic planes based on the Cayley numbers which have dimension 16.

The proofs of these results are based on deep results from theory of topological and Lie groups and cannot be given here. They are found in G . Section 53-55 and Tits [1] .

The last step in this direction would consist in determining all G-spaces with transitive groups of motion, but this is far beyond our present state of knowledge. The spaces with transitive abelian groups of motions are easily determined, either by using Pontrjagin's results as abelian topological groups or our present methods. They are the locally Minkowskian spaces which are topologically products of a finite number of straight lines and circles, see G pp.351, 352.

All two dimensional G-spaces with transitive groups of motions are known. The interest of this case lies in the fact that it revealed <u>a new hitherto unnoticed geometry</u>. Obvious examples of G-surfaces with transitive groups of motions are : the spherical and the elliptic geometries, the plane, cylinder and tori with Minkowskian metrics, and, of course, the hyperbolic plane. However, the latter possesses a generalization which in some respects is similar to the generalization of the euclidean plane to the Minkowskian. It can be defined as follows :

The transformations

(10.4) $\qquad x' = \alpha x + \beta \;, \quad y' = \alpha y \;, \quad \alpha > 0 \;, \quad \beta$ real

act transitively on the upper half plane $y > 0$. <u>The socalled quasi-hyperbolic geometries are the metrizations of $y > 0$ as G-spaces for which the transformations (10.4) are motions</u>.

The line element of such a geometry has the form

(10.5) $\qquad ds = F(dx, dy)/y \;,$

where $F(u,v)$ satisfies the conditions for a Minkowski metric. ($F(u,v) > 0$ for $(u,v) \neq (0,0)$. $F(ku, kv) = |k| \cdot F(u,v)$, $F(u,v)$ is convex). If we make differentiability hypotheses (10.5) is obvious. If $H(x,y,dx,dy)$ is invariant under (10.4), then applying $x' = x + \beta$, $y' = y$ we see that H does not depend on x . Applying $x' = \alpha x$, $y' = \alpha y$ we find $H(y,dx,dy) = H(\alpha y, adx, ady) = \alpha H(\alpha y, dx, dy)$. Differentiation with respect to α gives

$$0 = H(\alpha y, dx, dy) + \alpha y H_y(\alpha y, dx, dy)$$

hence for $\alpha = 1$

$$H_y(y,dx,dy)/H(y,dx,dy) = - 1/y \quad \text{or}$$

$$\log H = -\log y + \log F(dx,dy)$$

or (10.5). Not every $F(u,v)$ leads to a G-space. In contrast to the Minkowskian case, which would lead one to expect that $F(u,v)$ must be strictly convex, we find that <u>$F(u,v)$ must be differentiable for $(u,v) \neq (0,0)$ and strictly convex</u>, so to say, <u>only in one direction</u>: the supporting lines of the form $v = \text{const}$ of $C : F(u,v) = 1$ must touch C in only one point.

The space is then straight. The geodesics are found in a most interesting way which elucidates the wellknown hyperbolic case $ds = (dx^2 + dy^2)^{1/2}/ky$. One solves the following <u>isoperimetric problem</u> : in (u,v)-plane with the ordinary area given by $\int du dv$ find among all closed curves of a given Minkowski length $\int F(u',v')dt$ those which bound the maximal area. <u>These curves are homothetic to C only when perpendicularity is symmetric</u>.

In general they will be new closed convex curves (with centers) which are, of course, all homothetic. If we write one solution in the form $J(u,v) = 1$ with $J(ku,kv) = |k| J(u,v)$, then <u>the geodesics are the curves</u>

$$J(x - \beta , y) = k, \ y > 0 ,$$

<u>and in addition their semitangents at their intersections with the x-axis</u>. That these are parallel follows from the construction of $J(u,v) = 1$ and the fact that supporting lines of C parallel to the x-axis touches C in one point only.

A quasi-hyperbolic geometry does in general not have negative curvature, not are the loci C_T^{α} $(xT \leq \alpha)$ convex. <u>But there is</u> (under a very weak differentiability hypothesis) <u>exactly one angular</u>

measure so normalized that straight angles have measure π and invariant under (10.4) for which the Gauss-Bonnet Theorem holds. This means the following : the excess of a geodesic triangle is defined as the sum of the angles minus π , which leads to a finitely additive set function defined in polygonal regions. This set function can be extended to a completely additive set function defined on all Borel sets. For then the Gauss-Bonnet Theorem can be established in its usual form. References for quasi-hyperbolic geometry are G Sections 51, 52 and Busemann [3] .

11. DIFFERENTIABILITY.

Although the theory of G-spaces shows that a large part of differential geometry does not depend on differentiability, two questions arise naturally : 1) are there theorems which can be formulated and are false without, but correct with differentiability hypotheses, 2) can differentiability be formulated within the natural frame work of G-spaces, i.e. without postulating the existence of coordinates?

Normal coordinates on a differentiable Finsler manifold can be obtained in a purely geometric way as follows. Consider a sphere $S(p, \rho(p))$. For $0 < t < 1$ let $p_t = p$ and for $a \neq p$ let $(pa_t a)$ and $pa_t = tpa$. Then for any two points a, b in $S(p, \rho(p))$

$$(11.1) \qquad m_p(a,b) = \lim_{t \to 0+} a_t b_t / t$$

exists and is part of a Minkowski metric, the normal Minkowski metric at p . Obviously $m_p(p,a) = pa$, so that the $T(p,a)$ are segments both for ab and $m_p(a,b)$. Normal coordinates are affine coordinates $x^1,\dots,x^n$ in which $m_p(x,y)$ takes the standard form $F(y-x)$ with p corresponding to $(o,\dots,o)$.

In Section 4 we constructed an example where the plane is metrized as a G-space (with the ordinary lines as geodesics, but non-Desarguesian planes with this property also exist, see Busemann [5] where dilations from one point p exist and

$$ab = a_t b_t / t \quad \text{for all} \quad t > 0 .$$

Therefore the existence $\lim a_t b_t / t$ does not guarantee that $m_p(a,b)$ is Minkowskian. However, a slight strengthening of the condition will lead to the desired result. Consider three distinct points q, a, b in $S(p, \rho)$ and define a_t, b_t, by $(qa_t a)$, $tqa_t = qa$ and $(qb_t b)$, $tqb_t = qb$, $0 < t < 1$.

Differentiability means always a locally linear behaviour: We say therefore that <u>the space is continuously differentiable at p</u> if for any sequence of distinct points q^ν, a^ν, b^ν tending to p and $0 < t_\nu < 1$ the relation

$$\lim_{\nu\to} a^\nu_{t_\nu} b^\nu_{t_\nu} / t_\nu a^\nu b^\nu = 1 \tag{11.2}$$

holds.

Under these hypotheses it is easy to prove that the limit (11.1) exists and that the metric $m_p(a,b)$ approximates ab locally in the sense that

$$m_p(a^\nu, b^\nu)/a^\nu b^\nu \to 1 \quad \text{if} \quad a^\nu \to p,\ b^\nu \to p \text{ and } a^\nu \neq b^\nu. \tag{11.3}$$

Moreover $m_p(a,b)$ can be extended beyond $S(p, \rho(p))$ to a metric space in which our axioms I to IV hold.

If $m_p(a,b)$ also satisfies Axiom V , then the metric is Minkowskian. V expresses the <u>Legendre Condition</u> of the calculus of variation. Thus

(11.4) <u>If the G-space is continuously differentiable at P</u> (i.e. satisfies (11.2) <u>and</u> $m_p(a,b)$ <u>satisfies Axiom V , then</u> $m_p(a,b)$ <u>is part of a Minkowski metric which approximates ab in the sense of (11.3)</u>.

These and the following results are found in Busemann [5]. In principle at least we can now decide whether the given space is a smooth Finsler space. The space must be continuously differentiable everywhere. If $x^1,\ldots,x^n$ are normal coordinates at p and $y^1,\ldots, y^n$ are normal coordinates at q , then a point x which also lies in the neighborhood covered by the y is expressible as

$$x^i = f^i(y^1,\ldots,y^n)$$

and we can postulate that the x^i are of class C^k for $y \neq (0,\ldots,0)$ (i.e. $y \neq q$) and for $f(y) \neq 0$. It is known that in the latter cases we can, in general, only attain class C^1. If the given space is optimally, i.e. with a suitable coordinatization, is of class C^k, $k \geqslant 4$, then it is of class at least C^{k-2} in terms of normal coordinates (because introducing normal coordinates requires solving the differential equations of the extremals). Thus we reach <u>a partial decision for the smoothness of a given G-space for finite C^k and a complete decision for C^∞</u>.

At the same time we find a theorem formulable but false without but correct with differentiability hypotheses.

Call similarity of a G-space R a mapping α of R on itself satisfying

$$x\alpha\, y\alpha = kxy\ , \quad k > 0\ , \quad k \neq 1\ .$$

Then α^{-1} is also a similarity, so that we may assume $k < 1$. The relations $x\alpha^{\nu}\, y\alpha^{\nu} = k^{\nu}\, xy$ and $x\alpha^{\nu}\, x\alpha^{\nu-1} = k^{\nu}\, x\alpha x$ show that $x\alpha^{\nu}$ is a Cauchy sequence and hence converges to a point f , moreover $y\alpha^{\nu}$ f for any other point y . Therefore f is a fixed point of α and the only one. The space is straight. To see this we have to show that for any two distinct points x,y a point z with (xyz) exists. For large ν the points $x\alpha^{\nu}$ and $y\alpha^{\nu}$ lie in $S(p, \rho(f))$ and hence a point z_{ν} with $(x\alpha^{\nu}\, y\alpha^{\nu} z_{\nu})$ will exist. Then $(xyz_{\nu}\alpha^{-\nu})$ because all distances are multiplied by $k^{-\nu}$.

We know from the mentioned example that a G-space may possess similarity without being Minkowskian. However

(11.5) <u>If the G-space R possesses a similarity α and is continuously differentiable at the fixed point f of α, then R is Minkowskian.</u>

According to our results on spaces with non-positive curvature it suffices to prove that if a,b,c are distinct and b',c' are the midpoints of a,b and a,c respectively, the relation

$$2b'c' = bc$$

holds, i.e. that the curvature is zero.

Putting generally $x_{\nu} = x\alpha^{\nu}$ we have

$$e'_{\nu} c'_{\nu} \neq b_{\nu} c_{\nu} = k\, b'_{\nu-1} c'_{\nu-1} / k\, b_{\nu-1} c_{\nu-1} = b'_{\nu-1} c'_{\nu-1} / b_{\nu-1} c_{\nu-1}$$

Moreover b'_{ν}, c'_{ν} are the midpoints of a_{ν}, b_{ν} and a_{ν}, c_{ν} respectively. Therefore (11.2) gives with $t_{\nu} = 1/2$ that

$$b'_{\nu} c'_{\nu} / 2^{-1} b_{\nu} c_{\nu} \quad 1\ ,$$

and therefore $2b'c' = bc$.

The theorem (11.5) is certainly not a deep contribution to differential geometry, but it is the result of a long and arduous search for a theorem where differentiability does not enter the assertion, but is nevertheless necessary. The fact that it is apparently hard to find such theorems indicates <u>that much of differentiability is not in the same way thoroughly understood as other parts of mathematics are</u>. The theory of convex surfaces and more generally of surfaces with bounded curvature by A.D.Alexandrov [1], Pogorelov and others exhibits the same phenomena even for classical surface theory. Moreover, in both cases the synthetic methods have yielded results which seem inaccessible to the classical methods.

I hope that I have shown you that there is a very large field for future research.

REFERENCES

Alexandrov, A.D., [1] Die innere Geometrie der konvexen Flächen, Berlin 1955.

Auslander, L., [1] On curvature in Finsler geometry Trans. Am. Math. Soc.79 (1955), pp.378-388.

Blaschke, W., [1] Räumliche Variationsprobleme mit symmetrischer Transversalitätsbedingung, Ber.Sächs.Akad.Wiss. Leipzig 68 (1916) pp.50-55.

Busemann, H., [1] Local metric geometry, Trans.Math.Am.Soc. 56 (1944) pp.200-276.

[2] The geometry of geodesics, New York 1955, quoted as G .

[3] Quasihyperbolic geometry, Rend.Circ. Mat. Palermo Ser.II, 4 (1955) pp.1-14.

[4] Metrizations of projective spaces, Proc. Am. Math. Soc. 8 (1957), pp.387-390.

[5] Similarities and differentiability, Tohòku Math. J. Sec. Ser. 9 (1957), pp.56-67.

Busemann, H. and F.P.Pedersen [1] Tori with one-parameter groups of motions, Math.Scand. 3 (1955) pp.209-220.

Cartan, E., [1] Les espaces de Finsler, Paris 1934.

[2] Leçons sur la geometrie des Espaces de Riemann, 2nd ed., Paris 1951.

Hedlund, G.A. [1] Geodesics on a two-dimensional Riemann manifold with periodic coefficients, Ann.Math.33 (1932), pp.719-739.

Hopf, E., [1] Closed surfaces without conjugate points, Proc.Nat. Acad.Sci. U.S.A. 34 (1938) pp.47-51.

Kelly, P.J. and L.J.Paige [1] Symmetric perpendicularity in Hilbert geometries, Pacific J. Math. 2 (1952) pp.319-322.

Kelly, P.J. and E.G.Strauss [1] Curvature in Hilbert geometry, Pacific J. Math. 8 (1958) pp.119-126.

Morse, M. [1] A fundamental class of geodesics on any closed surface of genus greater than 1, Trans. Am. Math. Soc. 26 (1924) pp.25-60.

Pedersen, F.P. [1] On spaces with negative curvature, Mat. Tidsskift B 1952, pp.66-89.

Petty, C.M. [1] On the geometry of the Minkowski plane, Rivista Mat. Univ. Parma 6 (1955) pp.269-292.

Radon, J. [1] Über eine besondere Art ebener konvexer Kurven, Ber.Sächs.Akad.Wiss. Leipzig 68 (1916), pp.131-134.

Rhodes, F. [1] Homogeneity and isotropy in geodesic spaces, Quart. J. Math. Oxford (2) 9 (1958) pp.55-62.

Rund, H., [1] The differential geometry of Finsler spaces, Berlin-Göttingen-Heidelberg 1959.

Skornyakov, L.A. [1] Metrization of the projective plane in connection with a given system of curves (Russian) Izo. Akad. Nauk. SSSR, Ser. Mat. 19 (1955) pp.471-482.

Szenthe, J. [1] Über ein Problem von H.Busemann, Publ. Math. 7 (1960), pp.408-413

Tits, J. [1] Sur certaines classes d'espaces homogènes de groupes de Lie, Acad. roy. Belgique, Mem. Coll. in 8°, 39 fasc. 3, 1955.

Wang, H.C. [1] Two theorems on metric spaces, Pacific J. Math. 1 (1951), pp.473-480.

[2] Two-point homogeneous spaces, Ann. Math. (1952), pp.177-191.

Zaustinsky, E.M. [1] Spaces with non-symmetric distances, Mem. Am. Math. Soc. No 34 (1959).

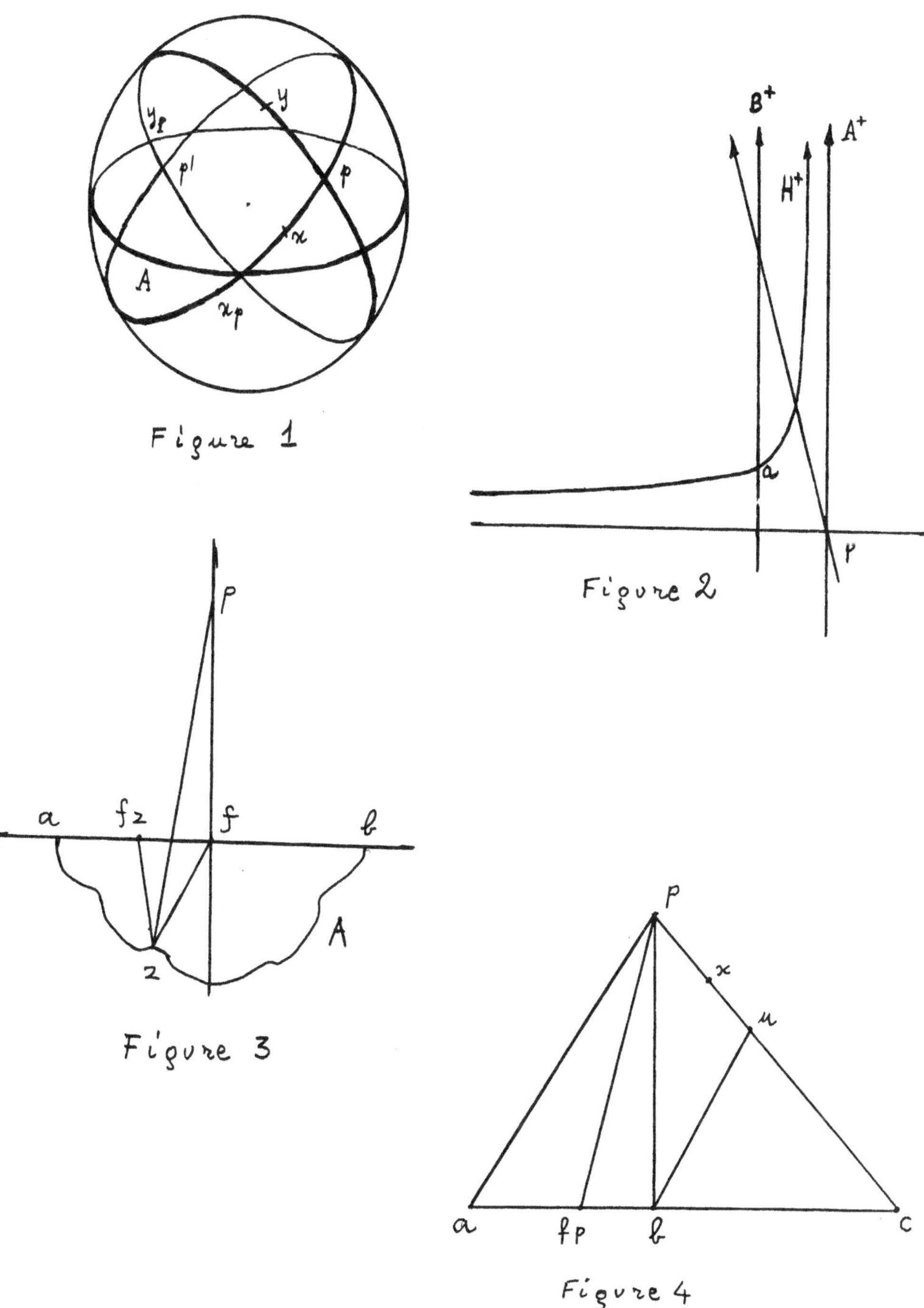

Figure 1

Figure 2

Figure 3

Figure 4

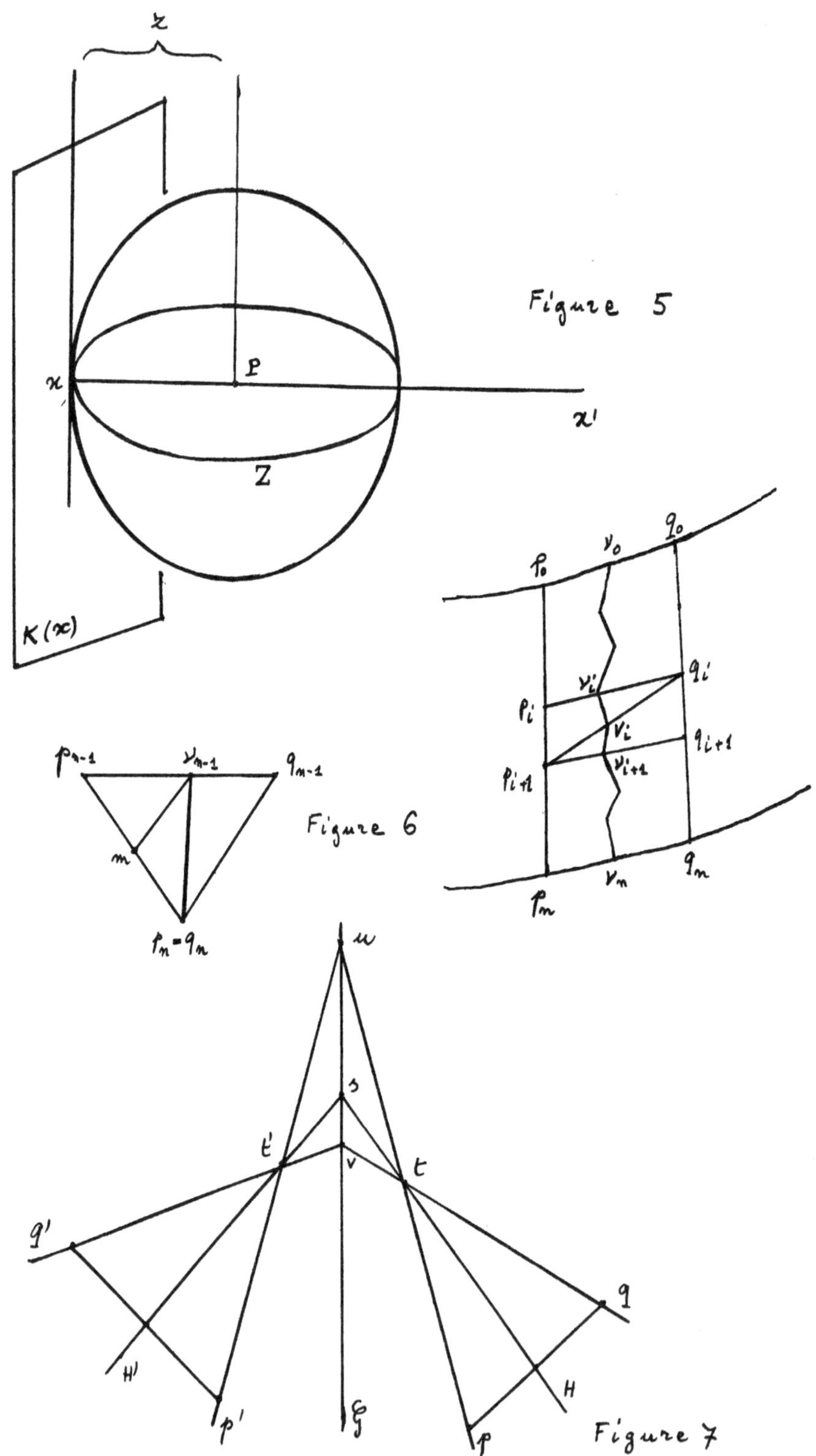

z
x
P
x'
Z
K(x)
Figure 5
p0
v0
q0
qi
vi'
pi
vi
qi+1
pi+1
vi+1
pn
vn
qn
pn-1
vn-1
qn-1
m
pn=qn
Figure 6
u
s
t'
v
t
q'
q
H'
H
p'
g
p
Figure 7

CENTRO INTERNAZIONALE MATEMATICO ESTIVO

(C.I.M.E.)

E. T. DAVIES

VEDUTE GENERALI SUGLI SPAZI VARIAZIONALI

ROMA - Istituto Matematico dell'Università

E.T.Davies

I LEZIONE

In queste lezioni vogliamo considerare alcuni concetti geometrici che si possono associare ai problemi di calcolo delle variazioni. Per questo scopo è necessario avere i problemi in forma parametrica. Per integrali semplici la teoria fu sviluppata da Weierstrass.

Cerchiamo le condizioni perchè un integrale sia indipendente da ogni trasformazione di parametro per un integrale della forma $\int_{t_1}^{t_2} L(t, x(t), \dot{x}(t))dt$ lungo una curva C le cui equazioni parametriche sono : $x^i = x^i(t)$.

Se facciamo una trasformazione $t \rightarrow \tau$ l'integrale corrispondente

$$\int_{\tau_1}^{\tau_2} L(\tau , x(\tau), \dot{x}(\tau))d\tau$$

deve essere uguale al primo per ogni trasformazione $t \rightarrow \tau$ per ogni posizione di un punto terminale per ogni curva regolare in cui L è di una data classe.

E' dimostrato nei libri classici di Calcolo delle variazioni che per la trasformazione semplice $t = \tau + c$ (c costante) l'uguaglianza non sussiste se t si trova esplicitamente nella funzione L . Un'altra trasformazione semplice $t = k\tau$ $(k > o)$ darà in modo analogo la condizione che L deve essere positivamente omogenea di primo grado nelle $\dot{x}^i(t)$, cioè

$$(1) \qquad \dot{x}^i \frac{\partial L}{\partial \dot{x}^i} = L \quad o \quad \dot{x}^i L_i = L .$$

Derivando rispetto alle $\dot{x}^j$ e badando all'omogeneità avremo

$$(2) \qquad \dot{x}^i \frac{\partial^2 L}{\partial \dot{x}^i \partial \dot{x}^j} = 0 \qquad \dot{x}^i L_{ij} = 0 \, ;$$

di qui segue che il determinante $|L_{ij}|$ deve essere di rango n-1 al massimo. Quando il rango è massimo diremo che si tratta di un problema <u>regolare</u> di Calcolo delle variazioni.

Il problema fondamentale di calcolo delle variazioni è di trovare fra tutti gli archi di curva C congiungenti due punti A e B , quello che rende minimo il valore dell'integrale $\int_A^B L(x, \dot{x})dt$.

Per ottenere le condizioni in forma analitica consideriamo un sistema $x^i = x^i(t, e)$ ad un parametro e, che ricopra univocamente una piccola regione nella vicinanza di una curva C_o, e tale che C_o sia la curva $x^i = x^i(t, o)$.

Vogliamo esprimere le condizioni perchè l'integrale

$$(3) \quad I(e) = \int_{t_1(e)}^{t_2(e)} L\left[x(t,e), \dot{x}(t,e)\right] dt \, , \qquad \dot{x} = \frac{\partial x}{\partial t}$$

con punti terminali variabili sia minimo relativo per e = 0, cioè tale che $I'(0) = 0$, $I''(0) > 0$.

Svolgendo il calcolo, dobbiamo avere

$$(4) \qquad I'(0) = \left[L \frac{dt}{de}\right]_{t_1}^{t_2} + \int_{t_1}^{t_2} \frac{\partial L}{\partial e} dt$$

con

$$(5) \qquad \frac{\partial L}{\partial e} = \frac{\partial x^i}{\partial e} \frac{\partial L}{\partial x^i} + \frac{\partial \dot{x}^i}{\partial e} \frac{\partial L}{\partial \dot{x}^i} \, .$$

Introducendo le notazioni

$$(6) \qquad E_i(L) \equiv \frac{\partial L}{\partial x^i} - \frac{d}{dt}\frac{\partial L}{\partial \dot{x}^i}\,, \quad \left(\frac{\partial x^i(t,e)}{\partial e}\right)_{e=0} = \eta^i(t) = \eta^i$$

possiamo scrivere

$$(7) \qquad I'(0) = \left[L\frac{dt}{de} + L_i\,\eta^i\right]_{t_1}^{t_2} + \int_{t_2}^{t_1} \eta^i E_i dt$$

Siccome alle estremità le $t_1(e)$, $t_2(e)$ sono funzioni di e, i punti A, B si muovono sulle curve $x^i = x^i(t_1(e),e)$ e $x^i = x^i(t_2(e),e)$. Quindi ai punti terminali variabili avremo

$$(8) \qquad \frac{dx^i}{de} = \frac{\partial x^i}{\partial e} + \frac{\partial x^i}{\partial t}\frac{dt}{de} = \eta^i + \dot{x}^i\frac{dt}{de}\,.$$

Sostituendo nell'equazione (7) e badando alla relazione $\dot{x}L_i = L$, si ottiene

$$(9) \qquad I'(0) = \left[L_i\frac{dx^i}{de}\right]_{t_1}^{t_2} + \int_{t_1}^{t_2} \eta^i E_i\, dt\,.$$

Se nei punti A, B la condizione $L_i\frac{dx^i}{de} = 0$ è valida, si dice che A e B si muovono su curve trasversali alla curva C_0. Se l'integrale nel secondo membro si deve annullare per ogni variazione η^i, dobbiamo avere le equazioni ben note (di Eulero):

$$(10) \qquad E_i \equiv \frac{\partial L}{\partial x^i} - \frac{\partial^2 L}{\partial \dot{x}^i \partial x^j}\dot{x}^j - \frac{\partial^2 L}{\partial \dot{x}^i \partial \dot{x}^j}\ddot{x}^j = 0\,.$$

Le n equazioni (10) non sono tutte indipendenti. Si può facilmente verificare l'identità

$$(11) \qquad \dot{x}^i E_i \equiv 0\,.$$

Osserviamo che nell'espressione (9) sotto l'integrale

la componente tangenziale del vettore η^i non contribuisce al valore dell'integrale in virtù dell'identità $\dot{x}^i E_i \equiv 0$.

Nella parte fuori dell'integrale della (9) invece si vede che non può sussistere la trasversalità se il vettore $\frac{dx^i}{de}$ ha una componente tangenziale alla curva.

Per la seconda variazione avremo :

$$I''(0) = \left[\frac{d}{de}\left(L\frac{dt}{de}\right) + \frac{\partial L}{\partial e}\frac{dt}{de}\right]_{t_1}^{t_2} + \int_{t_1}^{t_2} \frac{\partial^2 L}{\partial e^2}\, dt \tag{12}$$

dove si può mettere

$$\frac{\partial^2 L}{\partial e^2} = 2\Omega(\eta, \dot{\eta}) + E_i \frac{\partial \eta^i}{\partial e} + \frac{d}{dt}\left(\frac{\partial L}{\partial \dot{x}^i}\frac{\partial \eta^i}{\partial e}\right) \tag{13}$$

con

$$2\Omega(\eta,\dot{\eta}) = \frac{\partial^2 L}{\partial x^i \partial x^j}\eta^i\eta^j + 2\frac{\partial^2 L}{\partial x^i \partial \dot{x}^j}\eta^i\dot{\eta}^j +$$

$$+ \frac{\partial^2 L}{\partial \dot{x}^i \partial \dot{x}^j}\dot{\eta}^i\dot{\eta}^j = \eta^i\frac{\partial \Omega}{\partial \eta^i} + \dot{\eta}^i\frac{\partial \Omega}{\partial \dot{\eta}^i} =$$

$$= \eta^i\left(\frac{\partial \Omega}{\partial \eta^i} - \frac{d}{dt}\frac{\partial \Omega}{\partial \dot{\eta}^i}\right) + \frac{d}{dt}\left(\frac{\partial \Omega}{\partial \dot{\eta}^i}\eta^i\right).$$

Nel caso di punti terminali fissi e per una curva che soddisfi a $E_i = 0$, avremo

$$I''(0) = \int_{t_1}^{t_2} 2\Omega(\eta,\dot{\eta})\, dt . \tag{14}$$

Potremmo considerare il problema di trovare, nello spazio $\eta^1,\dots,\eta^n$, le condizioni perchè la curva $\eta^i = \eta^i(t,0)$ del sistema $\eta^i = \eta^i(t,e)$ renda minimo l'integrale (14). Le corrispondenti equazioni di Eulero sono :

$$J_i \equiv \frac{\partial \Omega}{\partial \eta^i} - \frac{d}{dt}\frac{\partial \Omega}{\partial \dot{\eta}^i} = 0 \tag{15}$$

Il problema di Calcolo delle variazioni associato alla funzione Ω è il cosidetto problema variazionale <u>accessorio</u> del problema L . Sia $x^i = \varphi^i(t,e)$ un sistema ad un parametro di soluzioni delle $E_i = 0$, in modo che se al posto di x , $\dot{x}$ nelle espressioni E_i mettiamo i valori corrispondenti a $\varphi^i(t,e)$, abbiamo $E_i \equiv 0$. Si può verificare facilmente che

(16) $$\frac{\partial}{\partial e}\left(\frac{\partial L}{\partial x^i}\right) \equiv \frac{\partial \Omega}{\partial \eta^i} , \frac{\partial}{\partial e}\left(\frac{\partial L}{\partial \dot{x}^i}\right) \equiv \frac{\partial \Omega}{\partial \dot{\eta}^i}$$ cioè

(17) $$J_i \equiv \frac{\partial E_i}{\partial e} \equiv 0$$ per un sistema di soluzioni di $E_i = 0$.

Cioè si verifica che le funzioni $\eta^i(t) = \left(\frac{\partial \varphi}{\partial e}\right)_{e=o}$ sono soluzioni delle cosidette "equazioni di Jacobi" $J_i = 0$.

Le equazioni $J_i = 0$ non sono tutte indipendenti. Lungo un estremale si può derivare l'identità $\dot{x}^i E_i \equiv 0$ rispetto al parametro e , ottenendo $\dot{x}^i J_i + \eta^i E_i \equiv 0$, cioè per $E_i = 0$, si ha

(18) $$\dot{x}^i J_i \equiv 0$$

E' ovvio che per punti terminali fissi la seconda variazione è data da

(19) $$I''(0) = \int_{t_1}^{t_2} \eta^i J_i \, dt$$

espressione che mette in evidenza l'importanza dell'espressione J_i per il segno della seconda variazione.

Sia $\psi^i(t)$ un sistema di soluzioni dell'equazione di Jacobi per cui $\psi^i(t_1) = \psi^i(\tau) = 0$ con $t_1 < \tau < t_2$. Diciamo in questo caso che il punto "τ" è "coniugato" al punto t_1.

Si può dimostrare che se esiste un tal punto nell'intervallo t_1 t_2 , la curva che soddisfa alle condizioni $E_i = 0$ non può essere una curva minimizzante per ogni variazione $\eta^i(t)$ possibile.

Secondo la teoria generale delle equazioni differenziali del 2° ordine, vi sarà un sistema ad (n-1) parametri di soluzioni del sistema $E_i = 0$ (di cui (n-1) sono indipendenti) passante per un punto dato A . La rappresentazione di questo sistema di soluzioni può essere allora

$$x^i = x^i(t, y^1, y^2, \ldots, y^{n-1}) \tag{20}$$

tale che in una regione che non include il punto A e nemmeno uno dei punti coniugati ad A , avremo un "campo" di estremali rappresentato dal sistema (20).

Per ogni punto della regione passerà una curva estremale ed una sola. Il "principio di Jacobi" ci insegna che, se l'estremale g è caratterizzata da $y^\alpha = 0$ ($\alpha = 1, \ldots, n-1$), allora le funzioni $\eta^i_\alpha(t) \equiv \left(\frac{\partial x^i}{\partial y^\alpha}\right)_{y=0}$ saranno soluzioni di $J_i = 0$ per $\alpha = 1, \ldots, n-1$. Se nel punto A il valore di t è t_0, e se consideriamo lo jacobiano

$$\frac{\partial(x^1, \ldots, x^n)}{\partial(t, y^1, \ldots, y^{n-1})} \tag{21}$$

possiamo dire che i punti coniugati del punto $t = t_0$ dell'estremale g sono i punti di g corrispondenti a valori di $t \neq t_0$ per cui lo jacobiano (21) sarà zero.

Si possono prendere $y^1, y^2, \ldots, y^{n-1}, y^n \equiv t$ come nuove coordinate in una regione per cui (21) non è zero. L'estremale g sarà l'asse y^n. Colla corrispondenza

$$(x^1, x^2, \ldots, x^n) \longrightarrow (y^1, \ldots, y^{n-1}, t) = (y^1, \ldots, y^n)$$

avremo

$$L(x,\dot{x}) = L(y^1, \ldots, y^n; \dot{y}^1, \ldots, \dot{y}^n) = \underset{\cdot}{L}(y^\alpha, t; \dot{y}^\alpha, 1) = f(t, y^\alpha, \dot{y}^\alpha)$$

cioè si passa dalla forma parametrica alla forma non parametrica dell'integrale. Si possono confrontare i risultati espressi nelle due forme. Se facciamo il calcolo delle condizioni di Eulero relativo al problema con f invece di L , avremo (n-1) equazioni indipendenti

$$E_\alpha \equiv \frac{\partial f}{\partial y^\alpha} - \frac{d}{dt} \frac{\partial f}{\partial \dot{y}^\alpha} = 0 \tag{22}$$

Abbiamo già fatto cenno alle condizioni di Eulero. Vi sono ancora due condizioni : quelle di Legendre e di Weierstrass, che si riferiscono ai punti di un estremale.

Secondo la condizione di Legendre dobbiamo avere

$$\frac{\partial^2 L}{\partial \dot{x}^i \partial \dot{x}^j} \lambda^i \lambda^j > 0 \qquad \text{per} \qquad \lambda^i \neq k\, \dot{x}^i \tag{23}$$

se l'arco g è minimizzante. La condizione di Legendre per la forma non parametrica si scrive

$$\frac{\partial^2 f}{\partial \dot{y}^\alpha \partial \dot{y}^\beta} w^\alpha w^\beta > 0 . \tag{24}$$

Abbiamo già detto che un problema di calcolo di variazione è regolare se il determinante $|L_{ij}|$ è di rango n-1. Una condizione equivalente è che la funzione $F_1 \neq 0$ deve soddisfare

(25) $$F_1(x, \dot{x}) = \frac{A^{ij}\, \dot{x}^i\, \dot{x}^j}{(\dot{x}^i\, \dot{x}^j)^2}$$

A^{ij} essendo il cofattore di L_{ij} nel determinante $|L_{ij}|$. La forma equivalente della condizione per la forma non parametrica è che il determinante

$$\frac{\partial^2 f}{\partial \dot{y}^\alpha \partial \dot{y}^\beta} \neq 0 .$$

La funzione eccesso di Weierstrass è la funzione

$$E(x, \dot{x}, p) = L(x,p) - L(x,\dot{x}) - (p^i - \dot{x}^i) \dot{\partial}_i L(x,\dot{x})$$

che si può scrivere, in vista dell'omogeneità, come

(26) $$E(x, \dot{x}, p) = L(x, p) - p^i \dot{\partial}_i L(x, \dot{x})$$

La condizione di Weierstrass è che $E(x, \dot{x}, p)$ sia positiva per ogni vettore $p \neq \dot{x}$ ai punti di un estremale.

La condizione di Jacobi è che non esiste un punto coniugato del punto iniziale A dell'arco g per un minimo valore dell'integrale.

Le condizioni di Legendre e di Weierstrass sono di natura strettamente "locale". E' la condizione di Jacobi che si presta a problemi di Calcolo delle variazioni in grande. Prendiamo il caso in cui la funzione $L(x, \dot{x})$ è la lunghezza d'arco di una curva $x^i = x^i(t)$ in uno spazio riemanniano V_n. Scriviamo

(27) $$2F = L^2 = \left(\frac{ds}{dt}\right)^2 = g_{ij}(x)\, \dot{x}^i\, \dot{x}^j$$

da cui deduciamo

(28) $\dot{\partial}_i F = F_i = L\, L_i = g_{ij}\, \dot{x}^j$

(29) $F_{ij} = L\, L_{ij} + L_i\, L_j = g_{ij}$

Scrivendo $l^i = \dfrac{\dot{x}^i}{L}$ per il vettore unitario nella direzione della tangente alla curva, avremo anche $L_i = g_{ij} l^j = l_i$. Segue da (29) che

(30) $L_{ij} = L^{-1}\,(g_{ij} - l_i\, l_j)$.

Sia λ^i un vettore qualunque indipendente da x ; allora la forma $L_{ij}\, \lambda^i\, \lambda^j$ che compare nella condizione di Legendre diviene, ricordando che le $l_i\, \lambda^i$ è $\lambda \cos\vartheta$, dove ϑ è l'angolo tra i vettori $\dot{x}^i$ e λ^i

(31) $L_{ij}\, \lambda^i \lambda^j = L^{-1}\; \lambda^2\, (1 - \cos^2\vartheta\,) > 0$.

In modo analogo la funzione eccesso di Weierstrass si esprime

(32) $E(x,\, \dot{x},\, \lambda\,) \equiv L(x,\, \lambda\,) - \lambda^i \dot{\partial}_i L(x,\, \dot{x}) = \lambda\,(1 - \cos\vartheta\,) > 0$

Per uno spazio riemanniano con una forma differenziale positiva definita le condizioni di Legendre e di Weierstrass sono soddisfatte.

Prendiamo ancora l'espressione (9) per la prima variazione di un integrale. Nel caso in cui $E_i = 0$ avremo

(33) $\delta I = \left[l_i\, \delta x^i \right]_2 - \left[l_i\, \delta x^i \right]_1 = |\delta x|_2 \cos\vartheta_2 - |\delta x|_1 \cos\vartheta_1$

Ciò esprime il fatto che se un arco geodetico AB diviene un arco geodetico A'B' infinitamente vicino, la prima variazione della lunghezza sarà

(34) $\delta I = BB' \cos(AB, BB') - AA' \cos(AB, AA')$.

Nel calcolo delle variazioni per un integrale semplice si passa da una funzione integranda $L(x, \dot{x})$ ad un sistema di equazioni differenziali ordinarie del 2° ordine che danno le curve estremali dell'integrale $\int L(x, \dot{x})dt$. Fin dal 1920, vi sono due sviluppi paralleli che si possono illustrare in un diagramma :

$\int ds$ nello spazio di Riemann con $ds^2 = g_{ij}(x)dx^i dx^j$ ——— geodetiche

Spazio di Weyl a connessione affine coi coefficienti dipendenti dalle sole x : $\Gamma^i_{jk}(x)$ ——— geometria dei cammini di Eisenhart e Veblen $\ddot{x}^i + \Gamma^i_{jk}(x)\dot{x}^j\dot{x}^k = 0$

Spazio di Finsler fondato sulla funzione $L(x, \dot{x})$ con una connessione dipendente da $\Gamma^i_{jk}(x, \dot{x})$ ——— spazio generale dei cammini di Douglas $\ddot{x}^i + 2G^i(x,\dot{x}) = 0$ con $\Gamma^i_{jk}(x,\dot{x})$

Spazio di Kawaguchi fondato sulla funzione $L(x, \dot{x}, \ldots, x^{(n)})$ ——— teoria geometrica delle equazion[i] $x^{(n)i} + H^i(x, \dot{x}, \ldots, x^{(n-1)}) = 0$

conducente alla teoria degli <u>estensori</u> di Craig.

Spazio fondato su un integrale multiplo colla funzione integranda $L\left(x^i, \frac{\partial x^i}{\partial t^\alpha}\right)$ ——— teoria dei "K-spreads" di Douglas con un sistema di equazioni $\frac{\partial^2 x^i}{\partial t^\alpha \partial t^\beta} + H^i_{\alpha\beta}\left(x^i, \frac{\partial x^i}{\partial t^\alpha}\right) = 0$

Un problema inverso di quello che abbiamo considerato in questo capitolo sarebbe quello di partire da un sistema di equazioni differenziali e di cercare un problema di calcolo delle variazioni di cui sono le soluzioni. E' il cosidetto problema inverso del Calcolo delle variazioni risoluto da Douglas in una memoria dei Transactuns of the American Mathematical Society 1941.

Nella geometria dei cammini si tratta di una teoria invariantiva di un sistema di equazioni come

$$\frac{d^2x^i}{dt^2} + 2\, G^i(x, \frac{dx}{dt}) = 0 \tag{35}$$

relativo a trasformazioni (a) delle coordinate $x \longrightarrow \bar{x}$(b) del parametro $t \longrightarrow \bar{t}$. Un sistema come (35) avrà come soluzione generale un sistema di curve a 2n-2 parametri $a_1, \ldots, a_{2n-2}$ della forma

$$x^i = f^i(t, a)\ . \tag{36}$$

Douglas ha studiato il sistema (36) in tutta generalità rispetto ai tre gruppi di trasformazioni

$$\text{(a)} \quad \bar{x} = \bar{x}\,(x) \quad ; \quad \text{(b)} \quad \bar{t} = \bar{t}\,(t,\, a) \quad ; \quad \text{(c)} \quad \bar{a} = \bar{a}\,(a).$$

Il terzo gruppo non viene in considerazione quando si studiano le equazioni differenziali dei cammini, e lo studio è ristretto per la maggior parte a quello relativo ad un gruppo G della classe (a) e un gruppo Γ della classe (b). Quando si tratta delle proprietà invarianti per G e Γ insieme, si parla della geometria (G, Γ) dei cammini.

Prendiamo il caso in cui G è arbitrario, Γ il gruppo

$\bar{t} = \alpha t + \beta$ dove α e β sono funzioni arbitrarie dei parametri a; siamo nel caso della geometria <u>affine</u> dei cammini.

Il gruppo Γ ($\bar{t} = \frac{\alpha t + \beta}{\gamma t + \delta}$ con α, β, γ, δ funzioni di a) conduce alla geometria <u>proiettiva</u> dei cammini.

Il caso particolare della geometria affine in cui $\alpha \equiv 1$ cioè Γ è il gruppo $\bar{t} = t + \beta$, conduce alla cosidetta geometria <u>metrica</u> dei cammini. Vogliamo far cenno ad una proprietà interessante di questa geometria.

Differenziando due volte le equazioni (36) rispetto a t, avremo

$$\dot{x}^i \equiv p^i = f'^i(t, a) \tag{37}$$

$$\ddot{x}^i = f''^i(t, a) \tag{38}$$

Le equazioni (36) e (37) insieme sono 2n equazioni che ci daranno i 2n-1 parametri t, a come funzioni di x, p, e di più daranno una relazione in x, p della forma

$$\theta(x, p) = 0 \tag{39}$$

dove $\theta(x,p)$ non è omogenea nelle p, altrimenti vi sarebbe una restrizione sui rapporti delle p, che contraddice al fatto che esiste un cammino per ogni punto in ogni direzione. Sostituendo nelle (38) i valori ottenuti per t ed a dalle (36) e (37) otterremo un sistema di equazioni differenziali della forma :

$$\ddot{x}^i + \psi^i(x, p) = 0 . \tag{40}$$

Si può dimostrare che quando sussiste (39), si possono

trovare funzioni $F(x,p)$ e $G^i(x,p)$ omogenee del secondo grado nelle p, e tali che

$$(41) \qquad F(x,p) \equiv F(x, \dot{x}) = 1$$

$$(42) \qquad \ddot{x}^i + 2\,G^i(x, p) = 0$$

Differenziando la (41) rispetto a t, e utilizzando (42) avremo :

$$(43) \qquad \frac{\partial F}{\partial x^i}\, p^i - 2\,\frac{\partial F}{\partial p^i}\, G^i = 0$$

relazione che significa che una qualunque delle n equazioni (42) è una conseguenza delle (n-1) altre insieme a (43).

Inversamente, dato un sistema di equazioni differenziali della forma (41) e (42), per cui sussiste (43), l'integrale generale di queste equazioni avrà la forma

$$x^i = f^i\,(t + \beta\,(a),\ a)$$

poichè il sistema è invariante per le trasformazioni $\bar{t} = t + \beta$. Supponiamo di avere nello spazio dei cammini una metrica definita da $ds^2 = F(x, dx)$; allora da (41) abbiamo che $dt^2 = F(x, dx)$, cioè $ds^2 = dt^2$, ossia $s = \pm\, t + \beta$, cioè il parametro t è la lunghezza presa da un punto arbitrario su ogni cammino nei due sensi.

Abbiamo visto allora che la funzione $F(x, \dot{x})$ che si presenta in modo naturale nella geometria <u>metrica</u> dei cammini può stare alla base di una geometria di Finsler, e che dal sistema dei cammini si arriva anche ad un problema di calcolo delle variazioni relativo alla funzione $L(x, \dot{x}) = (2F)^{1/2}$.

E.T.Davies

LEZIONE II

QUALCHE RISULTATO PER GLI SPAZI RIEMANNIANI

Prendiamo il caso in cui

$$(1) \qquad 2F = L^2 = \left(\frac{ds}{dt}\right)^2 = g_{ij}(x)\,\frac{dx^i}{dt}\,\frac{dx^j}{dt}$$

si verifica facilmente che sussiste

$$(2) \qquad E_i(F) = L\,E_i(L) - \frac{dL}{dt}\,L_i$$

in modo che $E_i(L) = 0$ può scriversi

$$(3) \qquad F_{ij}\,\ddot{x}^j + \left(\frac{\partial^2 F}{\partial \dot{x}^i \partial x^j} - \frac{\partial F}{\partial x^i}\right) - \frac{dL}{dt}\,\frac{\partial L}{\partial \dot{x}^i} = 0$$

Questa forma è più comoda perchè $F_{ij} = g_{ij}(x)$ il cui determinante non svanisce come quello di L_{ij}. Se mettiamo

$$(4) \qquad \omega^i = \ddot{x}^i + \left\{ {i \atop jk} \right\} \dot{x}^j\,\dot{x}^k$$

Un calcolo facile ci darà

$$(5) \qquad E_i(F) = -\,\omega^i \qquad \text{e} \qquad \frac{dL}{dt} = L_i\,\omega^i$$

Le equazioni di Eulero $E_i(L) = 0$ divengono

$$(6) \qquad (g_{ij} - l_i l_j)\,\omega^j = 0$$

colle equazioni corrispondenti $E^i(L) = 0$ nella forma

$$(7) \qquad \omega^i - \frac{d(\log L)}{dt}\,\dot{x}^i = 0$$

che si può scrivere anche nelle due forme equivalenti

(8) $\dot{x} \wedge \omega = 0 \qquad$ o $\qquad \dot{x}^{[i}\,\omega^{j]} = 0$

Siccome dalla prima lezione abbiamo $J_i = \dfrac{\partial E_i(L)}{\partial e}$ un calcolo diretto ci darà per $J_i = 0$ l'equazione

(9) $(g_{ij} - l_i l_j)\,\phi^j = 0$ dove

$$\phi^i = \frac{D^2\eta^i}{dt^2} + R^i_{j\ kl}\,\dot{x}^j\dot{x}^k\,\eta^l$$

Le equazioni corrispondenti $J^i = 0$ sono

(10) $(\delta^i_j - l^i l_j)\,\phi^j = 0$

o, in forma equivalente

(11) $\dot{x} \wedge \phi = 0$

Nel caso in cui il parametro t è la lunghezza d'arco s , abbiamo $L \equiv 1$ lungo la curva, allora le equazioni (8) e (10) divengono rispettivamente $\omega^i = 0$ e $\phi^i = 0$. Le equazioni $\phi^i = 0$ sono conosciute come le equazioni dello scostamento geodetico di Levi-Civita.

Se facciamo le ipotesi che (a)t = s, (b) la variazione η^i è ortogonale alla tangente alla curva, avremo

(12) $I''(0) = \displaystyle\int_{s_1}^{s_2} \eta_i\, J^i\, ds = -$

$$= - \int_{s_1}^{s_2} \left[\eta_i \frac{D^2\eta^i}{ds^2} + R^i_{j\ kl}\,\dot{x}^j\,\eta_i\,\dot{x}^k\,\eta^l\right] ds$$

dove il nostro tensore di Riemann-Christoffel è

$$R^i_{j\,kl} = \frac{\partial \left\{ {i \atop jk} \right\}}{\partial x^l} - \frac{\partial \left\{ {i \atop jl} \right\}}{\partial x^k} + \left\{ {i \atop ml} \right\} \left\{ {m \atop jk} \right\} - \left\{ {i \atop mk} \right\} \left\{ {m \atop jl} \right\}$$

con

$$R_{jikl} = g_{im} R^m_{j\,kl}$$

Se mettiamo $\eta^i = \eta \mu^i$, dove μ^i è il vettore unitario nella direzione della variazione, avremo

$$\eta_i \frac{D^2 \eta^i}{ds^2} = \eta \ddot{\eta} + \eta^2 \mu_i \frac{D^2 \mu^i}{ds^2}$$
$$= \eta \ddot{\eta} - \eta^2 \bar{\mu}^2$$

con

$$\bar{\mu}^2 = g_{ij} \frac{D\mu^i}{ds} \frac{D\mu^j}{ds}$$

mentre il secondo termine dell'espressione sotto l'integrale è

$$\eta^2 R_{jikl} \dot{x}^j \mu^i \dot{x}^k \mu^l = \eta^2 K(l, \mu)$$

dove $K(l,\mu)$ è la curvatura riemanniana nella giacitura determinata dai vettori l e μ. Allora possiamo scrivere

$$I''(0) = - \int_{s_1}^{s_2} \eta \left[\ddot{\eta} + \eta (K - \bar{\mu}^2) \right] ds \tag{13}$$

Siccome vogliamo considerare il segno di $I''(0)$ per η^i arbitrario, prendiamo il caso in cui $\bar{\mu} = 0$, nel qual caso

$$I''(0) = - \int_{s_1}^{s_2} \eta \left[\ddot{\eta} + \eta K \right] ds \tag{14}$$

Nel seguito avremo bisogno del teorema fondamentale per spazi <u>completi</u> che si può esprimere come segue : <u>Due punti qualunque di una varietà completa possono essere congiunti da una geo-</u>

detica minimizzante.

Vogliamo adesso dimostrare il teorema di Myers : - Se la curvatura riemanniana $K(P,\gamma)$ per ogni punto P e per ogni giacitura γ di una V_n completa, soddisfa alla condizione $K \geqslant K_0 > 0$ allora un arco geodetico di lunghezza maggiore di $\pi/\sqrt{K_0}$ non può essere un arco geodetico minimizzante.

Sia l la lunghezza dell'arco geodetico, si ha allora

$$I''(0) = -\int_0^l \eta\,[\ddot{\eta} + \eta K]\,ds \leq -\int_0^l \eta\,[\ddot{\eta} + \eta K_0]\,ds \tag{15}$$

Siccome η è ancora arbitrario possiamo scegliere

$\eta = \operatorname{sen} \frac{\pi s}{l}$, che ci dà

$$I''(0) \leqslant \int_0^l \eta^2 \left[\frac{\pi^2}{l^2} - K_0\right] ds < 0 \quad \text{per} \quad l > \frac{\pi}{\sqrt{K_0}} \tag{16}$$

Cioè l'arco non è minimizzante. Ma secondo il teorema fondamentale per spazi completi esiste sempre una geodetica minimizzante. Abbiamo dimostrato allora che la geodetica minimizzante (sempre esistente) dev'essere di lunghezza minore di $\pi/\sqrt{K_0}$ che sarà il "diametro" della varietà. Siccome la massima "distanza" tra due punti è finita, si conclude che la varietà è compatta e chiusa. Allora : - Una varietà completa a curvatura positiva $K \geqslant K_0 > 0$ è chiusa con un diametro $\pi/\sqrt{K_0}$.

Un altro risultato di Myers si riferisce alla curvatura media di Ricci di una V_n completa. Siano $\mu^i_{(\alpha)}$ ($\alpha = 1,2,\ldots,n-1$) vettori unitari normali alla curva geodetica, e tali che siano tutti trasportati per parallelismo lungo la curva. La curvatura media di Ricci è definita da $K_M = \sum_{\alpha=1}^{n-1} K(1,\mu_\alpha)$ e supponiamo che

$K_M \geqslant K_o > 0$. Nell'equazione (14) prendiamo $\eta = \operatorname{sen} s\sqrt{\left(\frac{K_0}{n-1}\right)}$ per ogni α che ci da

$$(17) \qquad I''(0) \equiv J(\eta_\alpha) = \int_0^1 \eta^2\left[\frac{K_0}{n-1} - K(1,\mu_\alpha)\right]ds$$

Prendendo la somma per $\alpha = 1,\dots, n-1$ abbiamo

$$(18) \qquad \sum_{\alpha=1}^{n-1} J(\eta_\alpha) = \int_0^1 \eta^2 \left[K_0 - K_M\right] ds$$

La condizione $K_M \geqslant K_o$ conduce al risultato che $\sum J(\eta_\alpha) \lessgtr 0$ e vi sarà almeno un valore di α per cui $J(\eta_\alpha) < 0$ e per cui il corrispondente $I''(0) < 0$. Esiste allora un punto coniugato sull'arco, e non può essere minimizzante. Ma secondo il teorema fondamentale esiste sempre un arco minimizzante che avrà allora una lunghezza minore di $\pi\sqrt{\frac{n-1}{K_0}}$. Il diametro della V_n sarà dunque $\pi\sqrt{\frac{n-1}{K_0}}$.

Vogliamo far cenno ad un risultato di natura topologica sulle varietà V_n che abbiamo preso in considerazione. Sia R uno spazio topologico connesso (cioè tale che due punti qualsivoglia di R possono unirsi con un cammino). Sia P la totalità dei cammini chiusi che escono da un punto qualunque p di R . P è divisa in classi di curve equivalenti o omotope. Queste classi A, B,.. sono gli elementi del gruppo fondamentale G dello spazio. Uno spazio connesso R è semplicemente connesso se il suo gruppo fondamentale contiene solamente l'identità. In questo caso ogni curva chiusa è omotopa a zero.

Uno spazio R è localmente semplicemente connesso se, per ogni punto p e per ogni intorno U di p esiste un intorno $V \subset U$

dello stesso punto tale che ogni curva chiusa di origine p , e contenuta in V , sia omotopa a zero.

Uno spazio R è <u>localmente connesso</u> se per ogni punto p e per ogni intorno U di p esiste un intorno $V \subset U$ dello stesso punto tale che per $x \in V$ esiste un cammino in U che congiunge p e x .

Vogliamo adesso introdurre la nozione di <u>spazio semplicemente connesso di ricoprimento</u> per gli spazi che sono <u>connessi</u>, <u>localmente semplicemente connessi</u>, e <u>localmente connessi</u>. Sia P l'insieme di tutti i cammini dello spazio R che cominciano nel punto p .

Dividiamo P in classi equivalenti. Sia S l'insieme delle <u>classi</u>. Esiste una corrispondenza tra i punti di R e i punti dell'insieme S che non è (1,1) eccetto nel caso in cui R è semplicemente connesso. Ad ogni punto di S corrisponde un punto solo di R , ma ad ogni punto di R corrispondono più punti di S : A_1, A_2,.. .., dove i punti A_i corrispondono ad elementi del gruppo fondamentale. Lo spazio S è un ricoprimento per lo spazio R . Si può dimostrare che S è semplicemente connesso. S è lo <u>spazio semplicemente connesso di ricoprimento</u> di R .

Si può <u>metrizzare</u> lo spazio S quando R è uno spazio riemanniano V_n che supponiamo anche completo. In questo caso per due punti qualsivoglia a e b di V_n esiste una geodetica minimizzante che congiunge a e b. La lunghezza dell'arco sarà la distanza d(a,b) in V_n . Si può definire la distanza tra due punti corrispondenti a_i e b_i come uguale a d(a,b).

Ad ogni curva chiusa di V_n del tipo di omotopia corrispondente ad un elemento A del gruppo fondamentale e passante per un punto qualunque x di V_n corrisponde una curva che congiunge due punti distinti x' e x" dello spazio semplicemente connesso di ricoprimento $\bar{V}_n$. Ad ogni elemento A($\neq$ e) del gruppo fondamentale corrisponde allora una trasformazione $x' \longrightarrow x'' = A(x')$ dello spazio $\bar{V}_n$ su se stessa. Questa trasformazione è una trasformazione senza punti fissi, ed i punti che corrispondono nella trasformazione corrispondono allo stesso punto x di V_n . Dal modo in cui abbiamo metrizzato $\bar{V}_n$ la trasformazione è una isometria. Ad ogni elemento A del gruppo fondamentale corrisponde allora una isometria che indicheremo anche con A .

La proprietà di essere completo sarà valida anche per $\bar{V}_n$. Siccome la geometria di uno spazio riemanniano è determinata completamente dalla metrica, si può concludere che se V_n è a curvatura positiva, la $\bar{V}_n$ sarà anche a curvatura positiva giacchè le curvature riemanniane in punti corrispondenti sono uguali. Possiamo senz'altro enunciare teoremi corrispondenti a quelli di Myers per lo spazio semplicemente connesso di ricoprimento, cioè che quando V_n è chiuso, $\bar{V}_n$ è anche chiuso. Ad ogni elemento del gruppo fondamentale di V_n corrisponde una regione fondamentale di $\bar{V}_n$. Lo spazio $\bar{V}_n$ essendo compatto, il numero di queste regioni fondamentali, ed allora il numero degli elementi del gruppo fondamentale, sarà finito. Si può enunciare il teorema di Myers allora sotto la forma : - <u>Il gruppo fondamentale di uno spazio completo V_n a curvatura positiva è finito</u>.

Consideriamo adesso in $\bar{V}_n$ i punti $x' \subset V'$ e $x'' \subset V''$ con

$x'' = A(x')$, dove V' e V'' sono due regioni fondamentali di $\bar{V}_n$, e consideriamo tutte le curve di $\bar{V}_n$ che congiungono x' e x''. Secondo il teorema fondamentale sugli spazi completi esiste in $\bar{V}_n$ una sola geodetica di lunghezza minima da x' a x''. In generale la curva "chiusa" corrispondente in V_n avrà le due tangenti nel punto x non coincidenti, cioè ci sarà un angolo $\neq \pi$ tra le due tangenti. Diremo <u>geodetica chiusa</u> solamente per il caso in cui le due tangenti sono di direzioni opposte. Adesso arriviamo al teorema :

Lo spazio V_n ammette sempre una geodetica chiusa (nel senso ora definito) appartenente alla classe A di curve (corrispondente all'elemento $A(\neq e)$ del gruppo fondamentale).

Per la dimostrazione consideriamo l'espressione per la prima variazione (vedi la prima lezione) nel caso di un arco geodetico che comincia in un punto x di V_n e riviene ad un angolo $\neq \pi$. Se si sposta il punto x in un punto y l'arco geodetico corrispondente avrà subito una variazione data da $xy(\cos\beta - \cos\alpha)$ dove β è l'angolo tra xy e la direzione finale della tangente, mentre α è l'angolo tra xy e la direzione iniziale. Se invece l'angolo β è come nel diagramma, avremo la variazione

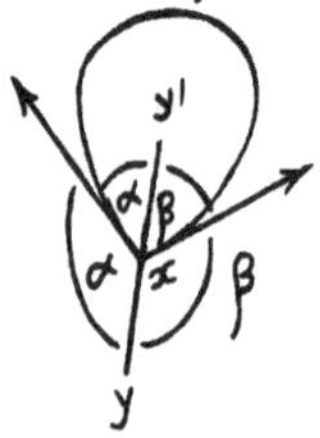

$$\delta I = - xy\ (\cos\alpha + \cos\beta)$$

Se la geodetica è chiusa, $\alpha + \beta = \pi$ ed allora la variazione sarà nulla. Se non è chiusa possiamo trovare punti y nella vicinanza di x tali che $\alpha > \pi/2$ e $\beta > \pi/2$ nel qual caso $\delta I > 0$. Possiamo anche trovare punti y' tali che $\alpha < \pi/2$ e $\beta < \pi/2$ nel qual caso $\delta I < 0$.

E' allora ovvio che $\delta I = 0$ solamente nel caso in cui $\cos\alpha + \cos\beta = 0$, cioè quando la geodetica è chiusa. La situazione corrispondente nello spazio $\bar{V}_n$ è che la distanza tra i punti x' e x" deve essere stazionaria. La funzione $f(x') \equiv d(x', x") \equiv \equiv d(x', A(x'))$ è una funzione continua che prende i suoi valori limiti. La distanza d(x', x") non può essere mai zero, allora f(x') prende il valore minimo in un punto ξ_1 di V' . La geodetica corrispondente nello spazio V_n sarà allora una geodetica chiusa g_1 . Ma in questo caso, siccome V_n è compatto, vi sarà anche un valore massimo di f(x') nel punto ξ_2 di V', e vi sarà una corrispondente geodetica chiusa g_2 .

E.T. Davies

III LEZIONE

SPAZI RIEMANNIANI OMEOMORFI AD UNA SFERA

In questa lezione vogliamo trattare di un problema posto per la prima volta da Rauch (1951). E' il problema di trovare i limiti della variabilità della curvatura riemanniana di uno spazio V_n a curvatura positiva tale che l'equivalenza topologica non sia perduta. E' un problema che è stato ripreso in considerazione recentemente da Klingenberg e da Berger. Il caso in cui la dimensione dello spazio riemanniano è pari presenta qualche semplificazione nelle dimostrazioni, e prendiamo questo caso nel seguito. Cominciamo con un teorema di Synge : <u>In uno spazio riemanniano orientabile completo a curvatura positiva e di dimensione pari non esiste una geodetica chiusa minimizzante</u>.

Supponiamo che g sia una tale geodetica chiusa. Scegliamo un punto p di g e un vettore μ ortogonale alla tangente a g che sarà trasportato per parallelismo lungo g , ritornando al punto p come un vettore unitario $'\mu$ che sarà in generale differente da μ . Siccome μ è preso arbitrariamente nello spazio vettoriale E_{n-1} ortogonale a g nel punto p, è ovvio che gli elementi di E_{n-1} subiscono una trasformazione ortogonale. Una trasformazione ortogonale di uno spazio euclideo lasciando invariante un punto p è una rotazione accompagnata o no da una simmetria. In questo caso prendiamo il caso in cui l'orientazione è conservata e si tratta di una rotazione. E' ben noto che in questo caso quando (n-1) è <u>dispari</u>, esiste un vettore μ_0 che non cambia nella trasformazione $\mu \longrightarrow {}'\mu$.

Prima di continuare avremo bisogno della seconda variazione della lunghezza d'arco geodetico quando i punti terminali si muovono lungo una varietà trasversale alla geodetica. L'espressione è

$$(1)\quad I''(0) = \left[l_i \frac{D^2x^i}{de^2} \right]_{s_1}^{s_2} + \int_{s_1}^{s_2} \left[g_{ij} \frac{D\eta^i}{ds} \frac{D\eta^j}{ds} - \eta^2 K(l,\mu) \right] ds$$

Consideriamo l'espressione fuori dell'integrale. Supponiamo che il punto s_1 sia fisso ma che il punto s_2 vari in una ipersuperficie della V_n data parametricamente da $x^i = x^i(v^1,\ldots,v^{n-1}) = x^i(v^\alpha)$ con $\alpha = 1, 2,\ldots, n-1$. La curva (di V_n) $x^i = x^i(s_2(e), e)$ è allora una curva $v^\alpha = v^\alpha(e)$ su V_{n-1}. Avremo

$$(2)\quad l_i \frac{D^2x^i}{de^2} = l_i\left(\frac{\partial^2 x^i}{\partial v^\alpha \partial v^\beta} + \left\{ {i \atop jk} \right\} \frac{\partial x^j}{\partial v^\alpha} \frac{\partial x^k}{\partial v^\beta} \right) \frac{dv^\alpha}{de} \frac{dv^\beta}{de} =$$

$$= a_{\alpha\beta} \frac{dv^\alpha}{de} \frac{dv^\beta}{de}$$

dove $a_{\alpha\beta}$ sono i coefficienti della seconda forma fondamentale di V_{n-1} in V_n . Se V_{n-1} è totalmente geodetica, il che possiamo supporre se il punto s_2 si muove lungo una geodetica (di V_n) ortogonale a g , allora il corrispondente termine fuori dell'integrale svanisce. Le espressioni per la prima e la seconda variazione della lunghezza d'arco coincidono allora con le espressioni per punti terminali fissi.

Prendiamo allora la geodetica chiusa g e spostiamo i punti di g lungo le geodetiche che escono da ogni punto di g nella direzione μ_0 definita in ogni punto di g . La seconda variazione della lunghezza l della geodetica chiusa g è allora data come nella seconda lezione da

$$(3) \qquad I''(0) = -\int_p^p \eta\left[\ddot{\eta} + \eta\, K\right] ds$$

Prendiamo $\eta = c$ (costante), nel qual caso la seconda variazione vale $-c^2\int Kds$ lungo g . Siccome $K > 0$ la geodetica è <u>massimante</u> piuttosto che <u>minimante</u>, il che dimostra il teorema di Synge. Di più Synge è arrivato al risultato che uno spazio V_n di cui trattiamo è <u>semplicemente connesso</u>, perchè abbiamo dimostrato nella seconda lezione che per ogni elemento $A(\neq e)$ del gruppo fondamentale esiste una geodetica chiusa <u>minimante</u>. Si conclude allora che non può esistere come altro elemento che l'identità del gruppo fondamentale, cioè lo spazio è semplicemente connesso.

Nelle considerazioni che seguono lo spazio V_n è completo, orientabile, di dimensione pari, di curvatura positiva $K \geqslant K_0 > 0$ in modo che valga il teorema di Synge che V_n è allora semplicemente connesso. Se pensiamo a tutte le geodetiche che escono da un punto p del V_n compatto, vi sarà un punto p' su ogni geodetica g con questa proprietà : Per un punto q sull'arco pp' di g la lunghezza d'arco pq è minimante mentre per q fuori dell'arco pp' questo fatto non vale più. Si può anche dire che, partendo da p lungo g , p' è l'ultimo punto per cui l'arco pq è minimante. Per un punto fisso p e g variabile, il punto p' giace su una sfera topologica che indichiamo con C(p). Per una superficie il luogo C(p) è stato studiato da Poincaré sotto il nome di "ligne de partage". Il luogo C(p) è stato ripreso in considerazione da Whitehead (1934) e da Myers (1935). Whitehead l'ha studiato per uno spazio di Finsler e lo chiama "cut-locus", mentre Myers l'ha studiato per il caso riemanniano a due dimensioni e lo chiama "minimum point locus". Il punto p' per cui la proprietà enunciata è soddisfatta può essere un punto coniugato. Allora cominciamo con qualche osservazione sui coniugati del punto p . Nel caso di una superficie V_2 , vi sarà una sola "equazione di Jacobi", alla quale si possono applicare i teoremi di confronto e di separazione di Sturm. Supponiamo che la curvatura gaussiana K della superficie soddisfi alla relazione $K_0 \leq K \leq K_1$ allora la distanza d(p,p') da p al suo primo punto coniugato p' sarà tale da soddisfare alla disuguaglianza

$$\frac{\pi}{\sqrt{K_1}} \leq d(p, p') \leq \frac{\pi}{\sqrt{K_0}} \tag{4}$$

Questi risultati sono generalizzati da Schoenberg e da Mosse ad un numero n di dimensioni. Il corrispondente teorema di confronto per un sistema di equazioni differenziali lineari del secondo ordine si può esprimere come segue : Nel caso del problema accessorio di un problema di Calcolo delle variazioni siano $J(\eta) = 2\int_{t_1}^{t_2}\Omega(t,\eta,\dot{\eta})dt$ e $J^*(\eta) = 2\int_{t_1}^{t_2}\Omega^*(t,\eta,\dot{\eta})dt$ le espressioni per la seconda variazione per due problemi I e I^*. Sia, nell'intervallo $t_1 \leq t \leq t_2$ la relazione $\Omega \leq \Omega^*$ soddisfatta. Allora il risultato è che i punti coniugati rispetto a $J(\eta)$ sono più vicini che i punti coniugati rispetto a $J^*(\eta)$. Più precisamente supponiamo che τ^* sia un punto coniugato di t_1 rispetto al problema J^*, allora se $\Omega \leq \Omega^*$ esiste un punto τ con $t_1 < \tau \leq \tau^*$ che sarà coniugato a t_1 rispetto al problema J . La dimostrazione è facile. Dalla definizione di τ^* vi sarà una soluzione $u^i(G)$ del sistema

$$J_i \equiv \frac{d}{dt}\frac{\partial \Omega^*}{\partial \dot{\eta}^i} - \frac{\partial \Omega^*}{\partial \eta^i} = 0 \quad \text{con} \begin{cases} u^i(t_1) = u^i(\bar{\tau}) = 0 \\ u^i(t) \neq 0 \,, \; t_1 < t < \bar{\tau} \end{cases}$$

da cui segue

$$(5) \qquad J^*(u) \equiv \int_{t_1}^{\tau^*} 2\,\Omega^* dt = \int_{t_1}^{\tau^*} u^i \left(\frac{\partial \Omega^*}{\partial u^i} - \frac{d}{dt}\frac{\partial \Omega^*}{\partial \dot{u}^i} \right) dt \equiv 0$$

ed allora per l'espressione corrispondente J avremo

$$(6) \qquad J(u) \leq J^*(u) \leq 0 \,. \qquad J(u) \leq J^*(u) \leq 0$$

Ma siccome siamo nel caso di un problema di calcolo delle variazioni in cui le condizioni di Weierstrass e di Legendre sono soddisfatte nel senso stretto, è ben noto che [Hadamard, § 288] : <u>se non esiste su</u> $t_1 \bar{\tau}$ <u>un punto coniugato a</u> t_1, <u>abbiamo sempre</u> $J(\eta) > 0$ <u>per ogni insieme</u> $\eta^i = \eta^i(t)$ <u>di funzioni di classe</u> D^1 <u>nell'intervallo</u> $t_1 \bar{\tau}$ (salvo nel caso banale $\eta^i \equiv 0$). Allora siccome abbiamo la disuguaglianza (6) deve esistere un punto τ coniugato a t_1 relativo al problema J nell'intervallo $t_1 < \tau \leq \bar{\tau}$. Nel caso della seconda variazione della lunghezza d'arco "l" in un V_n compatto, abbiamo

$$(7) \qquad J(\eta) = - \int_0^l \eta \left[\ddot{\eta} + \eta K \right] ds$$

e se $K_0 \leq K \leq K_1$ avremo $\overset{1}{J}(\eta) \leq J(\eta) \leq \overset{0}{J}(\eta)$ da cui deduciamo subito il risultato (4) per la distanza d(p,p'). Se pensiamo allora al <u>luogo</u> dei punti p' che sono coniugati ad un punto p per varie geodetiche g uscenti da p abbiamo subito che se, per ogni punto e per ogni giacitura in uno spazio sussiste $K_0 \leqslant K \leqslant K_1$ i punti coniugati sono tutti in una regione limitata dalle due sfere geodetiche di centro p e di raggi $\pi/\sqrt{K_1}$ e $\pi/\sqrt{K_0}$. Ma i punti del"cut locus" non sono necessariamente dei punti coniugati. Vogliamo enunciare alcune proprietà del C(p) che si trovano raccolte nella memoria di Klingeberg (Annals 1959) : (i) Sia p un punto qualunque del V_n di cui parliamo, e sia q un punto di C(p) che sta più vicino a p di tutti gli altri punti di C(p). Se q non coincide col punto p' coniugato a p rispetto alla geodetica minimizzante che congiunge p e q , allora q è il punto medio di una geodetica chiusa che comincia e che finisce nel punto p . (ii) Se q appartiene a C(p), allora p appartiene a C(q). (iii) La distanza d(p, C(p))è una funzione continua di p. (iv) Un limite inferiore della distanza d(p, C(p)) è dato dal minore dei due numeri seguenti (a) la minima distanza tra due punti coniugati qualunque di V_n (b) la metà della minima lunghezza di una geodetica chiusa di V_n . Sia p un punto di V_n e r un punto a distanza massima da p , allora r è un punto di C(p).

Facciamo l'osservazione che si può sempre normalizzare la metrica di uno spazio V_n in modo che il valore massimo K_1 della curvatura riemanniana sia l'unità. Scriviamo allora $0<\delta \leqslant K \leqslant 1$ invece di $0 < K_0 \leqslant K \leqslant K_1$. Enunciamo un teorema di Klingeberg.

Sia V_n una varietà riemanniana compatta, di dimensione pari, colla curvatura riemanniana soddisfacente alla disuguaglianza $0 < \delta \leq K \leq 1$. Allora sussistono (a) la lunghezza di una geodetica chiusa è almeno 2π (b) se due punti di V_n sono a distanza $< \pi$, esiste una geodetica minimizzante che li congiunge, ed allora si conclude che la distanza $d(p, C(p)) \geqslant \pi$ (c) il diametro $d(V_n)$ di V_n soddisfa alla relazione $\pi \leq d(V_n) \leq \pi/\sqrt{\delta}$

E' già stabilito nel teorema di Synge che la V_n di questo teorema è semplicemente connessa. Vogliamo adesso dimostrare che se $\delta \geqslant 1/4$, la V_n è omeomorfa alla sfera S_n. La dimostrazione è quella che si trova nella memoria di Berger (Annali di Pisa 1960). Avremo bisogno di un risultato di Topogonov [Doklady 1958] che è il seguente :

Siano p, q, r tre punti di una V_n di cui parliamo, e siano $\Gamma = \{\gamma(s)\}$ e $\Lambda = \{\lambda(s)\}$ due geodetiche di V_n tali che Γ congiunge p, q e Λ congiunge p, r , con $\gamma(0) = \lambda(0) = p$. Indichiamo con $S_2(\delta)$ la sfera a due dimensioni colla proprietà che tre punti $\hat{p}, \hat{q}, \hat{r}$ di $S_2(\delta)$ sono scelti con $\hat{d}(\hat{p}, \hat{q}) = d(p, q)$, $\hat{d}(\hat{p}, \hat{r}) = d(p, r)$ e tale che l'angolo $\hat{\alpha}$ in $\hat{p}$ della sfera verifica $\cos\hat{\alpha} = \langle \gamma'(0), \lambda'(0)\rangle$. Il risultato di Topogonov è che $d(q, r) \leq \hat{d}(\hat{q}, \hat{r})$. Segue come corollario che se $\delta > 1/4$, $d(p, q) \geqslant \pi$, $d(p, r) \geqslant \pi$ e $\cos\hat{\alpha} = \langle \gamma'(0), \lambda'(0)\rangle \geqslant 0$ allora $d(q, r) < \pi$.

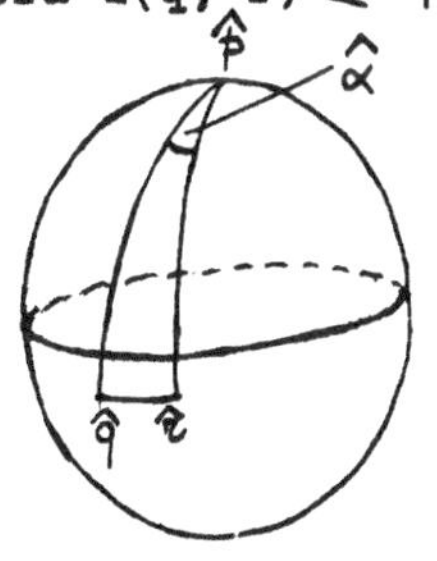

Con $\delta > 1/4$ il raggio di S_2 è $1/\sqrt{\delta}$ che sarà <2 . Il cerchio all'equatore è di circonferenza $2\pi/\sqrt{\delta}$ di cui la quarta parte sarà $1/2\ \pi/\sqrt{\delta}$ cioè $< \pi$ allora $d(q, r) \leq \hat{d}(\hat{q}, \hat{r}) < \pi$.

Nella dimostrazione del corollario abbiamo utilizzato il fatto che $\cos\hat{\alpha} > 0$. Berger ha dimostrato che si può sempre far l'ipotesi che $\cos\hat{\alpha} > 0$. Il lemma di Berger di cui avremo bisogno è il seguente : Siano p, q due punti tali che $d(p, q) = d(V_n)$, allora, per un vettore qualunque X dello spazio tangente T_p nel punto p , esiste una geodetica $\Lambda = \{\lambda(s)\}$ tale che Λ congiunge p e q , con $\lambda(0) = p$, e $\langle \lambda'(0), X \rangle \geqslant 0$. Un secondo lemma di Berger che ci occorre è il seguente : Per $\delta > 1/4$, e $d(V_n) \geqslant \pi$ siano p, q due punti tali che $d(p, q) = d(V_n)$. Allora, per un punto qualunque r di V_n avremo $d(r, p) < \pi$ e $d(r, q) < \pi$.

Per dimostrarlo prendiamo $\Gamma = \{\gamma(s)\}$ come la geodetica congiungente p ed r . Nel primo lemma di Berger sia $X = \gamma'(0)$ allora esiste una geodetica $\Lambda = \{\lambda(s)\}$ con $\lambda(0) = p$ e tale che $\langle \gamma'(0), \lambda'(0) \rangle \geqslant 0$. Se $d(r, p) < \pi$ il lemma è dimostrato. Se $d(r, p) \geqslant \pi$, siccome $d(p, q) = d(V_n) \geqslant \pi$ il corollario al teorema di Topogonov ci mostra che $d(r, q) < \pi$ come abbiamo voluto dimostrare.

Adesso abbiamo i mezzi necessari per dimostrare che la varietà V_n compatta, orientabile, di dimensione pari, colla curvatura riemanniana K soddisfacente a $0 < \delta \leq K \leq 1$, è omeomorfa ad una sfera S_n .

Prendiamo due punti p e q come nel secondo lemma di Berger, cioè con $d(p, q) = d(V_n)$. Su una geodetica qualunque $\Gamma = \{\gamma(s)\}$, $\gamma(0) = p$, uscente da p esiste sempre un sol punto m per cui $d(p, m) = d(m, q)$. Ciò si dimostra subito perchè se esiste un altro punto m' con $d(p, m') = d(m', q)$ supponiamo che m sia sul segmento pm'. Allora avremo un triangolo mm'q per cui $d(q, m') < d(q, m) + d(m, m')$, ma secondo la nostra ipotesi $d(q,m') = d(m',p) = d(pm) + d(mm') = d(q,m) + d(m,m')$. La supposizione che esista un secondo punto m' ci conduce allora ad un assurdo. Cioè il punto m è unico.

Sia C l'insieme di tutti i segmenti pm , e D l'insieme di tutti punti qm .

<u>La varietà</u> V_n <u>è coperta da</u> C <u>e</u> D , un fatto che si dimostra facilmente osservando che per un punto qualunque r di V_n, dalla proprietà fondamentale di varietà completa esistono segmenti geodetici congiungenti r con p e con q . Se $d(p, r) < d(q, r)$ esiste un arco geodetico dell'insieme pm passante per r con $d(pm) = d(qm) < \pi$, se $d(p,r) = d(q, r)$ allora $r = m$. Se $d(pr) > d(qr)$ allora r sta su di un segmento qn con $d(qn) = d(p,n)$. Abbiamo la conclusione che ogni punto di V_n appartiene a C o a D .

Sia S_n la sfera di raggio unità della dimensione n . Vogliamo definire un omeomorfismo φ di S su V_n . Facciamo prima un'applicazione del polo nord u di S sul punto p di V_n, e del polo sud v di S sul punto q di V_n . Avendo scelto un'applicazione lineare $d\varphi$ dello spazio tangente ad S nel punto u sullo spazio tangente a V_n nel punto p , è ovvio come si possa definire un omeomorfismo φ dell'emisfero nord di S sull'insieme C di V_n . Basta porre una corrispondenza tra gli archi di cerchi ue dal punto u (polo nord) ad un punto "e" sull'equatore e gli archi geodetici pm che costituiscono l'insieme C . Siccome i punti m del contorno di C sono anche i punti n del contorno di D , l'applicazione φ può estendersi ad un omeomorfismo dell'emisfero sud di S sull'insieme D di V_n , facendo corrispondere gli archi di cerchi vf della sfera e gli archi qn di D . Siccome V_n è semplicemente connesso, φ è un omeomorfismo globale.

E.T.Davies

IV LEZIONE

LO SPAZIO DI FINSLER

Tratteremo adesso di qualche concetto relativo alla teoria degli spazi di Finsler. Prendiamo (vedi il libro di Busemann) una varietà differenziale di dimensione n e di classe C^r $(r \geq 4)$. Sui vettori contravarianti $(x, \xi) = (x^1,\dots,x^n;\ \xi^1,\dots,\xi^n)$ sia definita una funzione $L(x, \xi)$ di classe C^4 che può determinare la lunghezza $\Lambda(\varphi)$ di una curva $\varphi(t) = (\varphi^1(t), \varphi^2(t),\dots,\varphi^n(t))$ di classe C^1 da $t_1 \leq t \leq t_2$ come

$$\Lambda(\varphi) = \int_{t_1}^{t_2} L\left[\varphi(t), \varphi'(t)\right] dt$$

che sarà la lunghezza nel senso della geometria riemanniana se

$$L^2(x, \xi) = g_{ij}(x)\, \xi^i \xi^j$$

Vogliamo che la funzione $L(x, \xi)$ soddisfi anche alle condizioni seguenti

(i) $L(x, \xi) > 0$ per $\xi \neq 0$

(ii) $L(x, k\xi) = |k|\, L(x, \xi)$

(iii) $L(x, \xi) + L(x, \eta) \geq L(x, \xi + \eta)$

(iv) La superficie $L(x_0, \xi) = 1$ nello spazio $\xi^1,\dots,\xi^n$ è a curvatura gaussiana positiva.

Possiamo definire la distanza $d(p,q)$ tra due punti come il numero

$$d(p,q) = \inf_{\varphi(t)} \Lambda(\varphi)$$

Con questa definizione della distanza lo spazio diviene uno spazio metrico. Prendiamo come un'altra condizione

(v) Lo spazio è completo rispetto alla metrica. Queste condizioni individuano lo spazio di Finsler.

Per sviluppare una geometria è necessario avere delle curve estremali o geodetiche i cui archi definiscono la distanza d(p,q) tra due punti in un certo intorno limitato. Le condizioni di Weierstrass e di Legendre corrispondono alle condizioni (iv) sopra indicate.

Nella seconda lezione abbiamo dato [II (3)] le condizioni $E_i(L) = 0$ in una forma in cui si può prendere la funzione $F = \frac{1}{2} L^2$ invece di L . Se scriviamo $g_{ij}(x, \dot{x}) = F_{ij}$ dove g_{ij} in questo caso è una funzione di x e di $\dot{x}$ che sarà omogenea del grado zero nelle $\dot{x}$, abbiamo che $g_{ij} \dot{x}^i \dot{x}^j = 2F$. Si può dire allora che la lunghezza del vettore $\dot{x}^i$ è $L(x, \dot{x})$. Sussistono le seguenti conseguenze della condizione di omogeneità :

$$\frac{\partial^2 F}{\partial \dot{x}^i \partial x^j} \dot{x}^i = 2 \frac{\partial F}{\partial x^j} \quad ; \quad \frac{\partial^3 F}{\partial \dot{x}^i \partial \dot{x}^j \partial x^k} \dot{x}^i = \frac{\partial^2 F}{\partial \dot{x}^j \partial x^k}$$

da cui segue che

$$\frac{\partial^2 F}{\partial \dot{x}^i \partial x^j} \dot{x}^i = \frac{\partial F_{ik}}{\partial x^j} \dot{x}^j \dot{x}^k = \frac{\partial g_{ik}}{\partial x^j} \dot{x}^j \dot{x}^k$$

Nello stesso modo

$$\frac{\partial F}{\partial x^i} = \frac{1}{2} \frac{\partial^2 F}{\partial \dot{x}^j \partial x^i} \dot{x}^j = \frac{1}{2} \frac{\partial g_{jk}}{\partial x^i} \dot{x}^j \dot{x}^k$$

Da queste relazioni si può facilmente verificare che le curve estremali per la funzione L come integranda saranno date da $E_i(L) = 0$ che si può scrivere nella forma

$$(1) \qquad \omega^i - \frac{d(\log L)}{dt} \dot{x}^i = 0$$

dove $\omega^i = \ddot{x}^i + \gamma^i_{jk} \dot{x}^j \dot{x}^k$ (γ^i_{jk} essendo i simboli di Christoffel) come nel caso dello spazio di Riemann. Una forma equivalente delle equazioni (1) è la seguente

$$(2) \qquad \omega^i \wedge \dot{x}^j = 0$$

Per il caso in cui $L(x, \dot{x}) = 1$, cioè in cui $t = s$ è la lunghezza d'arco, le equazioni (1) prendono la forma più semplice

$$(3) \qquad \omega^i \equiv \frac{d^2 x^i}{ds^2} + \gamma^i_{jk} \frac{dx^j}{ds} \frac{dx^k}{ds} = 0$$

che coincide colla forma ottenuta per lo spazio riemanniano. In questo caso però le funzioni γ sono funzioni delle $\dot{x}$ del grado zero, mentre nel caso riemanniano sono funzioni delle sole x . Nel caso riemanniano i simboli di Christoffel γ conducono ad una definizione di derivazione covariante perchè hanno la solita legge di trasformazione per coefficienti di connessione per una trasformazione $x^i \longrightarrow \bar{x}^\lambda$

$$(4) \qquad \frac{\partial^2 \bar{x}^\lambda}{\partial x^i \partial x^j} + \bar{\Gamma}^\lambda_{\mu\nu}(\bar{x}) \frac{\partial \bar{x}^\mu}{\partial x^i} \frac{\partial \bar{x}^\nu}{\partial x^j} = \Gamma^k_{ij}(x) \frac{\partial \bar{x}^\lambda}{\partial x^k}$$

La legge di trasformazione per $\gamma^i_{jk}(x, \dot{x})$ però è complicato, e ci resta il problema di trovare funzioni di x e $\dot{x}$ che

(i) avranno la trasformazione (4) in modo che da un campo di vettori $T^i(x,\dot{x})$ possiamo dedurre un tensore misto T^i_{ij} che sarà la derivata covariante di T^i, (ii) il corrispondente trasporto per parallelismo conserva la lunghezza di un vettore, il che è equivalente a richiedere che la derivata covariante del tensore fondamentale $g_{ij}(x, \dot{x})$ svanisca, (iii) le curve geodetiche che realizzano un valore estremale dell'integrale $\int Ldt$ sono anche curve autoparallele rispetto alla connessione definita dalle funzioni che cerchiamo.

Le tre condizioni accennate non bastano per determinare univocamente un sistema di funzioni che possono funzionare come coefficienti di connessione, ma saranno sufficienti per determinarle a meno di un tensore emisimmetrico del terzo ordine.

E.Cartan, nella sua monografia sullo spazio di Finsler, mette come base della sua teoria (i) l'espressione $g_{ij}(x,\dot{x})dx^i dx^j$ (ii) l'espressione $DT^i = dT^i + T^k(\Gamma^i_{kh}\, dx^h + C^i_{kh}\, d\dot{x}^h)$ dove Γ sono funzioni omogenee del grado zero, e C sono omogenee del grado -1 nelle $\dot{x}$. Per determinare le espressioni (i) e (ii) dalla funzione fondamentale $L(x, \dot{x})$ Cartan introduce quattro condizioni di natura intrinseca, che sono le seguenti

A. La lunghezza di un vettore ξ^i nella direzione del cosidetto "elemento d'appoggio" è $L(x, \xi)$.

B. Sia $\bar{D}X^i$ il differenziale assoluto di un vettore X^i quando le sue componenti X^i rimangono fisse, l'origine del vettore rimane fisso quando il suo elemento di appoggio subisce una piccola rotazione attorno al suo centro. Abbiamo per due vettori qualsiansi X^i e Y^i con lo stesso elemento di appoggio

$$\vec{X}.\overline{D\vec{Y}} = \vec{Y}.\overline{D\vec{X}}$$

C. Se il vettore X^i ha la stessa direzione del suo elemento d'appoggio $\overline{D\vec{X}} = 0$.

D. Se un vettore X^i si sposta in modo che il suo elemento di appoggio rimanga parallelo a se stesso e se $\overset{*}{\Gamma}{}^i_{jk}$ sono le funzioni che figurano nell'espressione

$$DX^i = dX^i + \overset{*}{\Gamma}{}^i_{kj} X^k dx^j$$

le $\overset{*}{\Gamma}$ sono simmetriche rispetto agli indici inferiori.

Siccome Cartan prende sempre il caso della connessione <u>euclidea</u>, cioè per cui $Dg_{ij} = 0$ le condizioni

$$(5) \quad \begin{cases} \dfrac{\partial g_{ij}}{\partial x^h} = g_{ik}\Gamma^k_{jh} + g_{jk}\Gamma^k_{ih} = \Gamma_{jih} + \Gamma_{ijh} \\[2ex] \dfrac{\partial g_{ij}}{\partial \dot{x}^h} = g_{ik}C^k_{jh} + g_{jk}C^k_{ih} = C_{jih} + C_{ijh} \end{cases}$$

sono sempre soddisfatte. La condizione B ci dà subito la simmetria delle C_{ijh} rispetto ai primi due indici, ed allora abbiamo subito $C_{ijh} = \frac{1}{2}\dfrac{\partial g_{ij}}{\partial \dot{x}^h}$. Le funzioni $\overset{*}{\Gamma}$ hanno l'espressione come funzioni di Γ e di C

$$(6) \qquad \overset{*}{\Gamma}{}^j_{ih} = \Gamma^j_{ih} - C^j_{ir}\dot{x}^k\Gamma^r_{kh}$$

e sono le funzioni che figurano nelle espressioni per la derivata covariante di un tensore. Queste funzioni $\overset{*}{\Gamma}$ coincidono coi simboli a tre indici di Christoffel per uno spazio riemanniano <u>osculatore</u> lungo una curva come è stato dimostrato da Varga. Per ottenerle cominciamo con un tensore $g_{ij}(x,\xi)$ dipendente dalle 2n variabili x^i, ξ^i e omogeneo del grado zero nelle ξ. Se le ξ^i sono funzioni di x in una regione avremo uno spazio riemanniano coi coefficienti $\gamma_{ij}(x) \equiv g_{ij}(x,\xi(x))$ del tensore metrico. Ma senza ridursi a questo caso speciale possiamo determinare le ξ^i lungo una curva arbitraria da una relazione <u>non-integrabile</u> tra i differenziali come

$$d\xi^i + \Gamma^i_j(x,\xi)dx^j = 0 \tag{7}$$

dove $\Gamma^i_j(x,\xi)$ sono funzioni arbitrarie di classe C^3 nelle 2n variabili x e ξ. Se indichiamo con $\xi[x]$ il fatto che ξ^i saranno funzioni di x solamente <u>lungo la curva</u> avremo

$$\frac{\partial \gamma_{ij}(x,\xi[x])}{\partial x^k} = \frac{\partial g_{ij}}{\partial x^k} + \frac{\partial g_{ij}}{\partial \xi^m}\frac{\partial \xi^m}{\partial x^k} = \tag{8}$$

$$= (\partial_k - \Gamma^m_k \dot{\partial}_m) g_{ij}$$

Se scriviamo

$$X_k \equiv \frac{\partial}{\partial x^k} - \Gamma^m_k \frac{\partial}{\partial \xi^m} \equiv \partial_k - \Gamma^m_k \dot{\partial}_m \tag{9}$$

i simboli di Christoffel di prima specie per lo spazio riemanniano osculatore corrispondente alle relazioni (7) sono

$$2\hat{\gamma}_{ijh} = X_h g_{ij} + X_i g_{jh} - X_j g_{ih} \qquad \text{con } \hat{\gamma}^k_{ih} = g^{kj}\hat{\gamma}_{ijh} \tag{10}$$

Per ogni sistema di funzioni Γ^i_j allora avremo uno spazio riemanniano osculatore, e le funzioni $\hat{\gamma}^k_{ih}$ corrispondenti avranno due delle proprietà necessarie (a) di avere la legge di trasformazione (4) (b) di dare coefficienti di una connessione euclidea nel senso che $g_{ij/k} = 0$. Vogliamo però che le funzioni Γ^i_j che sono arbitrarie siano determinate in modo che (c) si può dedurle dalla funzione fondamentale $L(x, \xi)$ (d) le curve autoparallele siano le curve estremali del problema $\int Ldt$.

Queste due condizioni sono soddisfatte se prendiamo

$$(11) \qquad \Gamma^i_j = \dot{\partial}_j G^i \qquad \text{dove} \qquad 2G^i = \gamma^i_{mn}(x, \xi)\, \xi^m \xi^n .$$

Le $\hat{\gamma}^i_{jk}$ corrispondenti a questa scelta delle Γ^i_j sono precisamente le funzioni $\overset{*}{\Gamma}{}^i_{jk}$ di Cartan. In ogni punto lungo una curva arbitraria esiste allora uno spazio riemanniano osculatore γ_{ij} coi simboli di Christoffel uguali alle $\overset{*}{\Gamma}$ di Cartan. In questo punto il relativo spazio riemanniano avrà una curvatura riemanniana che scriviamo $\overset{*}{R}_{ijkl}\, \lambda^i \mu^j \lambda^k \mu^l$ per una giacitura determinata dai vettori λ e μ che supponiamo unitari. La relazione tra il tensore $\overset{*}{R}{}^i_{jkl}$ e il tensore di curvatura R^i_{jkl} che appare nella monografia di Cartan è la seguente

$$(12) \qquad R^i_{jkh} = R^i_{jkh} + C^i_{jm}\, \xi^n R^m_{nkh}$$

con

$$(13) \qquad \overset{*}{R}{}^i_{jkh} = X_h \overset{*}{\Gamma}{}^i_{jk} - X_h \overset{*}{\Gamma}{}^i_{jh} + \overset{*}{\Gamma}{}^i_{mh} \overset{*}{\Gamma}{}^m_{jk} - \overset{*}{\Gamma}{}^i_{mk} \overset{*}{\Gamma}{}^m_{jh}$$

Se $l^i = \xi^i/L$, e se indichiamo con "o" una contrazione con l^i , possiamo dedurre da (12) che, siccome $C^i_{om} = 0$

$$(13') \qquad R^{i}_{okh} = \overset{*}{R}{}^{i}_{okh}$$

e si può anche scrivere (12) nella forma

$$(14) \qquad R^{i}_{jkh} = \overset{*}{R}{}^{i}_{jkh} + A^{i}_{jm} \overset{*}{R}{}^{m}_{okh} \qquad \text{con} \qquad A^{i}_{jm} = LC^{i}_{jm}$$

Per la curvatura riemanniana avremo bisogno di

$$(15) \qquad \overset{*}{R}_{jikh} = g_{in} \overset{*}{R}{}^{n}_{jkh} = R_{jikh} - A_{jim} \overset{*}{R}{}^{m}_{o\,kh}$$

e se i vettori che determinano la giacitura saranno l^{i} e un altro vettore indipendente μ^{i}, che può essere ortogonale al vettore l, avremo subito che

$$(16) \qquad \overset{*}{R}_{jikh} l^{j} \mu^{i} l^{k} \mu^{h} = R_{jikh} l^{j} \mu^{i} l^{k} \mu^{h} = R_{oioh} \mu^{i} \mu^{h}$$

$$= R^{i}_{o\,oh} \mu_{i} \mu^{h} = K(l, \mu) = \overset{*}{K}(l, \mu)$$

Dall'espressione che si ottiene per $\overset{*}{\Gamma}$ è evidente che le equazioni delle curve estremali sono anche le equazioni delle curve autoparallele. Basta osservare che per ogni scelta possibile delle Γ^{i}_{j} avremo $\hat{\gamma}^{i}_{oo} = \gamma^{i}_{oo}$. Le curve autoparallele di o-gni spazio riemanniano osculatore lungo le curve estremali relative al problema $\int L dt$ coincidono colle curve date dall'equazione (3). Si possono ora mettere queste considerazioni anche in relazione ai lavori di Myers sullo spazio riemanniano in grande. Se facciamo il calcolo della seconda variazione per l'integrale $\int L(x, \dot{x})dL$ con punti terminali fissi quando la prima variazione è zero avremo, come abbiamo visto nella prima lezione, l'espressione

$$(17)\qquad \int_{t_0}^{t_1} \eta^i J_i\, dt \equiv \int_{t_0}^{t_1} \eta_i J^i dt$$

Nella monografia di Cartan le equazioni per lo seostamento geodetico, che non sono altro che $J^i = 0$, sono

$$(18)\qquad \frac{D^2\eta^i}{ds^2} + R^{\;i}_{o\;oj}\,\eta^j = 0$$

dato che il vettore di variazione η^i è parallelo lungo l'estremale, e sempre normale alla geodetica. Sia, come nel caso riemanniano $\eta^i = \eta\,\mu^i$, dove μ^i è unitario.

Allora la seconda variazione (17) diviene

$$(19)\qquad \int_{s_0}^{s_1} \eta_i J^i = -\int_{s_0}^{s_1} \eta[\ddot{\eta} + \eta\, K(1,\mu)]\, ds = -\int_{s_0}^{s_1} \eta\, [\ddot{\eta} + \eta \overset{*}{K}(1,\mu)]\, ds$$

Sembra naturale chiamare $K(1,\mu)$ la curvatura riemanniana dello spazio di Finsler in un punto P nella giacitura determinata da 1 e μ. Si può allora enunciare un teorema corrispondente a quello di Myers : - Se per ogni punto P e per ogni giacitura γ di una F_n completa la curvatura riemanniana $K(P,\gamma)$ soddisfa alla condizione $K \geq K_o > 0$ allora un arco geodetico di lunghezza maggiore di $\pi/\sqrt{K_o}$ non può essere un arco geodetico minimizzante.

Nella teoria degli spazi di Riemann, come si vede nel libro di Cartan, si può utilizzare la nozione di spazio euclideo osculatore lungo una curva (espaces euclidien de raccordement le longue d'une ligne), e si può anche considerare (localmente) uno spazio di Riemann di dimensione n come uno spazio subordinato di uno spazio euclideo ad $N > n$ dimensioni. Nella teoria degli spazi di Finsler abbiamo già visto che vi sono molti punti di somiglianza tra gli spazi di Finsler a connessione euclidea e gli spazi riemanniani, in particolare che si può utilizzare la nozione di spazio riemanniano osculatore. Nel caso in cui il vettore $A_i = \partial \log \sqrt{g} / \partial x^i$ svanisce, il Cartan nella sua monografia fa cenno alla somiglianza ancora più forte. Ma Deicke dimostrò che nel caso $A_i = 0$, per $L > 0$, gli spazi di Finsler sono infatti spazi riemanniani. Possiamo adesso domandare se gli spazi di Finsler con $A_i \neq 0$ sono spazi veramente nuovi, o possono riguardarsi come spazi subordinati di spazi "puntuali" a 2n dimensioni. La risposta è che non possono riguardarsi come spazi riemanniani subordinati salvo nel caso in cui esiste un parallelismo assoluto degli elementi lineari di appoggio, caratterizzato dal fatto che il tensore $R_{ojkh} = 0$ nello spazio di Finsler. Nel caso generale si possono dedurre i tensori di F_n come quelli di una <u>distribuzione</u> (secondo la terminologia di Chevalley) o di una varietà anolonoma (secondo la terminologia più classica di Vranceanu, Schouten ed altri) in uno spazio a connessione euclidea con torsione, e di dimensione 2n.

Vogliamo dimostrare adesso che, dai dati fondamentali di uno spazio F_n di Finsler, possiamo costruire uno spazio V_{2n} a 2n dimensioni, con un tensore metrico $g_{ab}(x, \xi)$, $(a, b, c = 1, \ldots, 2n)$ e un tensore di torsione $S_{bc}^{\cdot\cdot a}(x, \xi)$ tale che si possono riguardare i tensori di F_n come quelli di una distribuzione in V_{2n}. In una varietà a 2n dimensioni con le coordinate x^a con x^i $(i\ ,\ j,\ k = 1,..,n)$ e con la convenzione che $\hat{i} = i + n$ che ci darà $x^{\hat{i}} = \xi^i$, siano due insiemi $p^{.a}_i$ e $q^{.a}_{\hat{i}}$ di vettori indipendenti che possono servire come base nella varietà a 2n dimensioni. Sia $(p^i_{.a}\ ,\ q^{\hat{i}}_{.a})$ la matrice reciproca tale che valgano le relazioni

$$(20)\qquad p^{.a}_i p^j_{.a} = \delta^j_i\ ,\ p^{.a}_i q^{\hat{j}}_{.a} = 0 = q^{.a}_{\hat{i}} p^j_{.a}\ ,\ q^{\hat{i}}_{.a} q^{.a}_{\hat{j}} = \delta^{\hat{i}}_{\hat{j}}$$

Se scriviamo

$$p^{.a}_i p^i_{.b} = \beta^a_b\ ,\qquad q^{.a}_{\hat{i}} q^{\hat{i}}_{.b} = \gamma^a_b$$

avremo

$$\beta^a_b + \gamma^a_b = \delta^a_b\ .$$

Sia 'V lo spazio vettoriale individuato dai vettori $p^{.a}_i$, e "V quello individuato dai vettori $q^{.a}_{\hat{i}}$. Un vettore X qualunque dello spazio a 2n dimensioni può decomporsi in due componenti 'X e "X utilizzando i "tensori di proiezione" β e γ come segue : $'X = \beta X$, $"X = \gamma X$. Le componenti $'X^i$ di 'X in 'V sono $p^i_{.a}X^a$, e in modo analogo $"X^{\hat{i}}$ avrà le componenti $q^{\hat{i}}_{.a}X^a$. Scriviamo $\omega^i = p^i_{.a}dx^a$, $\omega^{\hat{i}} = q^{\hat{i}}_{.a}dx^a$, allora in ogni punto della varietà a 2n dimensioni, le equazioni <u>non-integrabili</u>

$$(21)\qquad \omega^i = 0 \qquad e \qquad \omega^{\hat{i}} = 0$$

definiscono varietà subordinate anolonome o distribuzioni. Se mettiamo per le (eventuali) componenti del tensore metrico in V_{2n} le funzioni g_{ab} , possiamo scrivere

$$(22)\qquad ds^2 = g_{ij}\ \omega^i\omega^j + g_{\hat{i}\hat{j}}\ \omega^{\hat{i}}\omega^{\hat{j}}$$

in cui abbiamo messo

$$(23) \qquad g_{ij} = g_{ab} p_i^{.a} p_j^{.b} \quad ; \quad g_{\hat{i}\hat{j}} = g_{ab} q_{\hat{i}}^{.a} q_{\hat{j}}^{.b}$$

Se Γ^a_{bc} sono i coefficienti della connessione nella V_{2n}, vi saranno coefficienti come (i) Γ^i_{jk} che si riferiscono al trasporto di vettori di 'V per uno spostamento in 'V, (ii) $\Gamma^i_{j\hat{k}}$ che si riferiscono al trasporto di vettori di 'V per uno spostamento in "V (iii) $\Gamma^{\hat{i}}_{\hat{j}k}$ che si riferiscono al trasporto di vettori di "V per uno spostamento in "V , e (iv) $\Gamma^{\hat{i}}_{\hat{j}\hat{k}}$ per trasporto di vettori di "V per uno spostamento in "V .

I casi (i) e (ii) corrispondono ai coefficienti $\Gamma^i_{jk}(x,\xi)$ e $C^i_{jk}(x,\xi)$ dello spazio di Finsler. Dal libro di Schouten abbiamo in generale, per una geometria anolonoma

$$2\Gamma^a_{bc} = g^{ae}(X_b g_{ce} + X_c g_{be} - X_e g_{bc}) + \\ + (\Omega_{bc}^{..a} + \Omega^a_{.bc} + \Omega^a_{.cb}) + (T_{bc}^{..a} + T^a_{.bc} + T^a_{.cb})$$

dove $\Omega^a_{.bc} = g^{ad} g_{ce} \Omega_{db}^{..e}$

dove $T_{bc}^{..a}$ sono le componenti del tensore di torsione.

I coefficienti particolari che saranno necessari per il nostro scopo sono

$$(24) \quad 2\Gamma^i_{jk} = g^{im}(X_j g_{km} + X_k g_{jm} - X_m g_{jk}) + (\Omega_{jk}^{..i} + \Omega^i_{.jk} + \Omega^i_{.kj}) + \\ + (T_{jk}^{..i} + T^i_{.jk} + T^i_{.kj})$$

con $X_j = p_i^{.a} \partial_a$, $\Omega_{jk}^{..i} = p^i_{.a}(X_j p_k^{.a} - X_k p_i^{.a})$

(25) $$2\Gamma^{i}_{j\hat{k}} = g^{im}(X_j g_{\hat{k}m} + X_{\hat{k}} g_{jm} - X_m g_{j\hat{k}}) + \ldots + \ldots$$

con $$\Omega^{\cdot\cdot i}_{j\hat{k}} = p^{i}_{\cdot a}(X_j q^{\cdot a}_{k} - X_{\hat{k}} p^{\cdot a}_{j}) \quad , \quad X_{\hat{k}} = q^{\cdot a}_{k} \partial_a$$

Ritorniamo al problema di costruire uno spazio V_{2n} dai tensori di una F_n. Possiamo cominciare dal tensore $g_{ij}(x\xi)$ della metrica di F_n. Se mettiamo $g_{\hat{i}\hat{j}} = g_{ij}$ avremo determinato la metrica di V_{2n} secondo (22). Supponiamo naturalmente che un vettore $p^{\cdot a}_{i}$ sia sempre ortogonale ad un vettore $q^{\cdot a}_{\hat{i}}$, cioè prendiamo $g_{\hat{i}j}=0$. Siccome i vettori $p^{\cdot a}_{i}$, $q^{\cdot a}_{\hat{i}}$ sono da determinare, facciamo la scelta seguente

$$p^{\cdot a}_{i} = \delta^{j}_{i} \quad \text{per} \quad a = j \quad , \quad p^{\cdot a}_{j} = -\Gamma^{j}_{i} = -\Gamma^{j}_{mi}\xi^{m} = -\dot{\partial}_i G^{j} \quad \text{per } a = \hat{j}$$

(26) $$\begin{aligned} &q^{\cdot a}_{\hat{i}} = 0 \quad \text{per} \quad a = j \quad , \quad q^{\cdot a}_{\hat{i}} = \delta^{j}_{i} \quad \text{per} \quad a = \hat{j} \\ &p^{i}_{\cdot a} = \delta^{i}_{j} \quad \text{per} \quad a = j \quad , \quad p^{i}_{\cdot a} = 0 \quad \text{per} \quad a = \hat{j} \\ &q^{\hat{i}}_{\cdot a} = \Gamma^{i}_{j} \quad \text{per} \quad a = j \quad , \quad q^{\hat{i}}_{\cdot a} = \delta^{i}_{j} \quad \text{per} \quad a = \hat{j} \end{aligned}$$

Facciamo anche l'ipotesi che lo spazio V_{2n} che stiamo costruendo abbia la proprietà geometrica che le curve estremali siano anche curve autoparallele. Ciò avrà per conseguenza che $T^{a}_{\cdot bc} + T^{a}_{\cdot cb} = 0$ per ogni scelta di indici a,b,c. Con questa scelta avremo

(27) $$\Omega^{\cdot\cdot i}_{jk} = 0 \quad , \quad \Omega^{\cdot\cdot i}_{j\hat{k}} = 0 \quad , \quad \Omega^{\cdot\cdot \hat{i}}_{jk} = R^{i}_{ojk}$$

e le espressioni (24) e (25) divengono, tenendo conto del fatto che

$$2\overset{*}{\Gamma}{}^{i}_{jk} = g^{im}(X_j g_{km} + X_k g_{jm} - X_m g_{jk})$$

(28) $$2\Gamma^{i}_{jk} = 2\overset{*}{\Gamma}{}^{i}_{jk} + T^{..i}_{jk}$$

(29) $$2\Gamma^{i}_{j\hat{k}} = g^{im} X_{\hat{k}} g_{im} + \Omega^{i}_{.j\hat{k}} + T^{..i}_{j\hat{k}}$$

Dalla equazione (28) allora, siccome le funzioni $\overset{*}{\Gamma}$ sono note, avremo determinato le Γ^{i}_{jk} se le componenti del tensore di torsione del tipo $T^{..i}_{jk}$ svaniscono. L'equazione (29) conduce a $X_{\hat{k}} g_{im} = \dot{\partial}_k g_{im}$, ed allora $\Gamma^{i}_{j\hat{k}}$ sarà determinata come uguale a C^{i}_{jk} dello spazio di Finsler se sussiste

(30) $$\Omega^{i}_{.j\hat{k}} + T^{..i}_{j\hat{k}} = 0$$

Si dimostra facilmente che $\Omega^{i}_{.j\hat{k}} = g^{im} g_{\hat{n}\hat{k}} R^{n}_{o\,mj} = g^{im} R_{okmj}$ e allora bisogna prendere

(31) $$T^{..i}_{j\hat{k}} = - g^{im} R_{okmj}$$

Nella nostra ricerca di uno spazio V_{2n} a connessione euclidea con torsione tale che i coefficienti della metrica e della connessione di uno spazio F_n possano ottenersi con "proiezione" dai corrispondenti coefficienti dello spazio ambiente V_{2n}, abbiamo determinato tutti i coefficienti della metrica : g_{ij}, $g_{\hat{i}\hat{j}} = g_{ij}$, $g_{\hat{i}j} = 0$. Dal tensore di torsione però abbiamo determinato solamente i coefficienti dei tipi T_{ijk}, $T_{ij\hat{k}}$. Quelli come $T_{i\hat{j}\hat{k}}$ e $T_{\hat{i}\hat{j}\hat{k}}$ sono ancora arbitrari. Sarebbe possibile dare valore zero a tali coefficienti.

Dalla relazione (31) è evidente che lo spazio ambiente

può essere riemanniano solamente quando $R_{oijk} = 0$ cioè quando esiste un parallelismo assoluto di elementi lineari. Ma in questo caso anche $\Omega_{jk}^{..\hat{i}} = 0$ come segue da (27), ed allora la distribuzione $\omega^{\hat{i}} = 0$ è una varietà integrabile cioè si tratta di una V_n riemanniana immersa in una V_{2n} riemanniana.

E.T.Davies

V LEZIONE

GLI SPAZI VARIAZIONALI DI LICHNEROWICZ

Lichnerowicz si è occupato in una memoria (1945) di una generalizzazione degli spazi di Finsler suggerita da un esempio nello studio dei sistemi dinamici non conservativi. Se S rappresenta l'azione del sistema dinamico per un intervallo di tempo t_o, t_1 si può scrivere

$$(1) \qquad S = \int_{t_o}^{t_1} \left(T + \int_{t_o}^{t} Q_i \frac{dx^i}{d\tau} d\tau\right) dt$$

dove T è la forza viva e $Q_i dx^i$ il lavoro virtuale elementare delle forze applicate al sistema. L'integrale (1) è un caso particolare di integrale della forma

$$(2) \qquad J = \int_{t_o}^{t_1} H\left[F_{t_o}^{t}\left[x(\tau)\right], x(t), \dot{x}(t)\right] dt$$

calcolato lungo una curva C di equazioni parametriche $x^i = x^i(t)$, e dove

$$(3) \qquad F_{t_o}^{t} = \int_{t_o}^{t} \omega\left[x(\tau), \dot{x}(\tau)\right] d\tau .$$

Le funzioni H e ω sono omogenee del primo grado rispetto a $\dot{x}^i$. H è di classe C^3 nelle variabili F, x, $\dot{x}$; ω nelle variabili x, $\dot{x}$. In relazione ad un integrale come (2) Lichnerowicz dà il concetto di <u>sotto-variazione</u> come segue :

Sia $F_{k_o}^{t} \equiv F_{t_o}^{t}\left[x(t)\right]$ un funzionale tale che per una curva data, $F_{t_o}^{t}$ è un funzionale integrabile di t . Lichnerowicz definisce <u>sotto-differenziale</u> $\bar{\delta} F$ di F una funzione di x^i, $\dot{x}^i$, e di δx^i che sarà (a) lineare rispetto a δx^i (b) tale che $\lim_{\varepsilon \to 0} \frac{\Delta F - \bar{\delta} F}{\eta} = 0$ dove ΔF è l'incremento del funzionale F per $\delta x = \varepsilon \eta^i$. Relativamente all'integrale (3) in cui il funzionale F è incluso, Lichnerowicz chiama <u>sotto-variazione</u> $\bar{\delta} J$ un integrale della forma

$$\bar{\delta} J = \int_{t_0}^{t_1} K\left[x(t),\ \dot{x}(t),\ \delta x\right] dt$$

dove (a) K è lineare rispetto a δx (b) $\lim_{\varepsilon \to 0} \frac{\Delta J - \bar{\delta} J}{\varepsilon \eta} = 0$.

Un funzionale $F^t_{t_0}$ si dice <u>della classe regolare</u> quando (a) è derivabile rispetto a t (b) ammette derivate funzionali parziali rispetto alle x^i (c) per incrementi δx^i il differenziale di F può mettersi sotto la forma

$$\delta F = \int_{t_0}^{t} (F'_i\ \delta x^i) d\tau + a_i\left[x^i(t),\ \dot{x}^i(t)\right] \delta x^i$$

dove l'integrale nel secondo membro è la parte <u>regolare</u>, e dove l'altro termine dipendente da valori eccezionali è la parte <u>eccezionale</u>. Per un funzionale della classe regolare il sotto-differenziale $\bar{\delta} F$ è uguale alla parte eccezionale. Nel caso in cui F è un integrale come (3) per cui

$$\delta F = \int_{t_0}^{t_1} E_i(\omega)\, \delta x^i\, d\tau + \frac{\partial \omega}{\partial \dot{x}^i}\ \delta x^i(t)$$

avremo subito che $\bar{\delta} F = \frac{\partial \omega}{\partial \dot{x}^i}\ \delta x^i \equiv \omega_i\, \delta x^i$.

Con queste considerazioni Lichnerowicz ha dimostrato che la sotto-variazione dell'integrale (2) è data da

$$(4) \quad \bar{\delta} J = \int_{t_0}^{t_1} \left[\frac{\partial H(o)}{\partial F} \frac{\partial \omega}{\partial \dot{x}^i} + \frac{\partial H(o)}{\partial x^i} - \frac{d}{dt} \frac{\partial H(o)}{\partial \dot{x}^i} - \frac{\partial^2 H(o)}{\partial x^i\, \partial F}\, \omega \right] \delta x^i dt$$

Si chiama problema di calcolo delle variazioni anolonomo definito dall'integrale J il problema di trovare le curve lungo le quali $\bar{\delta} J \equiv 0$ per ogni incremento δx. L'ipotesi di omogeneità delle funzioni H e ω nelle $\dot{x}$ ci assicura un carattere intrinseco agli integrali F e J . Possiamo allora scrivere, se M_0, M, e M_1 sono punti sulla curva C presa

$$(5) \qquad F_{M_0}^{M} = \int_{M_0}^{M} \omega(x, dx) \qquad J = \int_{M_0}^{M_1} H\left[F_{M_0}^{M}, x(M), dx\right] .$$

Si chiama spazio variazionale generalizzato uno spazio fondato su un problema di calcolo delle variazioni anolonomo relativo ad un integrale (lungo una curva C) della forma J .

Nel secondo membro di (4) H(o) rappresenta la funzione H per il valore zero dell'argomento F . H(o) è allora una funzione di x e di $\dot{x}$, che è omogenea del primo grado nelle $\dot{x}$. Per indicare che si tratta del caso in cui $F \equiv 0$ nella funzione H , mettiamo $\overset{o}{H}(x, \dot{x})$ invece di $H(o, x, \dot{x})$, una funzione che può stare alla base di una geometria di Finsler colla metrica

$$(6) \qquad d\overset{o}{s}{}^2 = \overset{o}{H}{}^2(x,\dot{x}) = \overset{o}{g}_{ij}(x,\dot{x})dx^i dx^j$$

con

$$(7) \qquad \overset{o}{g}_{ij} = \frac{\partial^2(\frac{1}{2}\overset{o}{H}{}^2)}{\partial \dot{x}^i \partial \dot{x}^j} = \dot{\partial}_{ij}\left(\frac{1}{2}\overset{o}{H}{}^2\right) \quad \text{mettendo} \quad \dot{\partial}_i = \frac{\partial}{\partial \dot{x}^i}$$

con altri elementi necessari per stabilire una geometria. Scriveremo i coefficienti della connessione come

$$(8) \qquad \overset{o}{\Gamma}{}^i_{jk} = \overset{o}{\gamma}{}^i_{jk} + \overset{o}{g}{}^{ih}\left[\overset{o}{C}_{jkm}\dot{\partial}_h \overset{o}{G}{}^m - \overset{o}{C}_{hkm}\dot{\partial}_j \overset{o}{G}{}^m\right] ;$$

dove abbiamo messo, come nella IV Lezione

$$\overset{o}{G}{}^m \equiv \overset{o}{\gamma}{}^m_{rs} \dot{x}^r \dot{x}^s .$$

Prima di passare alla considerazione dell'integrale J nel caso generale, vogliamo fare qualche osservazione sugli spazi puntuali (a) di Weyl (b) a connessione euclidea con torsione. Nella teoria degli spazi W_n di Weyl sono dati due elementi fondamentali, cioè una forma quadratica differenziale $\overset{o}{g}_{ij}(x)dx^i dx^j$ e anche una forma differenziale lineare non-integrabile $\omega_i(x)dx^i$. La forma quadratica col tensore $\overset{o}{g}_{ij}(x)$ determina uno spazio di Riemann. Consideriamo adesso un integrale della forma $\int_M^m \omega_i(x)dx$ preso lungo una curva arbitraria che riunisce M col punto vicino m. Consideriamo ancora, nel punto m, la metrica definita da

$$(9) \qquad ds^2 = g_{ij}(x)dx^i dx^j \equiv e^{\int_M^m \omega_i dx^i} \overset{o}{g}_{ij}(x)dx^i dx^j = e^{\int_M^m \omega_i dx^i} \overset{o}{ds}{}^2$$

La metrica così definita può riguardarsi come una metrica "calibrata" di Weyl, e gli elementi fondamentali dello spazio di Weyl sono i coefficienti $g_{ij}(x)$ della metrica e quelli della connessione simmetrica Γ^i_{jk}. Dalla relazione (9) tra g_{ij} e $\overset{o}{g}_{ij}$ avremo per $m \rightarrow M$, <u>al punto M</u> , mettendo $\partial_k = \partial / \partial x^k$

$$(10) \qquad g_{ij} = \overset{o}{g}_{ij} \quad ; \quad \partial_k g_{ij} = \partial_k \overset{o}{g}_{ij} + \overset{o}{g}_{ij}\omega_k$$

ed i corrispondenti simboli di Christoffel saranno

$$(11) \qquad \gamma^i_{jk} = \overset{o}{\gamma}{}^i_{jk} + \frac{1}{2} g^{ih}\left[g_{jh}\omega_k + g_{kh}\omega_j - g_{jk}\omega_h\right]$$

le funzioni γ^i_{jk} così definite sono i coefficienti Γ^i_{jk} di connessione affine di Weyl.

Consideriamo ancora nel punto m due vettori collineari unitari $\overset{o}{l}{}^i$ e l^i , il primo essendo unitario rispetto alla metrica riemanniana $d\overset{o}{s}{}^2$ e l'altro rispetto alla metrica di Weyl ds^2. In virtù della (9) questi vettori sono legati dalla relazione

$$l^i = e^{-\frac{1}{2}\int_M^m \omega_i dx^i} \overset{o}{l}{}^i \ . \tag{12}$$

Segue immediatamente che nel punto M avremo

$$l^i = \overset{o}{l}{}^i \quad , \quad \partial_j l^i = \partial_j \overset{o}{l}{}^i - \frac{1}{2}\,\omega_j\, \overset{o}{l}{}^i \tag{13}$$

La derivata covariante di l^i nella geometria di Weyl sarà

$$\nabla_j l^i = \partial_j l^i + \gamma^i_{kj}\, l^k \tag{14}$$

Ma si può anche scrivere il secondo membro nella forma

$$\overset{o}{\nabla}_j \overset{o}{l}{}^i = \partial_j \overset{o}{l}{}^i + \Lambda^i_{kj}\, \overset{o}{l}{}^k \tag{15}$$ dove

$$\Lambda^i_{kj} = \overset{o}{\gamma}{}^i_{kj} + \frac{1}{2}\, g^{ih}\left[g_{jh}\,\omega_k - g_{jk}\,\omega_h\right] \tag{16}$$

la connessione con i coefficienti Λ^i_{kj} essendo emi-simmetrica nel senso di Schouten. Data allora una metrica $\overset{o}{g}_{ij}$ e un vettore covariante, possiamo riguardarli come gli elementi fondamentali di una W_n di Weyl, o di uno spazio a connessione euclidea che sarà emi-simmetrica.

Ritorniamo adesso alla considerazione dell'integrale J nel caso generale. Dall'espressione (4) per la sotto-variazione, e da ragionamenti classici del calcolo delle variazioni, siamo condotti alle equazioni differenziali seguenti per le curve che fanno $\bar{\delta} J \equiv 0$ per ogni variazione δx^i :

$$(17) \qquad \frac{d}{dt}\frac{\partial H(o)}{\partial \dot{x}^i} - \frac{\partial H(o)}{\partial x^i} = H'(o)\frac{\partial \omega}{\partial \dot{x}^i} - \omega \frac{\partial H'(o)}{\partial \dot{x}^i}$$

Una questione di interesse per la geometria differenziale è di sapere se le curve estremali di un tale problema di calcolo delle variazioni possono riguardarsi come curve autoparallele rispetto ad una "connessione" che si può definire in modo intrinseco, e dipendente solamente dalle funzioni H e ω .

Come nel caso degli spazi puntuali possiamo considerare, nella vicinanza di ogni punto M , un fascio di curve congiungenti M ad ogni punto m della vicinanza. Possiamo attribuire a questa vicinanza una metrica ds^2 definita al punto m dalla formula

$$(18) \qquad ds^2 = H^2(F, x, dx)$$

dove $F = \int_M^m \omega_i dx^i$ è calcolato lungo la curva scelta da M ad m. Mettiamo

$$(19) \qquad g_{ij}(F, x, \dot{x}) = \dot{\partial}_{ij}\left(\frac{1}{2} H^2(F)\right)$$

in modo che avremo, dalle condizioni di omogeneità

$$(20) \qquad ds^2 = g_{ij}(x) dx^i dx^j$$

Come nel caso degli spazi puntuali, abbiamo che quando $m \rightarrow M$ sussistono nel punto M le seguenti relazioni

(21) $\quad g_{ij} = \overset{o}{g}_{ij}$ per i coefficienti della metrica dello spazio di Finsler basato sulla funzione $\overset{o}{H}(x,\dot{x})$

(22) $$\dot{\partial}_k g_{ij} = \dot{\partial}_k \overset{o}{g}_{ij}$$

(23) $$\partial_k g_{ij} = \partial_k \overset{o}{g}_{ij} + g'_{ij}\omega_k \quad , \quad \text{con} \quad g'_{ij} = \frac{\partial g_{ij}}{\partial F} \quad , \quad \omega_k = \dot{\partial}_k \omega$$

I corrispondenti simboli di Christoffel saranno legati da

(24) $$\gamma^i_{jk} = \overset{o}{\gamma}{}^i_{jk} + \frac{1}{2} g^{ih}\left[g'_{hj}\omega_k + g'_{kh}\omega_j - g'_{jk}\omega_h\right]$$

Per il sistema di equazioni delle curve estremali avremo

(25) $$\dot{x}^i \wedge \Phi^j = 0 \quad , \quad \text{con} \quad \Phi^i = \ddot{x}^i + \gamma^i_{mn}\dot{x}^m \dot{x}^n$$

che si riducono subito alle equazioni delle geodetiche degli spazi di Finsler se $\overset{o}{g}_{ij}$ è preso invece di g_{ij}.

Abbiamo adesso quattro spazi da prendere in considerazione.

(i) lo spazio variazionale generalizzato associato alla funzione $H(F,x,\dot{x})$

(ii) lo spazio di Finsler associato alla funzione $\overset{o}{H}(x,\dot{x})$

(iii) lo spazio (di Weyl) ottenuto se le $\dot{x}$ sono funzioni di x lungo una curva, cioè si può parlare di uno spazio di Weyl osculatore.

(iv) lo spazio di Riemann che è osculatore allo spazio di Finsler lungo una curva.

Nel caso in cui il parametro t è uguale ad s definito dalla funzione $H(F, x, \dot{x})$ avremo come equazioni delle geodetiche

(26) $$\Phi^i = 0$$

Mettendo $2G^i = \gamma^i_{mn} \dot{x}^m \dot{x}^n$ con $\dot{x} = dx/ds$, possiamo dire senz'altro che i coefficienti della connessione dello spazio osculatore di Weyl saranno, scrivendo $X_i = (\partial_j - \dot{\partial}_j G^n \dot{\partial}_n)$

$$(27) \qquad \hat{\Gamma}^i_{jk} = \frac{1}{2} g^{ih} \left[X_j g_{hk} + X_k g_{jh} - X_h g_{jk} \right]$$

e se γ^i_{jk} sono i simboli di Christoffel calcolati con ∂_i , possiamo anche scrivere

$$(28) \qquad \hat{\Gamma}^i_{jk} = \gamma^i_{jk} - g^{ih} \left[C_{hkn} \dot{\partial}_j G^n + C_{jhn} \dot{\partial}_k G^n - C_{jkn} \dot{\partial}_h G^n \right]$$

dove si potrebbero anche esprimere le funzioni γ_{jk} in termini delle corrispondenti funzioni $\overset{o}{\gamma}$ secondo la relazione (24).

Le funzioni

$$(29) \qquad \bar{\Gamma}^i_{jk} = \hat{\Gamma}^i_{jk} + C^i_{jn} \dot{\partial}_k G^n = \gamma^i_{jk} + g^{ih} (C_{jkn} \dot{\partial}_h G^n - C_{hkn} \dot{\partial}_j G^n)$$

sono uguali alle funzioni che Lichnerowicz ha ottenuto come coefficienti della connessione "intermediaria". E' facile verificare che

$$(30) \qquad \bar{\Gamma}^i_{jk} \dot{x}^j \dot{x}^k = \gamma^i_{jk} \dot{x}^j \dot{x}^k$$

allora nelle equazioni (26) per la γ si può sostituire la $\bar{\Gamma}$.

Ciò dimostra che nella geometria per cui la connessione è caratterizzata dalla $\bar{\Gamma}$, le autoparallele coincidono con le curve estremali.

Come per gli spazi puntuali possiamo introdurre in un punto m vicino ad M due vettori $\overset{o}{l}{}^i$ e l^i, l'uno unitario rispetto alla metrica di $\overset{o}{H}$, e l'altro unitario rispetto a H(F). Esiste tra le due la relazione $l^i H(F) = \overset{o}{l}{}^i \overset{o}{H} = \overset{o}{l}{}^i H(o)$. Nel punto M avremo $\overset{o}{l}{}^i = l^i$

poichè nel punto M , F = O. Per le derivate parziali però avremo

$$(31)\qquad \partial_j l^i = \overset{o}{\partial}_j l^i - \frac{H'(o)}{H(o)}\,\omega_j l^i = \overset{o}{\partial}_j l^i - (\log H)'\omega_j l^i$$

Se scriviamo il differenziale assoluto per il vettore unitario l^i avremo, tenendo conto del fatto che $l^i C_{ijk} = 0$ come per lo spazio di Finsler

$$(32)\qquad Dl^i = dl^i + \bar{\Gamma}^i_{jk}\, l^j dx^k = \overset{o}{d}l^i + (\bar{\Gamma}^i_{jk} - (\log H)'\omega_k \delta^i_j) l^j dx^k$$

Le funzioni

$$(33)\qquad \Gamma^i_{jk} = \bar{\Gamma}^i_{jk} - (\log H)'\omega_k\, \delta^i_j$$

sono i coefficienti della prima connessione ottenuta da Lichnerowicz.

Per un campo vettoriale qualunque $V^i(x, \dot{x})$ scriviamo

$$(34)\qquad DV^i = dV^i + (\Gamma^i_{jk}\, dx^k + C^i_{jk} d\dot{x}^k) V^j$$

Dall'espressione corrispondente per l^i che scriviamo

$$(35)\qquad \Pi^i = Dl^i$$

avremo per $d\dot{x}^i$ l'espressione

$$(36)\qquad d\dot{x}^i = H(o)\Pi^i + \dot{x}^i \frac{dH(o)}{H(o)} - \Gamma^i_{jk}\, \dot{x}^j dx^k$$

in modo che si possa modificare (34) nella forma

$$(37)\qquad DV^i = V^i_{\ ,k}\, dx^k + (V^i\|_k + A^i_{jk} V^j)\, \Pi^k$$

dove abbiamo scritto $V^i_{\ ,j}$ per la derivata covariante del vettore

V^i. L'espressione definitiva per $V^i_{,j}$ sarà

(38) $$V^i_{,j} = \partial_j V^i - \Gamma^m_{nj} \dot{x}^n \dot{\partial}_m V^i + \overset{*}{\Gamma}{}^i_{jk} V^j$$

dove

(39) $$\overset{*}{\Gamma}{}^i_{jk} = \Gamma^i_{jk} - C^i_{jm} \Gamma^m_{nk} \dot{x}^n = \Gamma^i_{jk} - A^i_{jm} \Gamma^m_{ok}$$

Si può anche scrivere (38) nella forma

(40) $$V^i_{,j} = X_j V^i + \overset{*}{\Gamma}{}^i_{jk} V^j$$

perchè si può verificare facilmente che per questo caso, come nel caso degli spàzi di Finsler

(41) $$\dot{x}^n \Gamma^i_{nj} = \dot{\partial}_j G^i$$

Le funzioni Γ^i_{jk} e $\overset{*}{\Gamma}{}^i_{jk}$ sono completamente determinate dalle funzioni fondamentali H(o) e ω e dalle relazioni (33), (29) e (24). Possiamo procedere all'introduzione della nozione di curvatura nel caso particolare di Finsler.

VI LEZIONE

GEOMETRIA FONDATA SU UN INTEGRALE DELL'ORDINE (n-1) IN UNO SPAZIO DI DIMENSIONE n (SPAZI DI CARTAN)

Sia L una funzione derivabile delle variabili $x^i (i=1,..,n)$ di t^α ($\alpha, \beta, \dots = 1,..,m < n$) delle derivate prime, seconde, etc. delle x rispetto a t , che scriviamo

$$(1) \qquad p^i_\alpha = \frac{\partial x^i}{\partial t^\alpha} \quad , \quad p^i_{\alpha\beta} = \frac{\partial^2 x^i}{\partial t^\alpha \partial t^\beta}, \dots$$

Cerchiamo le condizioni perchè l'integrale

$$(2) \qquad \int_{(m)} L(t, x, p^i_\alpha, \dots) dt^1 \dots dt^m$$

sia invariante per trasformazioni $t \longrightarrow \underline{t}$ del parametro, cioè che

$$(3) \qquad \int_{(\bar{m})} L(\underline{t}, x, \underline{p}^i_\alpha, \dots) d\underline{t}^1 \dots d\underline{t}^m$$

sia identicamente uguale a (2) per ogni trasformazione di t .

Si può subito trasformare l'integrale (3) in un integrale sulla regione (m) e la condizione per l'invarianza diviene

$$(4) \qquad L(\underline{t}, x, \underline{p}^i_\alpha \dots)\left(\frac{\partial \underline{t}}{\partial t}\right) \equiv L(t, x, p^i_\alpha, \dots)$$

Siccome (4) deve sussistere per ogni trasformazione, prendiamo il caso semplice $t^\alpha = \underline{t}^\alpha + c$, nel qual caso (4) mostra subito che L non può contenere t esplicitamente. Il metodo generale che si trova nei lavori di de Donder ed altri sugli integrali multipli è quello di derivare l'identità (4) rispetto a $\underline{t}^\alpha_\beta \equiv \partial t^\alpha / \partial \underline{t}^\beta$, e poi di mettere $t^{\alpha}_{\beta} = \delta^{\alpha}_{\beta}$, un procedimento che darà, scrivendo $,i$ per $\partial/\partial x^i$, $;^{\alpha}_{i}$ per $\partial/\partial p^i_{\alpha}$, etc.

$$L;^{\alpha}_{i}\, p^i_{\beta} = \delta^{\alpha}_{\beta} L \tag{5}$$

$$L;^{\alpha\beta}_{i}\, p^i_{\gamma} = 0 \qquad \text{per derivate superiori} \tag{6}$$

Si può dedurre anche per la seconda derivata rispetto a p^j_{β} :

$$L;^{\alpha\beta}_{i\,j}\, p^j_{\gamma} = 0 \tag{7}$$

da cui segue che il determinante di ordine nm

$$\left| L;^{\alpha\beta}_{i\,j} \right| = 0 \tag{8}$$

Cerchiamo la prima e la seconda variazione di un integrale.

$$\int_{(m)} L(x, \frac{\partial x}{\partial t})\, dt \qquad dt \equiv dt^1 \ldots dt^m \tag{9}$$

per una superficie $\underset{o}{V_m}$ di dimensione m

$$x^i = x^i(t^1, \ldots, t^m) \tag{10}$$

che appartiene ad un sistema ad un parametro di V_m

$$x^i = x^i(t^1, \ldots, t^m; e) \tag{11}$$

e supponiamo che per $e = 0$ ($\underset{o}{V_m}$) l'integrale (9) abbia valore estremale.

Più in generale parleremo di una V_m vicino a $\underset{o}{V_m}$ come <u>ammissibile</u> se i punti al contorno variabile V_{m-1} sono dati dalle equazioni

$$t^{\alpha} = t^{\alpha}(y^1, \ldots, y^r) \quad , \quad x^i = x^i(y^1, \ldots, y^r) \qquad m \leq r \leq n \tag{12}$$

per valori di (y) vicino a (o), dove supponiamo che le funzioni nelle (12) siano di classe C^2 e che per (y) = (o) diano i punti del contorno di $\underset{o}{V_m}$. Avremo allora un sistema ad r-parametri di V_m se mettiamo

$$(13) \qquad x^i = \varphi^i(t^1,\dots,t^m;\, y^1,\dots,y^r)$$

dove φ^i sono (a) di classe C^2 (b) prendono i valori (10) per (y) = 0, e (c) prendono i valori (12) sul contorno V_{m-1}.

Siano $y^1(e),\dots,y^r(e)$ funzioni di classe C^2 per e vicino a e = 0 con (y(0)) = (0) e mettiamo

$$(14) \qquad \varphi^i(t;\, y(e)) \equiv x^i(t^1,\dots,t^m;\, e)$$

che ci darà il sistema (11) ad un parametro. Al contorno avremo l'identità

$$(15) \qquad x^i\,[y(e)] \;\equiv\; x^i\,[t(e);e]$$

ed allora per le derivate rispetto a e al contorno avremo

$$(16) \qquad \frac{dx^i}{de} = \frac{\partial x^i}{\partial e} + \frac{\partial x^i}{\partial t^\alpha}\,\frac{dt^\alpha}{de} \equiv \eta^i + p^i_\alpha\, T^\alpha$$

Si può riguardare l'integrale (9) come una funzione I(e) della sola variabile e . Avremo I(0) come valore estremale se I'(0) = 0 . Si può calcolare facilmente il valore

$$(17) \qquad I'(0) = \int_{(m)} \frac{\partial L}{\partial e}\, dt + \int_{(m)} \frac{d}{dt^\alpha}\,(LT^\alpha)^{dt}$$

Se il contorno V_{m-1} è parametrizzato da $\varrho^1,\dots,\varrho^{m-1}$, si può scrivere

$$(-1)^{\alpha+1} p_\alpha = \frac{\partial(t^1,\ldots,t^{\alpha-1} t^{\alpha+1},\ldots,t^m)}{\partial(\rho^1,\ldots,\rho^{m-1})}$$

e si può modificare (17) nella forma

$$(18) \qquad I'(0) = \int_{(m)} \frac{\partial L}{\partial e} dt + \int_{(m-1)} p_\alpha (LT^\alpha) d\rho$$

Possiamo dare anche la seconda variazione nella forma

$$(19) \qquad I''(0) = \int_{(m)} \frac{\partial^2 L}{\partial e^2} dt + \int_{(m-1)} p_\alpha \Big[(\frac{\partial L}{\partial e} T^\alpha) + \frac{d}{de}(LT^\alpha) + LT^\alpha \frac{\partial T^\beta}{\partial t^\beta} - LT^\beta \frac{\partial T^\alpha}{\partial t^\beta} \Big] d\rho$$

Si può esprimere la prima variazione I'(0) in una forma più espressiva introducendo il vettore di Eulero

$$(20) \qquad E_i = \partial_i L - \partial_\alpha (L;{}^\alpha_i)$$

che darà

$$(21) \qquad I'(0) = \int_{(m)} \eta^i E_i dt + \int_{(m-1)} p_\alpha (L;{}^\alpha_i \frac{dx^i}{de}) d\rho$$

La prima variazione sarà zero allora per ogni variazione η^i se sono soddisfatte le condizioni sufficienti

$$(22) \qquad (a) \quad E_i = 0 \quad \text{lungo la varietà } \underset{0}{V_m}$$

$$(23) \qquad (b) \quad p_\alpha L;{}^\alpha_i \frac{dx^i}{de} = 0 \quad \text{lungo il contorno } V_{m-1} .$$

Come nel caso m = 1, le equazioni $E_i = 0$ non sono tutte indipendenti perchè si può facilmente verificare che

$$(24) \qquad p^i_\alpha E_i = 0$$

vi saranno allora (n-m) equazioni indipendenti.

Prendiamo per il momento il caso di un contorno fisso. Si ottiene, per la seconda variazione, l'espressione

(25) $$I''(0) = \int_{(m)} \eta^i J_i \, dt$$

lungo la V_m estremale. Come per m=1 , $J_i \equiv \partial E/\partial e$ e le r equazioni $J_i = 0$ corrispondono alle equazioni di Jacobi per m=1 . Non sono tutte indipendenti, ma soddisfano alle identità

(26) $$p^i_\alpha J_i \equiv 0$$

Prendiamo il caso in cui l'integrale rappresenta il volume di una regione dello spazio riemanniano col tensore metrico g_{ij} , e tale che la varietà subordinata abbia la metrica $g_{\alpha\beta} = g_{ij} p^i_\alpha p^j_\beta$. La funzione $\bar{L}$ sarà $\sqrt{|g_{\alpha\beta}|}$.

Se scriviamo

(27) $$\omega^i = g^{\alpha\beta} (p^i_{\alpha\beta} + \left\{ {i \atop jk} \right\} p^i_\alpha p^k_\beta) = g^{\alpha\beta} \omega^i_{\alpha\beta}$$

(28) $$p^\gamma_i = g^{\gamma\delta} g_{ij} p^j_\delta \ , \quad \beta^i_j = p^i_\alpha p^\alpha_j \ , \quad \gamma^i_j = \delta^i_j - \beta^i_j$$

avremo per $E_i = 0$ l'espressione

(29) $$g_{ij} \gamma^j_k \omega^k = 0$$

Le equazioni corrispondenti per $J_i = 0$ sono date prima da Enea Bortolotti come generalizzazioni delle equazioni dello scostamento geodetico di Levi-Civita.

Nel caso speciale in cui

(a) m = n-1 (b) la variazione $\eta^i = l^i \eta$ dove l^i è l'unico vettore normale unitario, avremo

(30) $$I''(0) = - \int_{(n-1)} \eta \left[g^{\alpha\beta} \partial_\alpha \partial_\beta \eta + U \eta \right] dt$$

dove

(31) $$\begin{aligned} U &= M_1 - (n-1)(n-2)K_r \\ &= - R_{ijkl} p^j_\alpha p^l_\beta g^{\alpha\beta} l^i l^k + g_{ij} g^{\alpha\gamma} g^{\beta\delta} \omega^i_{\alpha\beta} \omega^j_{\gamma\delta} \end{aligned}$$

in cui M_1 è la curvatura media di Ricci per la direzione l^i, e K_r è la curvatura relativa di V_{n-1} in V_n.

Ritorniamo all'integrale (9) per il caso m = n-1. Fu dimostrato da Radon (1911) che se poniamo $(-1)^{i+1} p_i$ per il determinante ottenuto dalla matrice $\| p^i_\alpha \|$ sopprimendo la colonna "i", una funzione di x^i e di p^i_α che soddisfi alla condizione di omogeneità (cioè alla condizione per l'invarianza del parametro) può mettersi nella forma L(x,p) e le condizioni d'omogeneità sono

(32) $$\begin{cases} p_i \dfrac{\partial L}{\partial p_i} \equiv p_i \partial^i L \equiv p_i L^i = L \\[2ex] p_i \dfrac{\partial^2 L}{\partial p_i \partial p_j} \equiv p_i \partial^{ij} L \equiv p_i L^{ij} = 0 \end{cases}$$

E' importante l'osservazione che p_i sono le componenti di una densità vettoriale di peso -1 , mentre le componenti

(33) $$a^{ij} = \partial^j \partial^i \left(\frac{1}{2} L^2\right) = LL^{ij} + L^i L^j$$

si trasformano come una densità tensoriale di peso 2 da cui possiamo ottenere un tensore a due indici dalla relazione

(34) $$g^{ij} = a^{-\frac{1}{n-1}} a^{ij} \qquad \text{con} \qquad a = \left| a^{ij} \right| > 0$$

Dalle equazioni (33) e (34) avremo subito

(35) $$g g^{ik} p_k = L L^i \qquad g g^{ij} p_i p_j = L^2$$

che suggerisce l'introduzione di un vettore unitario con le componenti

(36) $$l_i = \frac{\sqrt{g}}{L} p_i \qquad \text{con} \qquad l^i = \frac{1}{\sqrt{g}} L^i$$

Siccome abbiamo la relazione evidente $p_i p^i_\alpha = 0$ il vettore l^i è normale all'elemento iperpiano individuato da p_i.

Abbiamo allora una funzione $L(x,p)$ sulle densità vettoriali covarianti di peso -1 , da cui abbiamo ottenuto un tensore a due indici g^{ij}, g_{ij} che è omogenea del grado zero nelle p , e che può servire come tensore metrico. Allora possiamo mettere alla base di una geometria una funzione $L(x,p)$ di classe C^4 nelle variabili x,p per le quali sussistono le condizioni

(a) il caso $p_1 = p_2 = \dots = p_n$ è escluso

(b) L è positivamente omognenea di primo grado in $p_1,\dots,p_n$ cioè per $k > 0$ avremo $L(x,kp) = kL(x,p)$

(c) $L > 0$

(d) $F_1 > 0$ dove $F_1 p_i p_k$ è il complemento algebrico di L^{ij} nel determinante $\left| L^{ij} \right|$.

Avendo introdotto già il tensore metrico, la seconda nozione fondamentale di cui avremo bisogno è quella di una connessione, che ci permetta di introdurre un procedimento di differenziazione assoluta per un vettore $T^i(x,p)$

(37) $$DT^i = dT^i + (\Gamma^i_{jk}(x,p)dx^k + C_j^{ih}(x,p)dp_h)X^j$$

e tale che la connessione sia <u>euclidea</u>, cioè $Dg_{ij} = 0$ che si può esprimere nella forma

(38) (a) $$\partial_k g_{ij} = \Gamma_{ijk} + \Gamma_{jik} \qquad \text{(b)} \quad \partial^k g_{ij} \equiv \frac{\partial g_{ij}}{\partial p_k} = C_{ij}^{\ \ k} + C_{ji}^{\ \ k}$$

Cartan ha introdotto delle convenzioni di natura intrinseca del tutto analoghe a quelle che abbiamo introdotto nel caso degli spazi di Finsler. Per la determinazione delle $C_{ij}^{\ \ k}$ ha introdotto la condizione che si traduce analiticamente con la simmetria delle C negli indici inferiori, che dà subito

(39) $$C_{ij}^{\ \ k} = \frac{1}{2}\partial^k g_{ij}$$

Vogliamo sostituire per le funzioni $C_{ij}^{\ \ k}$ (che sono del grado -1 nelle p) le funzioni $A_{ij}^{\ \ k} = \frac{L}{\sqrt{g}} C_{ij}^{\ \ k}$. Introducendo il differenziale assoluto del vettore l, e scrivendo $\omega_i = Dl_i$, possiamo modificare (37) nella forma

(40) $$DT^i = dT^i + (\overset{*}{\Gamma}{}^i_{jk}\, dx^k + A_j^{\ ih}\,\omega_h)T^j$$

dove

(41) $$\overset{*}{\Gamma}{}^i_{jk} = \Gamma^i_{jk} + A_j^{\ ir}\,\Gamma_{rok} \quad ; \quad \Gamma_{rok} = \Gamma_r{}^m{}_k\, l_m$$

La condizione di simmetria delle funzioni $\overset{*}{\Gamma}$ negli indici inferiori, insieme alle condizioni (38a) sono n^3 equazioni nelle n^3 funzioni Γ. Ma per il calcolo delle funzioni Γ è necessario che il tensore

$$(42) \qquad H^{ij} = g^{ij} + A_k A^{kij} \quad ; \qquad A_k = g_{ki} A^i ; \quad A^i = A_k^{\ ki}$$

abbia un determinante $|H^{ij}| \neq 0$. Gli spazi per cui il determinante non è zero sono gli spazi di Cartan <u>regolari</u>. In uno spazio di Cartan regolare si può riguardare un'ipersuperficie a dimensione n-1 come il luogo dei suoi elementi iperpiani. Per tale sottospazio a dimensione (n-1) si possono introdurre le due forme fondamentali come nello spazio riemanniano. Se le equazioni parametriche son $x^i = x^i(t^1, \ldots, t^{n-1})$ la prima forma fondamentale sarà

$$(43) \qquad ds^2 = g_{\alpha\beta}\, dt^\alpha\, dt^\beta \qquad \text{dove} \qquad g_{\alpha\beta} = g_{ij} p^i_\alpha\, p^j_\beta$$

La seconda forma fondamentale può definirsi come $l_i D^2 x^i = - dx^i D l_i$ che si può anche esprimere nella forma

$$(44) \qquad a_{\alpha\beta}\, dt^\alpha\, dt^\beta$$

L'invariante che è la generalizzazione della curvatura media nella geometria riemanniana è scritto

$$(45) \qquad M = g^{\alpha\beta} a_{\alpha\beta}$$

E' stato dimostrato da Cartan e anche da Berwald che le superficie ad (n-1) dimensioni che sono minimali per l'integrale (9) sono caratterizzate da $M = 0$.

Una teoria di curve è anche possibile, badando sempre al fatto che si tratta di un luogo di <u>elementi</u> piuttosto che di punti.

Per sviluppare una teoria di curve, allora bisogna associare ad ogni punto di una curva l'elemento <u>normale</u> alla curva. Le curve autoparallele sono le curve per cui $\omega^i \equiv Dl^i = 0$. Le curve che realizzano un valore estremale dell'integrale $\int ds$ però, sono caratterizzate da

$$DA^i + (H^{ij} + \frac{L^i}{\sqrt{g}} A^j)\omega_j = 0 \quad . \tag{46}$$

Vediamo allora dalle (46) che le curve estremali e le curve autoparallele non coincidono nella geometria di Cartan. In una memoria del 1948 Allardin si è posto il problema di trovare una connessione, che egli chiama connessione G-E, in cui le due classi di curve coincidono. Nella sua teoria un tensore simmetrico

$$K^{ij} = \frac{L}{\sqrt{g}} \partial^j A^i + 2A_k A^{kij} + \frac{L^i}{\sqrt{g}} A^j + g^{ij}$$

rientra nelle considerazioni invece del tensore H^{ij} di Cartan. Allardin ha dimostrato che se $|K^{ij}| \neq 0$ si possono calcolare i coefficienti di una connessione G-E.

Quest'anno è stato dimostrato da Aassi che si può ottenere una connessione quando il determinante

$$\left|(1-t)H^{ij} + t\, K^{ij}\right| \neq 0$$

per ogni valore di t . Per $t = 0$ si ottiene la connessione di Cartan, e per $t = 1$ si ottiene quella di Allardin.

E.T.Davies

VII LEZIONE

SPAZI AREALI

Abbiamo considerato gli spazi di Riemann, di Finsler e di Cartan. Nei tre casi siamo arrivati senza difficoltà ad un tensore a due indici che può funzionare come tensore metrico. Per il caso $1 < m < n-1$ però vi sono delle difficoltà che non si presentano negli spazi "modelli" di Riemann, Finsler e Cartan. Ci occuperemo di questo caso nella presente lezione.

Il calcolo delle variazioni per integrali multipli è stato relativamente poco studiato. Come per integrali semplici, gli autori hanno tentato di introdurre delle condizioni che generalizzano quelle di Legendre e Weierstrass, ma la generalizzazione non è semplice. Per il nostro scopo vogliamo solamente far cenno ad una memoria fondamentale di Caratheodory (Acta Szeged 1929) dove si trovano generalizzazioni delle variabili canoniche, campi geodetici, della funzione eccesso di Weierstrass, e della condizione di Legendre. Se l'integrale da considerare è

$$(1) \qquad \int_{(m)} F(x^i, t^\alpha, p^i) dt^1 \dots dt^m$$

e se scriviamo $F_{;i}^{\alpha} = \partial_i^\alpha F = \partial F / \partial p_\alpha^i$, $\quad p_i^\alpha = (\log F)_{;i}^{\alpha}$

$$\bar{F} = F(x, t, \bar{p}) \ , \qquad \text{dove} \qquad \bar{p}_\alpha^i \neq p_\alpha^i$$

la funzione eccesso di Weierstrass si scrive

$$(2) \qquad E = \bar{F} - F^{1-m} \left| \delta_\beta^\alpha F + F_{;i}^{\alpha} (\bar{p}_\beta^i - p_\beta^i) \right|$$

Lo sviluppo secondo potenze di $\bar{p}^i_\alpha - p^i_\alpha$ del determinante (20) darà un termine corrispondente a quello che ci da la condizione di Legendre per m=1. Se mettiamo

$$(3) \qquad L^{\alpha\beta}_{ij} = F^{-1} F_{;ij}^{\alpha\beta} - p^\alpha_i p^\beta_j + p^\alpha_j p^\beta_i$$

la condizione di Legendre per integrali multipli si scrive

$$(4) \qquad L^{\alpha\beta}_{ij} (\bar{p}^i_\alpha - p^i_\alpha)(\bar{p}^j_\beta - p^j_\beta) > 0$$

Siccome si tratta in questa lezione della possibilità di fondare una geometria fondata su un integrale multiplo come (1), sussistono naturalmente le condizioni d'invarianza rispetto a cambiamenti dei parametri t . Allora le t non entrano esplicitamente, e sussistono anche le condizioni

$$(5) \qquad F_{;i}^{\alpha} p^i_\beta = \delta^\alpha_\beta F \qquad o \qquad p^\alpha_i p^i_\beta = \delta^\alpha_\beta$$

La funzione F(x,p) come funzione delle n + nm variabili indipendenti x,p sarà di classe C^4 in tutte le variabili, e soddisfa anche alle condizioni

(a) $F(x,p_\alpha) > 0$ per p_α linearmente indipendenti

(b) $F(x, \lambda^\alpha_{\alpha'} p_\alpha) = |\lambda^\alpha_{\alpha'}| F(x,p_\alpha)$ per $|\lambda^\alpha_{\alpha'}| > 0$

Sotto tali condizioni possiamo definire come <u>area</u> m-dimensionale di una regione Ω detto sotto-spazio definito dalle equazioni parametriche

$$(6) \qquad x^i = x^i(t^1,\dots,t^m)$$

come il valore dell'integrale

$$(7) \qquad S = \int_\Omega F(x, \frac{\partial x}{\partial t}) dt$$

e tratteremo della possibilità di definire la "misura" di un m-vettore, la lunghezza di un vettore, e di introdurre le nozioni geometriche di connessione, di curvatura etc.

La funzione F che soddisfa alle condizioni annunciate può esprimersi come una funzione positivamente omogenea nelle componenti di un m-vettore semplice

$$(8) \qquad p^I \equiv m!\, p^{i_1}_{[1} \cdots p^{i_m}_{m]}$$

cioè, esiste una funzione f di x e di p^I tale che

$$(9) \qquad f(x, p^I) \equiv F(x, p^i_\alpha)$$

Siccome le componenti p^I non sono tutte indipendenti, ma soddisfano alle identità di Plücker, non si possono formare le derivate parziali rispetto a p^I nel senso ordinario, ma un'operazione che possiede qualche proprietà di una derivata è stata introdotta da Iwamoto che si può definire come segue

$$(10) \qquad f_I = \frac{m!}{f^{m-1}} F^{[1}_{;i_1} \cdots F^{m]}_{;i_m} = m!\, F^{[1}_{;i_1} p^{2}_{i_2} \cdots p^{m]}_{i_m}$$

La seconda "derivata" f_{IJ} è molto più complicata e si devono consultare le memorie di Iwamoto, Kawaguchi e Barthel per l'espressione. Conosciuta l'espressione di f_{IJ}, un tensore a 2m indici si può definire con la formula

$$(11) \qquad g_{IJ} = (\tfrac{1}{2} f^2)_{IJ} = f_I f_J + f\, f_{IJ}$$

Segue subito dalla definizione di queste "derivate" generalizzate che g_{IJ} è omogeneo del grado zero nelle p^I, e che

$$(12) \qquad g_{IJ}p^J = f\, f_I \quad , \quad g_{IJ}\, p^I p^J = f^2 = F^2$$

Si può dire allora che $f(x,p^I)$ è la "misura" di p^I nello stesso modo che per lo spazio di Finsler abbiamo definito come la "lunghezza" di ξ il numero $L(x,\xi)$. Nel caso di uno spazio di Riemann

$$(13) \qquad g_{IJ} \equiv g_{i_1 \ldots i_m j_1 \ldots j_m} = \begin{vmatrix} g_{i_1 j_1} & \ldots & g_{i_1 j_m} \\ \cdot & \cdot\ \cdot\ \cdot & \cdot \\ g_{i_m j_1} & \ldots & g_{i_m j_m} \end{vmatrix}$$

Relazione dello stesso tipo sussistono anche per gli spazi di Finsler e di Cartan. Nel caso generale in cui $1 < m < n-1$ però non esiste in generale un modo di ottenere un teorema a <u>due</u> indici che può servire come un tensore per ottenere la lunghezza di vettori. Kawaguchi dimostrò che se il determinante di ordine $\binom{n}{m}$ delle componenti di g_{IJ} non svanisce, e se m e n sono numeri senza fattore comune, un tensore a due indici può sempre ottenersi dalla funzione F con operazioni algebriche, ma in questo caso la relazione (13) non vale. Infatti Tandai ha dimostrato che sussiste la relazione (13) solamente nei casi di Riemann, Finsler (m=1) e Cartan (m=n-1).

Per il caso di uno spazio di Riemann in cui la funzione F è l'elemento di area m-dimensionale, la "forma di Legendre" (3) prende una forma semplice, come si può dimostrare. Poniamo

$$(14) \qquad g_{\alpha\beta} = g_{ij} p^i_\alpha p^j_\beta \quad , \quad p^\alpha_i = g^{\alpha\beta} g_{ij}\, p^j_\beta \quad \text{cioè} \quad p^\alpha_i\, p^i_\beta = \delta^\alpha_\beta$$

$$(15) \qquad \beta^i_j = p^i_\alpha\, p^\alpha_j \quad , \quad \gamma^i_j = \delta^i_j - \beta^i_j$$

(16) $$F^2 = |g_{\alpha\beta}| = g_{IJ}p^I p^J$$

In questo caso avremo

(17) $$F_{;ij}^{\alpha\beta} = F\,(g^{\alpha\beta}g_{ij} - g^{\alpha\beta}\,g_{ik}\beta^k_j + p^\alpha_i\,p^\beta_j - p^\alpha_j\,p^\beta_i)$$

che dà

(18) $$L^{\alpha\beta}_{ij} = g^{\alpha\beta}\,g_{ik}\,\gamma^k_j$$

La forma di Legendre prende la forma (18) anche nei casi di Finsler e di Cartan. Il fatto che la forma di Legendre possa esprimersi secondo (18) è equivalente all'esistenza di un tensore a due indici per cui sussiste (13). Gli autori chiamano "metric" tali casi. Il caso in cui si può dedurre un tensore a due indici senza che valga (13) è chiamato il caso "submetric".

Vogliamo adesso dare un'espressione per la forma di Legendre nel caso "submetric". Sia $\bar{g}_{ij}$ il tensore ottenuto da F , e sia $\bar{g}_{\alpha\beta} = \bar{g}_{ij}\,p^i_\alpha p^j_\beta$. Possiamo introdurre un fattore ρ tale che se $g_{\alpha\beta} = \varphi\,\bar{g}_{\alpha\beta}$ sussiste $|g_{\alpha\beta}| = F^2$. Dato $g_{\alpha\beta}$ possiamo dedurre il cosidetto "tensore metrico normalizzato" dalla formula

(19) $$g_{ij} = (\frac{1}{m}\,L^{\alpha\beta}_{ij} + p^\alpha_i\,p^\beta_j)g_{\alpha\beta}$$

Segue subito dalla proprietà $L^{\alpha\beta}_{ij}\,p^j_\gamma = 0$ che

(20) $$g_{\alpha\beta} = g_{ij}p^i_\alpha\,p^j_\beta \quad \text{e} \quad p^\alpha_i = g^{\alpha\beta}\,g_{ij}p^j_\beta$$

Non è difficile ottenere l'espressione

(21) $$L^{\alpha\beta}_{ij} = g^{\alpha\delta}\,g_{mn;j}^{\;\;\beta}\,\gamma^m_i\,p^n_\delta + g^{\alpha\beta}\,g_{ik}\,\gamma^k_j = \overset{*}{L}{}^{\alpha\beta}_{ij} + \hat{L}^{\alpha\beta}_{ij}$$

dove $\hat{L}$ corrisponde all'espressione per $L^{\alpha\beta}_{ij}$ nel caso di uno degli spazi "modelli" di Riemann, Finsler e Cartan, e $\overset{*}{L}$ è la parte di L che svanisce per il caso "metric", e che Kawaguchi chiama il tensore "ecmetric". Recentemente Tandai ha studiato un caso "semimetric" per cui $\overset{*}{L}{}^{(\alpha\beta)}_{ij} = 0$. Si può dimostrare un fatto geometrico interessante che in questo caso il vettore di Eulero per i sottospazi estremali ci da

$$(22) \qquad \hat{L}^{\alpha\beta}_{ij} \left(\partial_\beta p^j_\alpha + \left\{ {j \atop mn} \right\} p^m_\alpha p^n_\beta \right) = 0$$

Il problema centrale in una geometria è di trovare una legge di trasporto, e di ottenere le funzioni che definiscono la connessione della funzione fondamentale F . Vogliamo dimostrare adesso come si può determinare anche in questo caso (submetric) di $1 < m < n-1$ una tale connessione. Sia $g_{ij}(x,\dot{p})$ il tensore metrico normalizzato nel caso "submetric" di cui trattiamo. Se in una certa regione le p^i_α sono funzioni delle x , vi sarà un campo di m-vettori nella regione, e le $g_{ij}(x,p(x)) \equiv \gamma_{ij}(x)$ definiscono uno spazio riemanniano. Senza ridursi a questo caso speciale possiamo determinare un trasporto <u>lungo una curva</u> da una relazione <u>non integrabile</u> della forma

$$(23) \qquad dp^i_\alpha + G^i_{j\alpha}(x,p)dx^j = 0$$

dove per il momento le funzioni G sono arbitrarie salvo per l'omogeneità di grado uno nelle p , e di classe C^2. Per ogni scelta di G abbiamo un determinato spazio osculatore lungo la curva $x^i = x^i(t)$ con la metrica $\gamma_{ij}(x) = g_{ij}(x,p[x])$ dove $p[x]$ significa che non si tratta di una funzione come nel caso in cui (23) è integrabile.

Per le derivate parziali delle γ_{ij} avremo

(24) $$\partial_k \gamma_{ij} = \partial_k g_{ij} - G^m_{k\alpha} \partial^\alpha_m g_{ij} = X_k g_{ij}$$

dove scriviamo $X_k = \partial_k - G^m_{k\alpha}\partial^\alpha_m$. I simboli di Christoffel per le γ_{ij} sono allora utilizzabili per definire un differenziale assoluto. Scriviamo

(25) $$2\overset{o}{\Gamma}_{ijk} = X_k g_{ij} + X_i g_{jk} - X_j g_{ik} \quad \text{con} \quad \overset{o}{\Gamma}{}^h_{ik} = g^{hj}\overset{o}{\Gamma}_{ijk}$$

che si possono prendere come coefficienti di connessione nello spazio areale. Soddisferanno alle condizioni (i) di avere la ben nota legge di trasformazione (ii) di determinare una connessione euclidea, cioè tale che la derivata covariante di g_{ij} sia zero. Ma dobbiamo trovare una funzione G tale che le corrispondenti funzioni $\overset{o}{\Gamma}$ siano (iii) dedotte dalla funzione fondamentale F cioè dalle g_{ij}, (iv) tale che per m=1 si riducono alle funzioni $\overset{*}{\Gamma}$ di Cartan per lo spazio di Finsler. Se scriviamo

$$2\gamma_{ijk} = \partial_k g_{ij} + \partial_i g_{jk} - \partial_j g_{ik} \quad \text{con} \quad \gamma^h_{ik} = g^{hj}\gamma_{ijk}$$

vi sono varie espressioni che si possono prendere per G^i_{jk} tali che $G^i_{j\alpha} \equiv G^i_{jk}p^k_\alpha$ soddisfino alle condizioni necessarie. Per esempio possiamo prendere

(26) $$G^i_{jk} = \gamma^i_{jk} - C^{i\,\gamma}_{j,m}\gamma^m_{nk}p^n_\gamma \quad \text{con} \quad C^{i\gamma}_{j,m} = g^{ih}C^{\ \gamma}_{jh,m}$$

e dove

(27) $$2C_{ij,k}^{\ \alpha} = g_{ij;k}^{\ \alpha} + g_{mn;k}^{\ \alpha}(\gamma^m_i\beta^n_j - \gamma^m_j\beta^n_i)$$

E' ovvio che le funzioni $G^{i}_{j\alpha}$ siano dedotte dalla metrica e che $C^{i\,\alpha}_{j',k}$ si riducano alle funzioni C^{i}_{jk} di Cartan per il caso m=1, cioè le $\overset{o}{\Gamma}$ divengano le $\overset{*}{\Gamma}$ di Cartan. Le condizioni (i), (ii), (iii) e (iv) sono allora soddisfatte dalle funzioni $\overset{o}{\Gamma}{}^{i}_{jk}$.

Un calcolo della prima variazione dell'integrale (7) darà come condizione per uno spazio subordinato di area minima l'equazione

$$(28) \qquad L^{\alpha\beta}_{ij} (\partial_\beta p^{i}_{\alpha} + \overset{o}{\Gamma}{}^{i}_{mn} p^{m}_{\alpha} p^{n}_{\beta}) + \beta^{m}_{n} \overset{o}{\Gamma}{}^{n}_{mk;j}{}^{\alpha} p^{k}_{\alpha} = 0$$

Dobbiamo allora cercare funzioni $\overset{*}{\Gamma}$ che soddisfino alle condizioni già ottenute per le $\overset{o}{\Gamma}$, ma che soddisfino anche alla relazione

$$(29) \qquad \beta^{m}_{n} \overset{*}{\Gamma}{}^{n}_{mk;j}{}^{\alpha} p^{k}_{\alpha} = 0$$

Per ottenere tali funzioni scriviamo prima

$$(30) \qquad \Gamma^{h}_{ij} = \gamma^{h}_{ij} - \frac{1}{2} g^{hn} \left[G^{m}_{i\varepsilon}\, g_{jn;m}^{\ \ \cdot\,\varepsilon} - G^{m}_{n\varepsilon}\, g_{ij;m}^{\ \ \cdot\,\varepsilon} \right]$$

e poi

$$(31) \qquad \overset{*}{\Gamma}{}^{h}_{ij} = \Gamma^{h}_{ij} - C^{h\,\alpha}_{i,m}\, \Gamma^{m}_{\alpha j}$$

Queste funzioni soddisfano alla (29), il che dimostra allora che la condizione (v) di avere le superficie estremali caratterizzate da

$$(32) \qquad L^{\alpha\beta}_{ij} (\partial_\beta p^{i}_{\alpha} + \overset{*}{\Gamma}{}^{i}_{mn} p^{m}_{\alpha} p^{n}_{\beta}) = 0$$

è soddisfatta.

E.T.Davies

VIII LEZIONE

CALCOLO TENSORIALE DI CONTATTO

Abbiamo considerato delle geometrie in cui i tensori fondamentali sono funzioni delle coordinate x e anche di un ente geometrico come un vettore per gli spazi di Finsler, o una densità nel caso degli spazi di Cartan. Abbiamo però considerato solamente le trasformazioni delle coordinate.

In questo capitolo vogliamo considerare la possibilità di estendere i procedimenti del calcolo tensoriale classico (rispetto a trasformazioni di coordinate x) ad un calcolo rispetto a trasformazioni più generali tra 2n variabili $(x^i, p_i) \longrightarrow (x^\lambda, p_\lambda)$. In questo capitolo gli indici α, β, γ,... , ad anche a, b, c,... corrono da 1 a 2n ; i, j, k, ..., e λ, μ, ν,... da 1 ad n , mentre il simbolo $\hat{i}$, o $\hat{\mu}$ sarà sempre i + n o μ + n.

Una trasformazione tra due sistemi di indici $(x^i, p_i) \longrightarrow (x^\lambda, p_\lambda)$ di classe C^2 sarà una trasformazione <u>canonica</u> se esiste una funzione $\psi\,(x^i, p_i)$ tale che

$$p_\lambda\, dx^\lambda \equiv p_i\, dx^i + d\psi\,(x,p) \tag{1}$$

una condizione che è equivalente alle equazioni

$$p_\lambda \frac{\partial x^\lambda}{\partial x^i} = p_i + \frac{\partial \psi}{\partial x^i} \quad , \qquad p_\lambda \frac{\partial x^\lambda}{\partial p_i} = \frac{\partial \psi}{\partial p_i} \tag{2}$$

Derivando le (2) ed eliminando le derivate seconde, abbiamo

$$(3)\qquad \begin{aligned} &\frac{\partial x^\lambda}{\partial x^i}\frac{\partial p_\lambda}{\partial x^j} - i|j = 0 \\ &\frac{\partial x^\lambda}{\partial x^i}\frac{\partial p_\lambda}{\partial p_j} - i|j = \delta_i^j \\ &\frac{\partial x^\lambda}{\partial p_i}\frac{\partial p_\lambda}{\partial p_j} - i|j = 0 \end{aligned}$$

e si può verificare facilmente che il determinante funzionale

$$\left[\frac{\partial(x^i,\ p_i)}{\partial(x^\lambda,\ p_\lambda)}\right]^2 = 1$$

Nei trattati classici si trovano anche le relazioni

$$(4)\qquad \frac{\partial p_\lambda}{\partial p_i} = \frac{\partial x^i}{\partial x^\lambda}\ ,\quad \frac{\partial x^i}{\partial p_\lambda} + \frac{\partial x^\lambda}{\partial p_i} = 0\ ,\quad \frac{\partial p_\lambda}{\partial x^i} + \frac{\partial p_i}{\partial x^\lambda} = 0$$

Un caso speciale importante è il caso $d\psi \equiv 0$, quando le (2) divengono

$$(5)\qquad p_\lambda \frac{\partial x^\lambda}{\partial x^i} = p_i \quad ; \quad p_\lambda \frac{\partial x^\lambda}{\partial p_i} = 0$$

che si può modificare, dalle equazioni (4), nella forma

$$(6)\qquad p_\lambda \frac{\partial p_i}{\partial p_\lambda} = p_i \quad ; \quad p_\lambda \frac{\partial x^i}{\partial p_\lambda} = 0$$

che conducono alla conclusione che le funzioni $x^i(x^\lambda, p_\lambda)$, $p_i = p_i(x^\lambda, p_\lambda)$ sono omogenee rispettivamente del grado 0 e 1 nelle p_λ . Le corrispondenti trasformazioni canoniche si chiamano <u>trasformazioni canoniche omogenee</u>.

Siano x^i, p_i, z (2n+1) variabili. Una trasformazione $x,p,z \longrightarrow \bar{x},\bar{p},\bar{z}$ tale che

$$d\bar{z} - \bar{p}_i d\bar{x}^i = \rho\,(x,p,z)(dz - p_i dx^i)$$

è una trasformazione di contatto. E' dimostrato da Caratheodory [Satz.I, p.108] che la più generale trasformazione di contatto nello spazio ad (n+1) dimensioni e la più generale trasformazione canonica omogenea in (2n+2) variabili, sono identiche come concetti. Possiamo allora parlare indifferentemente di trasformazioni per cui le equazioni (4) e (5) sono valide come trasformazioni canoniche o trasformazioni di contatto. Vogliamo dare qualche indicazione della possibilità di sviluppare un calcolo tensoriale "di contatto".

Sia V_{2n} una varietà dotata di una metrica col tensore fondamentale g_{ab} e un tensore di torsione $S_{bc}^{\cdot\cdot a}$ come nella lezione IV. Siano $(p_i^{\cdot a},\ q_{\hat{i}}^{\cdot a})(p_{\cdot a}^{i},\ q_{\cdot a}^{\hat{i}})$ due matrici reciproche come al IV. I tensori β e γ sono proiettori nel solito senso della teoria degli spazi vettoriali con la proprietà

$$\beta^2 = \beta \qquad , \qquad \gamma^2 = \gamma \ .$$

Se v è un vettore qualunque di V_{2n} , βv e γv appartengono rispettivamente agli spazi vettoriali caratterizzati dai vettori $p_i^{\cdot a}$ e $q_{\hat{i}}^{\cdot a}$ rispettivamente.

Sia v^i un vettore uscente da un punto qualunque x^a di V_{2n} che appartiene allo spazio vettoriale delle $p_i^{\cdot a}$ nel punto. Si può riguardare v^i anche come vettore di V_{2n} con le componenti $w^a = p_i^{\cdot a} v^i$. Si può definire una derivata assoluta subordinata nello spazio anolonomo V_{2n}^{n} delle $p_i^{\cdot a}$ mettendo

$$Dv^i = p_{\cdot a}^{i}\, Dw^a$$

che servirà a definire coefficienti di connessione $\Gamma_{jk}^{i}(x)$ in ogni punto di V_{2n} per vettori di V^n_{2n}. Se $\omega^{\hat{i}} = q^{\hat{i}}_a dx^a = 0$ è una distribuzione integrabile, avremo una varietà V_n olonoma che si può parametrizzare con parametri u^i. In questo caso v^i sono vettori dello spazio V_n ed hanno, per trasformazioni di parametri u una legge di trasformazione

$$v^i = v^\lambda \frac{\partial u^i}{\partial u^\lambda}$$

Nel caso in cui $\omega^{\hat{i}} = 0$ non è integrabile possiamo considerare trasformazioni lineari più generali

$$v^i = v^i B^i_\lambda \qquad \text{con} \quad |B^i_\lambda| \neq 0$$

Le funzioni Γ^i_{jk} avranno una legge di trasformazione che si può dedurre dalla condizione che Dv^i deve avere la stessa trasformazione di v^i, ed avremo subito

(7) $$X_j B^\lambda_i + \Gamma^\lambda_{\mu\nu} B^\mu_i B^\nu_j - \Gamma^k_{ij} B^\lambda_k = 0$$

come relazione tra le $\Gamma^\lambda_{\mu\nu}$ e le Γ^i_{jk}, dove abbiamo messo

(8) $$X_j = p^{\cdot b}_j \partial_b$$

come una derivata pfaffiana.

Per una funzione f(x) qualunque si può mettere

$$df = \partial_a f\, dx^a = X_i f \cdot \omega^i + X_{\hat{i}} f\, \omega^{\hat{i}}$$

Cerchiamo adesso le forme speciali che debbono avere le matrici B^λ_i $C^{\hat{\lambda}}_{\hat{i}}$ con $|B^\lambda_i| \neq 0$ $|C^{\hat{\lambda}}_{\hat{i}}| \neq 0$ in modo che, per una trasformazione di variabili nello spazio generale V_{2n} da $x^a \longrightarrow x^\alpha$ avremo

(9) $$X_i f = B^\lambda_i X_\lambda f \qquad\qquad X_{\hat{i}} f = C^{\hat{\lambda}}_{\hat{i}} X_{\hat{\lambda}} f$$

dove

$$(10) \qquad X_\lambda = p^{\cdot\alpha}_{\lambda}\partial_\alpha \qquad\qquad X_{\hat\lambda} = q^{\cdot\alpha}_{\hat\lambda}\partial_\alpha$$

Le relazioni (9) debbono essere valide per ogni funzione f , allora dobbiamo avere

$$(11) \quad (a)\ \ p_i^{\cdot a} = B_i^\lambda\, p_\lambda^{\cdot\alpha}\frac{\partial x^a}{\partial x^\alpha} = B_i^\lambda\, X_\lambda\, x^a \qquad (b)\ \ q_{\hat i}^{\cdot a} = C_{\hat i}^{\hat\lambda}\, X_{\hat\lambda}\, x^a$$

Prendiamo adesso il caso particolare in cui le matrici (p,q) hanno la forma speciale

$$(12) \quad \begin{cases} p_i^{\cdot a} = \delta_i^j \quad \text{per} \quad a = j & \qquad q_{\hat i}^{\cdot a} = \delta_{\hat i}^{\hat j} \quad \text{per} \quad a = \hat j \\ p^{i}_{\cdot a} = \delta_j^i \quad \text{per} \quad a = j & \qquad q^{\hat i}_{\cdot a} = \delta_{\hat j}^{\hat i} \quad \text{per} \quad a = \hat j \end{cases}$$

Le relazioni di reciprocità delle matrici ci daranno

$$(13) \quad \begin{cases} p_i^{\cdot\hat j} + q^{\hat j}_{\cdot i} = 0 & \qquad p^{i}_{\cdot\hat j} + q_{\hat j}^{\cdot i} = 0 \\ p_i^{\cdot\hat j}\, p^{k}_{\cdot\hat j} = 0 & \qquad q_{\hat j}^{\cdot i}\, q^{\hat j}_{\cdot k} = 0 \end{cases}$$

Se mettiamo a = j nella 11(a) e a = $\hat j$ in 11(b) avremo allora

$$(a)\ \ \delta_i^j = B_i^\lambda\, X_\lambda\, x^j \qquad\qquad (b)\ \ \delta_{\hat i}^{\hat j} = C_{\hat i}^{\hat\lambda}\, X_{\hat\lambda}\, x^{\hat j}$$

che possiamo ritenere nella forma

$$(14) \qquad (a)\ \ B_i^\lambda = X_i x^\lambda \qquad\qquad (b)\ \ C_{\hat i}^{\hat\lambda} = X_{\hat i} x^{\hat\lambda}$$

che danno allora le forme che debbono avere B_i^λ , $C_{\hat i}^{\hat\lambda}$ perchè sussistano le (9).

Siccome dobbiamo avere anche

$$(15)\quad (a)\quad \omega^{i} X_{i} f = \omega^{\lambda} X_{\lambda} f \qquad (b)\quad \omega^{\hat{i}} X_{\hat{i}} f = \omega^{\hat{\lambda}} X_{\hat{\lambda}} f$$

saranno valide

$$(16)\quad (a)\quad \omega^{i} = \omega^{\lambda} X_{\lambda}\, x^{i} \qquad (b)\quad \omega^{\hat{i}} = \omega^{\hat{\lambda}} X_{\hat{\lambda}}\, x^{\hat{i}}$$

Dalla definizione di ω^{i} possiamo dedurre

$$(a)\quad p^{i}_{\cdot a} \frac{\partial x^{a}}{\partial x^{\alpha}} = p^{\lambda}_{\cdot\alpha} X_{\lambda}\, x^{i} \qquad (b)\quad q^{\hat{i}}_{\cdot a} \frac{\partial x^{a}}{\partial x^{\alpha}} = q^{\hat{\lambda}}_{\cdot\alpha} X_{\hat{\lambda}}\, x^{\hat{i}}$$

di qui, per $\alpha = \mu$, otteniamo facilmente

$$(17)\quad (a)\quad p_{\mu}^{\cdot\hat{\lambda}} \partial_{\hat{\lambda}} x^{i} = -q^{\cdot i}_{\hat{j}} \partial_{\mu} x^{\hat{j}} \qquad (b)\quad p_{j}^{\cdot i} \partial_{\hat{\mu}}\, x^{j} = -q^{\cdot\lambda}_{\hat{\mu}}\, \partial_{\lambda}\, x^{\hat{i}}$$

Mettiamo finalmente

$$\begin{cases} a = \hat{j} \\ a = j \end{cases} \quad \text{nella} \quad \begin{cases} (11)(a) \\ (11)(b) \end{cases} \quad \text{ed avremo}$$

$$(18)\quad (a)\quad p_{i}^{\cdot\hat{j}} = X_{i} x^{\lambda}\, X_{\lambda}\, x^{\hat{j}} \qquad (b)\quad q_{\hat{i}}^{\cdot j} = X_{\hat{i}} x^{\hat{\lambda}}\, X_{\hat{\lambda}}\, x^{j}$$

che saranno ritenute nella forma

$$(19)\quad (a)\quad p_{i}^{\cdot\hat{j}} X_{\mu} x^{i} = X_{\mu} x^{\hat{j}} \qquad (b)\quad q_{\hat{i}}^{\cdot j} X_{\hat{\mu}} x^{\hat{i}} = X_{\hat{\mu}}\, x^{j}$$

Ritorniamo adesso alle trasformazioni di contatto, che sono trasformazioni di 2n variabili $(x^{i}, x^{\hat{i}})$ scritte nella forma (x^{i}, p_{i}). In conformità a questa convenzione scriviamo

$$(20)\qquad x^{\hat{i}} = p_{i}\ ,\quad p_{i}^{\cdot\hat{j}} = \Gamma_{ij}\ ,\quad q_{\hat{i}}^{\cdot j} = \Pi^{ij}\ ,\qquad \partial_{\hat{i}} = \partial \big| \partial p_{i} = \partial^{i}$$

la relazione (14)(b) diviene

(21) $$C^{\hat{\lambda}}_{\hat{i}} = X^i_{p_\lambda} = \frac{\partial p_\lambda}{\partial p_i} + \Pi^{ij} \frac{\partial p_\lambda}{\partial x^j} = \partial^i p_\lambda + \Pi^{ij} \partial_j p_\lambda$$

Le (17) divengono

(22) (a) $$\Gamma_{\mu\lambda} \partial^\lambda x^i = -\Pi^{ji} \partial_\mu p_j$$

(b) $$\Gamma_{ji} \partial_\mu x^j = -\Pi^{\mu\lambda} \partial_\lambda p_i$$

e finalmente le (19) divengono

(23) (a) $$\Gamma_{ij} X_\mu x^i = X_\mu p_j \qquad \text{(b)} \quad \Pi^{ij} X^\mu p_i = X^\mu x^j$$

Le funzioni Γ_{ij}, Π^{ij} con le loro trasformate $\Gamma_{\lambda\mu}$, $\Pi^{\lambda\mu}$ sono chiamate "contactframes" da Eisenhart.

Le relazioni (22) insieme alle relazioni fondamentali che definiscono una trasformazione di contatto (4) ci danno subito la relazione fondamentale

(24) $$X^i p_\lambda = X_\lambda x^i$$

Abbiamo adesso gli elementi necessari per un calcolo tensoriale "di contatto". I tensori in questo senso saranno sistemi di funzioni $T^{i_1 \dots i_r}_{j_1 \dots j_s}$ di x e di p. Con una legge di trasformazione

(25) $$T^{\lambda_1 \dots \lambda_r}_{\mu_1 \dots \mu_s} = T^{i_1 \dots i_r}_{j_1 \dots j_s} X^{\lambda_1}_{i_1} \dots\dots X^{\lambda_r}_{i_r} X^{j_1}_{\mu_1} \dots X^{j_s}_{\mu_s}$$

per una trasformazione $(x^i, p_i) \longrightarrow (x^\lambda, p_\lambda)$ che lascia invariante la forma esterna

(26) $$dx \wedge dp \equiv dx^i \delta p_i - \delta x^i dp_i$$

Si può dedurre da un tensore qualunque un tensore con un

indice di covarianza nuova utilizzando funzioni $\Gamma^i_{jk}(x,p)$ con una legge di trasformazione corrispondente alle (7) in cui $B^\lambda_i = X_i x^\lambda$ etc. Una derivazione covariante rispetto ad un indice j corrispondente ad una derivata parziale rispetto a x^j allora si può scrivere, per un vettore T^i qualunque, come

$$(27)\qquad \nabla_j T^i \equiv T^i{}_{,j} = X_j T^i + \Gamma^i_{kj}\, T^k$$

Ma in questo calcolo vi sarà un'altra derivazione che conduce ad un tensore con un indice nuovo di contravarianza. E' una derivazione che si può chiamare derivazione contravariante, che corrisponde alla derivata parziale rispetto alla variabile p_i. Si può scriverla come segue

$$(28)\qquad \nabla^j T^i \equiv T^{i,j} = X^j T^i + C^{ij}_k\, T^k$$

Le funzioni Γ^i_{kj} e C^{ij}_k possono ottenersi con un procedimento di "proiezione" dalle funzioni Γ^a_{hc} definite nella V_{2n} originale. Basta dare l'espressione per C^{ij}_k nella forma

$$(29)\qquad C^{ij}_k = \Gamma^i_{k\hat{j}} = \flat^i_{.a}\,(X_{\hat{j}} p^{.a}_k + \Gamma^a_{bc}\, p^{.b}_k q^{.c}_j)$$

che avrà allora una legge di trasformazione

$$(30)\qquad X^j X_i x^\lambda + C^{\lambda\nu}_\mu\, X_i x^\mu X_\nu\, x^j - C^{kj}_i X_k x^\lambda = 0$$

I procedimenti usuali di alternazione rispetto a queste derivate covarianti e contravarianti ci conducono a tre tipi di tensori di curvatura.

Se esiste un sistema di coordinate per cui le funzioni

$\Pi^{ij} = 0$, le condizioni

$$p_i \Pi^{ij} = 0 \quad , \quad x^k \Pi^{ij} - x^j \Pi^{ik} = 0 \tag{31}$$

debbono essere soddisfatte. Le sole trasformazioni che conservano le equazioni $\Pi^{ij} = 0$ sono le trasformazioni del tipo $x^i = x^i(x^\lambda)$ $p_i = p_\lambda \dfrac{\partial x^\lambda}{\partial x^i}$.

Siamo allora nel caso delle varie geometrie, di Finsler, di Cartan etc. per cui le trasformazioni sono trasformazioni puntuali. Consideriamo in particolare una geometria fondata su una funzione $L(x, p)$ omogenea del primo grado nelle p. Si può facilmente trovare la metrica e la connessione per una tale geometria. Abbiamo subito un tensore metrico

$$g^{ij} = \partial^2(\tfrac{1}{2} L^2)/\partial p_i \partial p_j \quad , \quad \text{con } |g^{ij}| \neq 0 \tag{32}$$

Le funzioni necessarie per definire una connessione saranno $\Gamma^i_{jk}(x,p)$ di grado zero nelle p, e $C^{ik}_j(x,p)$ di grado -1 nelle p, e sono tali che

$$\begin{cases} \partial_k g_{ij} = g_{im} \Gamma^m_{jk} + g_{mj} \Gamma^m_{ik} = \Gamma_{jik} + \Gamma_{ijk} \\ \partial^k g_{ij} = g_{im} C^{mk}_j + g_{mj} C^{mk}_i = C_{ji}{}^k + C_{ij}{}^k \end{cases} \tag{33}$$

Con convenzioni del tutto analoghe a quelle utilizzate per lo spazio di Finsler, avremo per $C_{ij}{}^k$ i valori $\frac{1}{2} \partial^k g_{ij}$. Per determinare le funzioni Γ sarà sufficiente domandare la simmetria negli indici inferiori delle funzioni

$$\overset{*}{\Gamma}{}^i_{jk} = \Gamma^i_{jk} + C^{in}_j \Gamma_{nok} \tag{34}$$

che darà

$$(35) \qquad \overset{*}{\Gamma}_{ijh} = \frac{1}{2}(X_h g_{ij} + X_i g_{jh} - X_j g_{ih})$$

dove

$$(36) \qquad X_h g_{ij} = \partial_h g_{ij} + p_n \Gamma_h{}^n{}_r \dot\partial^r g_{ij} = \partial_h g_{ij} + \Gamma_{hor} \dot\partial^r g_{ij}$$

allora

$$(37) \qquad \overset{*}{\Gamma}_{ijh} = \gamma_{ijh} + C_{ij}{}^r \Gamma_{hor} + C_{jh}{}^r \Gamma_{ior} - C_{ih}{}^r \Gamma_{jor}$$

utilizzando le identità ovvie, avremo

$$(38) \qquad \overset{*}{\Gamma}_{ioh} = \gamma_{ioh} - C_{ih}{}^r \overset{*}{\Gamma}_{oor} \;, \qquad \overset{*}{\Gamma}_{ooh} = \gamma_{ooh}$$

che determina completamente le funzioni.

Vogliamo dimostrare che, da uno spazio riemanniano, si può sempre derivare, con una trasformazione di contatto, uno spazio fondato su una funzione fondamentale $L(x,p)$ sui vettori covarianti. Siano $g_{\lambda\mu}(x^1,\ldots,x^n)$ i coefficienti del tensore metrico di uno spazio riemanniano, e sia $|g_{\lambda\mu}| \neq 0$ in modo che esista un tensore reciproco $g^{\mu\nu}$. Prendiamo $\mathcal{F} = \frac{1}{2}\mathcal{L}^2 = \frac{1}{2} g^{\lambda\mu} p_\lambda p_\mu$ dove p_λ è un vettore covariante. Siano le funzioni

$$x^\lambda = x^\lambda (x^i, p_i)$$
$$p_\lambda = p_\lambda (x^i, p_i)$$

omogenee nelle p_i di grado zero ed uno rispettivamente, e tali che definiscano una trasformazione di contatto.

La funzione $\mathcal{F}$ si trasformerà in una funzione $F(x^i, p_i)$ che sarà del secondo grado nelle p.

La seconda derivata di F rispetto a p_i e p_j sarà allora un tensore che sarà omogeneo del grado zero nelle p, e tale che $g^{ij}p_ip_j = g^{\lambda\mu}p_\lambda p_\mu$, dove $g^{\lambda\mu}$ è una funzione di x solamente, mentre $g^{ij}(x,p)$ è una funzione delle 2n variabili. Siccome prendiamo il caso in cui $|g^{ij}| \neq 0$, si può trovare un tensore reciproco $g_{ij}(x,p)$ che può stare alla base di una geometria come abbiamo fatto nell'ultima sezione. Abbiamo visto allora che una funzione nello spazio di Riemànn V_n può trasformarsi in una funzione nello spazio basato su L(x,p) che **chiamiamo C_n. Se Vogliamo una trasforma**zione di tensori di V_n in tensori di C_n, avremo bisogno di funzioni che possano funzionare come "contact frames". Possiamo seguire un procedimento di Eisenhart. Cominciamo con l'osservazione che $Y^i \equiv \partial F/\partial p_i$ è un vettore contravariante in C_n. Cercheremo la sua legge di trasformazione per una trasformazione di contatto. Abbiamo

$$(39)\qquad Y^i \equiv \frac{\partial F}{\partial p_i} = \frac{\partial \mathcal{F}}{\partial p_\lambda}\frac{\partial p_\lambda}{\partial p_i} + \frac{\partial \mathcal{F}}{\partial x^\lambda}\frac{\partial x^\lambda}{\partial p_i} = \frac{\partial \mathcal{F}}{\partial p_\lambda}\frac{\partial x^i}{\partial x^\lambda} - \frac{\partial \mathcal{F}}{\partial x^\lambda}\frac{\partial x^i}{\partial p_\lambda}$$

Possiamo subito verificare che nello spazio di Riemann V_n

$$(40)\qquad \frac{\partial \mathcal{F}}{\partial x^\lambda} = -\frac{\partial \mathcal{F}}{\partial p^\nu}\, p_\mu \left\{ {\mu \atop \nu\lambda} \right\} = -Y^\nu \Gamma_{\nu\mu}$$

dove abbiamo posto, come definizione

$$(41)\qquad Y^\lambda \equiv \frac{\partial \mathcal{F}}{\partial p^\lambda}, \qquad \Gamma_{\nu\mu} = p_\lambda \left\{ {\lambda \atop \nu\mu} \right\}$$

Allora la legge di trasformazione del vettore Y^i è

$$(42)\qquad Y^i = Y^\lambda\left(\frac{\partial x^i}{\partial x^\lambda} + \Gamma_{\lambda\mu}\frac{\partial x^i}{\partial p_\mu}\right) = Y^\lambda X_\lambda\, x^i$$

Nello stesso modo avremo

$$Y^\lambda \equiv \frac{\partial \mathcal{F}}{\partial p_\lambda} = \frac{\partial F}{\partial p_i}\frac{\partial p_i}{\partial p_\lambda} + \frac{\partial F}{\partial x^i}\frac{\partial x^i}{\partial p_\lambda} = \frac{\partial F}{\partial p_i}\frac{\partial x^\lambda}{\partial x^i} - \frac{\partial F}{\partial x^i}\frac{\partial x^\lambda}{\partial p_i}$$

$$\frac{\partial F}{\partial x^i} = \frac{1}{2}\frac{\partial g^{mn}}{\partial x^i} p_m p_n = -g^{mj}\left\{ {n \atop ij} \right\} p_m p_n = - Y^j \Gamma_{ji}$$

dove abbiamo messo

$$Y^j = g^{mj} p_m \quad , \quad \Gamma_{ji} = \left\{ {n \atop ji} \right\} p_n \tag{43}$$

Allora possiamo scrivere

$$Y^\lambda = Y^j X_j x^\lambda \tag{44}$$

E' evidente che abbiamo, come derivate parziali ordinarie

$$X_j x^\lambda X_\mu x^j = \delta^\lambda_\mu \qquad X_j x^\lambda X_\lambda x^i = \delta^i_j \tag{45}$$

Finalmente, per le funzioni che si possono prendere come "contact frames" abbiamo

$$\frac{\partial F}{\partial x^i} = - Y^j \Gamma_{ji} = \frac{\partial \mathcal{F}}{\partial p_\lambda}\frac{\partial p_\lambda}{\partial x^i} + \frac{\partial \mathcal{F}}{\partial x^\mu}\frac{\partial x^\mu}{\partial x^i} = - Y^\lambda\left(\frac{\partial p_i}{\partial x^\lambda} + \Gamma_{\lambda\mu}\frac{\partial p_i}{\partial p_\mu}\right)$$

e da (44) si può dedurre la legge di trasformazione per le funzioni Γ_{ji} che è

$$\Gamma_{ji} X_\lambda x^j = X_\lambda p_i \tag{46}$$

Vediamo allora che le funzioni che abbiamo preso per "contact frames" si trasformano secondo (23).

Abbiamo determinato gli elementi necessari completamente dalla struttura dello spazio riemanniano originale.

E.T.Davies

B I B L I O G R A F I A

AKBAR-ZADEH, H. : Les espaces de Finsler Thèse, Paris, 1961.

ALARDIN, F. : L'autoparallélisme des courbes extrémales dans les espaces métriques fondés sur la notion d'aire, J. Math.pur.appl. IX, 27, 255-336 (1948).

BARTHEL, W. : Über die minimal flächen in gefaserten Finsler raümen, Annali di Mat. IV, 36, 1-33 (1954).

" " : Über das Verhältnis der Vektorübertragung zu den Variationsproblemen in Cartanschen Raümen, Rend.Circolo Mat.di Palermo II, 3, 270-281 (1954).

" " : Über eine Parallelverschiebung mit Langeninvarianz in lokal Minkowskischen Raümen, Arch.Math. 4, 346-365 (1953).

" " : Extremalprobleme in der Finslerschen Inhaltsgeometrie, Ann.Univ.Saraviensis - Scientia 4, 171-183 (1955).

" " : Über metrische Differentialgeometrie begrundet auf dem Begriff eines p - dimensionalen Areals, Math.Annalen, Bd. 137, 42-63 (1959).

BERGER, M. : Sur certaines variétés riemanniennes à courbure positive, Comptes Rendus de l'Acad. des Sci. (Paris) 247 (1958), 1165-1168.

" " : Les variétés riemanniennes à courbure positive (Bull.Soc.Math.Belgique, t 10, 1958).

" " : Sur quelques variétés riemanniennes suffisamment pincées, (Bull.Soc.Math.France, t 90, 1960).

" " : Les variétés riemanniennes (V 4) - pincées, (Annali Scuola Normale Pisa, Ser III, Vol XIV (1960)).

BERWALD, L. : Untersuchung der Krümmung allgemeiner metrischer Raümen, Math.Zeitschrift, 25, 40-71 (1926).

" " : Über Finslersche und Cartansche geometrie I, Matematica, Cluj, 16 (1940).

" " : Über Finslersche und Cartansche geometrie II, Compositio Math., 7 (1939).

BERWALD, L. : Über die n-dimensionalen Cartanschen Räumen, Acta Math., 71, 191-248 (1939).

BLISS, G.A. : Lectures on the Calculus of Variations, Chicago, 1946.

BOLZA, O. : Variazionsrechnung, Teubner 1909.

BOMPIANI, E. : Sulle connessioni affini non-posizionali, Arch. d.Math., 3, 183-186 (1952).

BORTOLOTTI, E. : Scostamento geodetico e sue generalizzazioni : Giornale di Mat. 66 (1928).

BOTT, R. : Bull.Soc.Math.d.France, 84 (1956) 251-281.

BUSEMANN, H. : The geometry of Finsler spaces, Bull.Amer.Math. Soc., 56, 5-16 (1950).

" " : The Geometry of Geodesics (Academic Press, 1955).

CARATHEODORY, C. : Variationsrechnung und frartielle Differentialgleichnuger ester Ordnung. 1935.

" " : Über die Variationsrechnung bei mehrfachen Integralen (Acta Szeged 4 (1929)).

CARTAN, E. : Les espaces métriques fondés sur la notion d'aire, Paris Hermann, 1933.

" " : Les espaces de Finsler, Paris, Hermann 1934.

COHN-WOSSEN, S. : Vollständige Riemannsche Räume, C.R.Acad des Sciences de l'U.R.S.S., IV (1935), 387-389.

" " : Kürzeste Wege und Total Krummung auf Flächen, Compositio Mathematica, 2, 69-133.

DAVIES, E.T. : On metric spaces based on a vector density, Proc. London Math.Soc. II, 49, 241-259 (1947).

" " : On the invariant Theory of Contact Transformations Math.Zeitschrift, 57, 415-427 (1953).

" " : Subspaces of a Finsler space, Proc.London Math. Soc. II, 49, 19-39 (1947).

" " : On the second variation of a simple integral with movable end-points, (Journal London Math.Soc. 24, (1949)).

DAVIES, E.T. : On the first and second variations of the volume integral (Quarterly Journal Math. (Oxford), 13 (1942).

" " : On the second variation of the volume integral when the boundary is variable (Quarterly Journal Math. (Oxford) 1950).

DEICKE, A. : Finsler spaces as non-holonomic subspaces of Riemannian spaces, Journal London Math.Soc. 30, 53-58, (1955).

DOYLE, T.C. : Tensor decomposition with applications (Annals of Math., 42 (1941)).

EISENHART, L.P. : Finsler spaces derived from Riemann spaces by contact transformations (Annals of Math. 49 (1948)).

GALVANI, O. : La réalisation des connexions euclidiennes d'éléments linéaires et des espaces de Finsler, (Annals Inst.Fourier, Grenoble, 2, 1950).

HOPF, H. : Differentialgeometrie und topologische Gestalt, Jahresbericht der D.M.V., Bd.41, 209-229 (1932).

HOPF, H. e RINOW, W. : Über den Begriff der Vollständigen differentialgeometrischen Fläche, Commentarii Mathematici Helvetici, 3 (1931).

IWAMOTO, H. : On Geometries associated with multiple integrals, Math.Jap., 1, 74-91 (1948).

KAWAGUCHI, A. : On Areal spaces, I-V, Tensor, N.S. (1950-1954).

KLINGENBERG, W. : Proc.Nat.Acad.Sci., 44 (1958), 586-588.

" " : Contributions to Riemannian Geometry in the large (Annals of Math. 69 (1959)).

LAUGWITZ, D. : Zur geometrischen Begründung der Parallelverschiebung in Finslerschen Räumen, Arch. der Math., 6, 448-453.

" " : Die Vektorübertragung in der Finslerschen Geometrie und die Wege geometrie, Indagagiones Math., 18, 21-28 (1956).

LEVI-CIVITA, T. : Sur l'écart géodesique, Math.Annalen, 97 (1926).

LICHNEROWICZ, A. : Les espaces variationnels généralisés (Annales École Normale Supérieure 1945).

MOOR, A. : Die oskulierenden Riemannschen Räume regularer Cartanscher Räume (Acta Math.Acad.1c. Hungaricae V (1954)).

" " : Metrische Dualitat der allgemeinen Räume (Acta Scient. Math. Szeged, XVI (1955)).

MORSE, M. : The Calculus of Variations in the large (American Math.Soc.Coll.Publ.1934).

MYERS, S.B. : Riemannian manifolds in the large, Duke Math. Journal, I, (1935).

" " : Riemannian manifolds with positive mean Curvature (Duke Math.Journal 8 (1941)).

" " : Connections between Differential Geometry and Topology (Duke Math.Journal I (1935)).

PONTRYAGIN, L. : Topological groups, Princeton (1941).

RAUCH, H.E. : On differential geometry in the large, Proc. Nat.Acad.Sci., 39 (1953), 440-443.

" " : A contribution to differential geometry in the large, Annals of Math., 54 (1951).

" " : Geodesics, symmetric spaces and differential geometry in the large, Commentarii Mathematici Helvetici, 27 (1953).

RINOW, W. : Über Zusammenhänge der Differentialgeometrie in grossen und in Kleinen, Math. Zeitschrift, 35 (1932).

RUND, H. : Eine Krümmungstheorie der Finslerschen Räume, Math.Annalen, 125, 1-18 (1952).

" " : Curvature tensors in Finsler spaces, Math. Annalen, 127, 82-104 (1954).

" " : The Differential Geometry of Finsler spaces, Springer Verlag, 1959.

SCHOENBERG, J.M. : Applications of the Calculus of Variations to Riemannian Geometry, Annals of Math. II, 33, 485-495 (1931).

SYNGE, J.L. : The first and second variations of the length integral in Riemannian space, Proc.London Math. Soc., 25 (1926).

" " : On the neighbourhood of a geodesic, Duke Math. Journal I (1925), 527-537.

" " : On the connectivity of spaces of positive curvature, Quarterly Journal of Math. (Oxford), 7, 316-320 (1936).

TANDAI, K. : Areal spaces VI, VII, VIII Tensor N.S. (1953-61).

VARGA, O. : Zur Herleitung des invarianten Differentials in Finslerschen Räumen, Monatshefte für Math., 50 (1942).

VRANCEANU, G. : Studio geometrico dei sistemi anolonomi, Annali di Mat. IV, 4, 9-43 (1928).

" " : Les espaces non-holonomes, Memorial des Sci.Math. (1936).

YANO, K. e DAVIES, E.T. : Contact Tensor Calculus, Annali di Mat., IV, 37 (1954).

CENTRO INTERNAZIONALE MATEMATICO ESTIVO

(C.I.M.E.)

D. LAUGWITZ

GEOMETRICAL METHODS IN THE DIFFERENTIAL GEOMETRY OF FINSLER SPACES

Roma - Istituto Matematico dell'Università

D. Laugwitz

INTRODUCTION

The treatment of metric spaces (Finsler spaces) by the methods of differential geometry involves a lot of geometric objects (tensors, objects of connection etc.), the geometrical background of which is in most cases not obvious. In some cases, these objects are more or less formal generalizations of corresponding objects of Riemannian geometry; in other instances, corresponding Riemannian objects do not exist. In Riemannian geometry, all of the objects needed have a significant geometrical background, either from elementary surface theory or from intrinsic properties. The situation in the more general metric spaces is apparently far more complicated. For instance, geometrical reasons for the choice of the tensors $C_{j\,k}^{\;i}$, $A_{j\,k}^{\;i}$ in the theory of Finsler spaces (CARTAN [1]) and for the choice of the different connection coefficients introduced by various authors are far from being evident. In fact, there are lots of connections for Finsler spaces [1)] which are essentially different from each other, each of them having its special advantages for certain problems and all of them generalizing Levi-Civita's parallelism for Riemannian geometry. The method for introducing a connection in Finsler geometry has in general been the setting up of a number of postulates which lead to a certain object of connection. All of these sets of postulates take a few of the properties of Levi-Civita's connection, which are said to be the essential geometrical properties of this connection; looked at without any bias, any choice of such "fundamental" properties

(1) SYNGE [1], TAYLOR [1], BERWALD [3] and [4], CARTAN [1], RUND [2], BARTHEL [1]; further references will be found in RUND [1].

will seem arbitrary. This state of basing Finsler geometry on objects which are apparently derived by only formal deductions, may look very unsatisfactory. One may ask, if any geometrical procedures might be found to furnish the fundamental geometric objects of Finsler geometry with a geometrical background. Such procedures should, moreover, decide for or against one or the other formal possibility to generalize an object of Riemannian geometry to Finsler spaces. The treatment of such geometrical procedures will be the subject of these lectures.

In the two first parts of these lectures I shall be concerned with

I.) punctal properties,

II.) infinitesimal properties

of Finsler spaces. The term "punctal" will apply to all objects or properties which may be defined by considering the tangent space of a single point. The word "infinitesimal" will be used for objects involving "infinitesimally distant" points of the manifold and their tangent spaces; more precisely, "infinitesimal" applies to all quantities which are expressed by differential equations involving spatial derivatives , $\frac{\partial}{\partial x^i}$ etc. In the elementary theory of surfaces in threedimensional euclidean space, the first and second fundamental forms are punctal objects, Levi-Civita's parallelism, geodesic curvature, and Gauss's curvature are infinitesimal objects. (The term "local", which is often used where I prefer "punctal", will be restricted to properties of finite neighbourhoods for instance, "locally euclidean" means a space of vanishing Riemann curvature. This terminology is in accordance to the use of modern topology.)

In the first part of my lectures I shall give a geometrical background to the punctal properties and objects of Finsler geometry. This background will be furnished by the affine geometry of the indicatrix and figuratrix surfaces. The second part will consist of the exposition and application of two different methods for the deduction of connections. The first method which is a geometrical one will use osculating Riemannian spaces; the other method might be called variational, using the equations of geodesics.

A third part will be concerned with characterizations of euclidean and Riemannian spaces. Apparently this is a different subject of interest. Nevertheless, it is closely related to the other parts of these lectures, not only by the use of the same methods as before, but also from intrinsic reasons. If a geometrical property is characteristic for Riemannian spaces, it can have no geometrical meaning in general Finsler geometry. This will show in which cases geometrical properties of Riemannian geometry cannot be generalized to Finsler spaces. This last part will contain a very short proof of (the punctal part of) the famous Helmholtz-Lie theorem on the characterization of Riemannian spaces and of a related theorem of H.Weyl.

References to the lecture notes prepared by E.BOMPIANI for the present cyclus of C.I.M.E. will be indicated by BOMPIANI p....

I. Punctal properties of Finsler spaces and the affine geometry of the Minkowski indicatrix.

I.1. Osculating quadrics of the indicatrix surface.

Considering the tangent space of a single point (x) of a Finsler space, we shall presently deal with a n-dimensional vector space which is furnished by

$$F(x;\xi) \underset{\text{def}}{=} F(\xi)$$

with a Minkowski metric. All functions will presently depend only on the vector variable ξ of this tangent space.

The surface $F(\xi) = 1$ is called the indicatrix S . By introducing

$$g_{ik}(\xi) \underset{\text{def}}{=} \frac{1}{2} \frac{\partial^2 F^2(\xi)}{\partial \xi^i \partial \xi^k}$$

we may easily describe the osculating quadric

$$g_{ik}(\xi)\,\eta^i \eta^k = 1$$

(Cf. BOMPIANI, p.7).

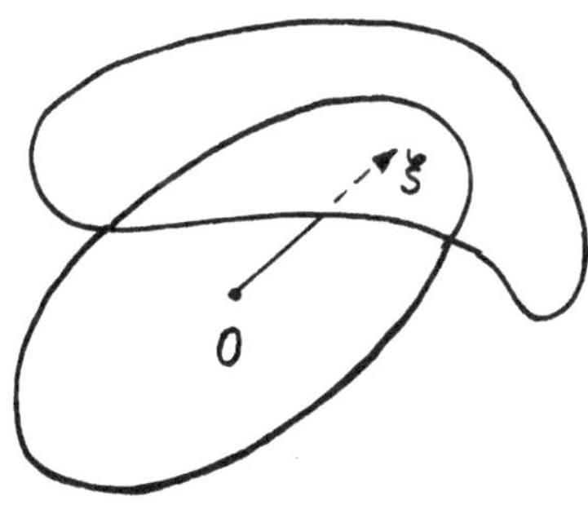

Figure 1

The quadric having the origin O as its center and osculating the surface S in the endpoint of ξ is uniquely determined. The relation between the two surfaces is obviously invariant under the group of all linear homogeneous transformations. This group is precisely the group of all coordinate transformations of the tangent space, induced by general differentiable coordinate transformations of the differentiable manifold. Hence, we have good reason to hope that the geometry of this group (called centro-affine geometry) will have a notable significance for Minkowski and Finsler geometry. We shall begin with some introductory remarks on centro-affine hypersurface theory.

I.2. First fundamental form for hypersurfaces in centro-affine spaces.

We give a definition for the measure of lengths in the surface S in a centro-affinely invariant manner :

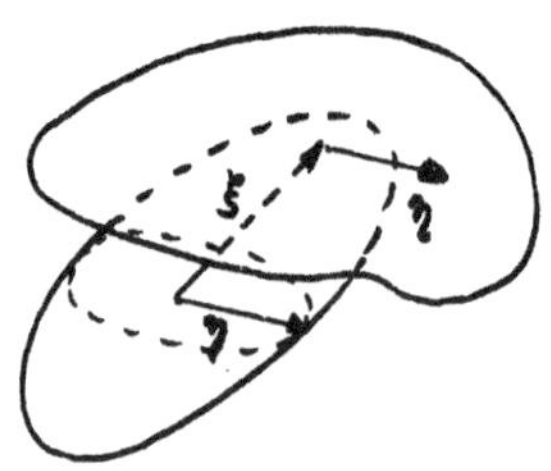

Figure 2

A tangent vector η of S in ξ will have unit length if after a parallel displacement into the origin O its endpoint lies on the quadric which osculates S in ξ and has center 0 . Obviously, this definition may be generalized to a vector η starting from any point P , coordinates ξ , of our vector space : Draw the straight line PO, which will meet S in a point $\tilde{\xi}$; η will by definition be of unit length if after parallel displacement into the origin its endpoint lies on the central quadric osculating S in ξ .

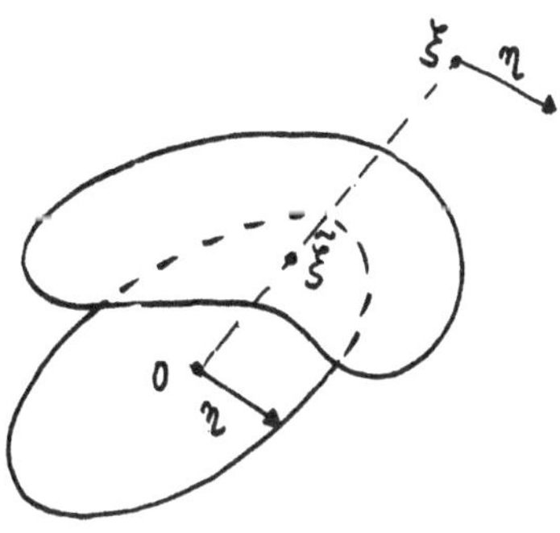

Figure 3

By this construction the whole space (more precisely: a conic section of the space, which is spanned by S) has been furnished with a Riemannian metric. Since

$$g_{ik}(\tilde{\xi}) = g_{ik}(\xi)$$

because of the homogeneity of order zero, we have for this Riemannian line element

$$ds^2 = g_{ik}(\xi)d\eta^i d\eta^k .$$

The metric in the surface S will be (S will have parameters u^α, $\xi^i = \xi^i(u^\alpha)$, Greek indices running from 1 to n-1):

$$d\sigma^2 = \gamma_{\alpha\beta}(u)du^\alpha du^\beta = g_{ik}(\xi(u)) \frac{\partial \xi^i}{\partial u^\alpha} \frac{\partial \xi^k}{\partial u^\beta} du^\alpha du^\beta$$

Here

$$\gamma_{\alpha\beta} \underset{\text{def}}{=} g_{ik} \frac{\partial \xi^i}{\partial u^\alpha} \frac{\partial \xi^k}{\partial u^\beta}$$

is the fundamental tensor of a Riemannian metric in the hypersurface S . This metric is invariant under centro-affine transformations.

The idea of furnishing Minkowski space with a Riemannian metric goes back to a paper by E.R.LORCH [1] . It has been developed in other papers : LORCH & LAUGWITZ [1] , LAUGWITZ [7] , [8] [12] ; earlier, VAGNER [1] published related ideas. The metric is regular, either definite or indefinite, if and only if no tangent plane of S meets O and S contains no parabolic points. This regularity assumptions will be made tacitly throughout these lectures.

In 1935, E.SALKOWSKI [1] and O.MAYER [1] defined a centro-affine metric in a surface in threedimensional space in a different manner; its tensor is (our definition covers all finite numbers of dimension) :

$$\sigma_{\alpha\beta} = \frac{[\xi^i_{\alpha\beta}, \xi^i_1, \dots, \xi^i_{(n-1)}]}{[\xi^i, \xi^i_1, \dots, \xi^i_{(n-1)}]} \qquad \xi^i_\alpha \underset{\text{def}}{=} \frac{\partial \xi^i}{\partial u^\alpha} \quad \text{etc.}$$

(The denominator does not vanish if we assume that no tangent plane of S meets the origin O). This tensor is rather complicated to manage.

We shall show now that it coincides (up to its sign) with $\gamma_{\alpha\beta}$ (LAUGWITZ [8]) :

Theorem :

Proof : We consider two linear forms,

$$l_j y^j \underset{\text{def}}{=} \frac{\left[y^i, \xi^i_1, \ldots, \xi^i_{(n-1)} \right]}{\left[\xi^i, \xi^i_1, \ldots, \xi^i_{(n-1)} \right]}$$

and

$$m_j y^j \underset{\text{def}}{=} g_{ij}(\xi)\, \xi^i y^j$$

From the equation

$$F^2(\xi) = g_{ij}(\xi)\, \xi^i \xi^j = 1$$

we have by the Euler relation

$$\frac{\partial g_{ij}}{\partial \xi} \xi^i = 0$$

that

$$m_j \xi^j = 1 \;, \quad m_j \xi^j_\alpha = 0$$

On the other hand,

$$l_j \xi^j = 1 \quad , \quad l_j \xi^j_\alpha = 0$$

Since the linear forms $l_j y^j$, $m_j y^j$ coincide for n linearly independent vectors $y = \xi$, ξ_α we have identically

$$l_j y^j = m_j y^j \;,$$

and the proposition follows from

$$\sigma_{\alpha\beta} = l_j \xi^j_{\alpha\beta} = m_j \xi^j_{\alpha\beta} = g_{ij}(\xi)\, \xi^i \xi^j_{\alpha\beta} = -\gamma_{\alpha\beta}$$

which holds since

$$g_{ij}(\xi)\, \xi^i \xi^j_{\alpha} = 0$$

yields by differentiation and by virtue of the Euler relation

$$g_{ij}(\xi)\, \xi^i_{\alpha} \xi^j_{\beta} + g_{ij}(\xi)\, \xi^i \xi^j_{\alpha\beta} = 0 .$$

This theorem is the foundation for many relations between the affine geometry of the indicatrix surface and Minkowski geometry (or punctal Finsler geometry).

Since S is now imbedded in a Riemannian space with metrical tensor

$$g_{ik}(\xi) = \frac{1}{2} \frac{\partial^2 F^2(\xi)}{\partial \xi^i \partial \xi^k}$$

wa may use Riemannian differential geometry for the deduction of affinly invariant properties.

Above all, the affine metric of the surface S has a notable bearing on Finsler geometry : It is nothing but the angular metric introduced by LANDSBERG [1] (cf. LAUGWITZ [7] , [12]). That this angular metric, considered as a Riemannian metric in the indicatrix, may serve as a foundation for some tensors of Cartan's theory, was also indicated by VARGA [2] .

Because of

$$g_{ij}(\xi)\, \xi^i \xi^j_{\alpha} = 0 \quad \text{and} \quad g_{ij}(\xi)\, \xi^i \xi^j = 1$$

ξ is the unit normal vector of the surface S . The affine connection of the embedding space has the components

$$\left\{ {i \atop jk} \right\} = \frac{1}{2} \, g^{ir}(\xi) \, \frac{\partial g_{jk}(\xi)}{\partial \xi^r} = C^{\ i}_{j\ k}$$

since

$$\frac{\partial g_{jk}}{\partial \xi^r} = \frac{1}{2} \, \frac{\partial^3 F^2}{\partial \xi^r \partial \xi^j \partial \xi^k} .$$

They coincide with the C's of Cartan's euclidean connection. The second fundamental form of a hypersurface in Riemannian space is defined by

$$l_{\alpha\beta} = - \xi^k_\beta \, \xi^i_\alpha \, \nabla_i \, n_k$$

In our case, this yields

$$l_{\alpha\beta} = - \xi^k_\beta \, \xi^j_\alpha \, \nabla_i \, \xi_k = - \xi^k_\beta \, \xi^i_\alpha \nabla_i \, g_{kj}(\xi) \, \xi^j$$

$$= - \xi^k_\beta \, \xi^i_\alpha \left(\frac{\partial}{\partial \xi^i} (g_{kj}(\xi) \, \xi^j) - \left\{ {r \atop ik} \right\} g_{rj} \, \xi^j \right.$$

$$= - \xi^k_\beta \, \xi^i_\alpha \left(g_{kj} \, \delta^k_i - \frac{1}{2} \, \frac{\partial g_{ik}}{\partial \xi^j} \, \xi^j \right) = - \gamma_{\alpha\beta} .$$

There are striking resemblances to the geometry of the unit sphere in euclidean space: All of the normal vectors point to the center O , and all surface points are umbilics. This analogy to the geometry of the sphere in euclidean space may be pursued in many ways. I shall prove, that the Riemannian curvature of the embedding space is always 0 if the dimension of space is 2. By the way, this does not hold for dimensions n=3, as has been shown by examples (KAWAGUCHI & LAUGWITZ [1]). Let n be equal to 2 ; we shall choose a special type of coordinate system. Take the indicatrix vector ξ as the first basis vector of a basis in the vector space and take the remaining (n-1) basis vectors in such a way that the equation of the osculating quadric becomes

$$\xi_1^2 + \sum_{k=2}^{n} \varepsilon_k \xi_k^2 = 1 , \quad \varepsilon_k^2 = 1$$

where

$$g_{ik}(\xi) = \varepsilon_k \delta_{ik}$$

and Euler's relation

$$g_{ikj}(\xi) \xi^j = 0$$

gives

$$g_{ik1}(\xi) = 0$$

Now, a direct calculation yields

$$R_{ijkl} = \frac{1}{4} g^{rs} (g_{irk} g_{jsl} - g_{irl} g_{jsk})$$

(LORCH & LAUGWITZ [1]).

For n=2 the only essential component of R is R_{1212}, and each term on the right side vanishes because og $g_{ij1} = 0$ and because of the symmetry of g_{ijk} in its three indices.

The curvature tensor R_{ijkl} is closely related to Cartan's tensor S_{ijkl}. By direct calculation,

$$S_{ijkl} = \frac{1}{F^2} R_{ijkl} .$$

The other two curvature tensors of Cartan's theory have not a purely local character. Hence it is impossible to relate them to our method of the affine geometry of the indicatrix.

General references for I.1 and I.2 :
BERWALD [1] , BLASCHKE [1] , CARTAN [1] , DELENS [1] , LAUGWITZ [7] , [11] , [12] , SALKOWSKI [1] .

I.3. The second or cubic fundamental form, the affine normal vector, and related Finsler quantities.

Since $\gamma_{\alpha\beta}$ determines both the metrical and the second quadratic fundamental form, another essential tensor must exist to form a complete system of invariants for the centro-affine theory of surfaces. We shall find this second fundamental form by writing down the equations of structure, presently with undetermined coefficients :

$$\xi^i_{\alpha\beta} = A^\rho_{\alpha\beta}\, \xi^i_\rho + L_{\alpha\beta}\, \xi^i$$

(These equations correspond to the Gauss equations of elementary surface theory. The equivalents to Weingarten's equations assume the trivial shape $\xi^i_\alpha = \xi^i_\alpha$).

By multiplying with $g_{ik}(\xi)\,\xi^k$ we receive

$$g_{ik}(\xi)\xi^k \xi^i_{\alpha\beta} = 0 + L_{\alpha\beta} \cdot 1$$

or

$$L_{\alpha\beta} = - \gamma_{\alpha\beta} \quad .$$

To determine the coefficients $A^{\rho}_{\alpha\beta}$, we multiply by $g_{ik}(\xi)\xi^k_{\gamma}$ which gives the equations

$$\text{(A)} \qquad g_{ik}\, \xi^i_{\alpha\beta}\, \xi^k_{\gamma} = A^{\rho}_{\alpha\beta}\, \gamma_{\rho\gamma} + 0 \quad .$$

Now differentiation of

$$g_{ik}\, \xi^i_{\alpha}\, \xi^k_{\gamma} = \gamma_{\alpha\gamma}$$

yields

$$\frac{\partial \gamma_{\alpha\gamma}}{\partial u^{\beta}} = \frac{\partial g_{ik}}{\partial \xi^j}\, \xi^j_{\beta}\, \xi^i_{\alpha}\, \xi^k_{\gamma} + g_{ik}\, \xi^i_{\alpha\beta}\, \xi^k_{\gamma} + g_{ik}\, \xi^i_{\alpha}\, \xi^k_{\gamma\beta}$$

or, for the Christoffel symbols

$$\text{(B)} \quad \{\alpha\beta,\gamma\} = \frac{1}{2}\, \frac{\partial g_{ik}}{\partial \xi^j}\, \xi^j_{\beta}\, \xi^i_{\alpha}\, \xi^k_{\gamma} + g_{ik}\, \xi^i_{\alpha\beta}\, \xi^k_{\gamma} \quad .$$

If we take

$$a_{\alpha\beta\gamma} \underset{\mathrm{def}}{=} \frac{1}{4}\, \frac{\partial^3 F^2}{\partial \xi^i\, \partial \xi^k\, \partial \xi^j}\, \xi^i_{\alpha}\, \xi^k_{\gamma}\, \xi^j_{\beta}$$

as the coefficients of a (cubic) second fundamental form, we receive by virtue of (A) and (B) :

$$A^{\rho}_{\alpha\beta}\, \gamma_{\rho\gamma} = \{\alpha\beta,\gamma\} - a_{\alpha\beta\gamma}$$

or

$$A^{\rho}_{\alpha\beta} = \left\{ {\rho \atop \alpha\beta} \right\} - a^{\rho}_{\alpha\beta}$$

Hence, the equations of structure are

$$\xi^i_{\alpha\beta} = (\left\{ {\rho \atop \alpha\beta} \right\} - a^\rho_{\alpha\beta})\, \xi^i_\rho - \gamma_{\alpha\beta}\, \xi^i$$

By routine calculations integrability conditions are deduced, and a well-known method shows that $\gamma_{\alpha\beta}$ and $a_{\alpha\beta\gamma}$ determine a surface up to homogeneous affinities (LAUGWITZ [7]).

Four our present purposes, we shall be interested in the affine normal vector. By using Beltrami's differential parameter, we define

$$N^i \underset{\text{def}}{=} \frac{1}{n-1} \gamma^{\alpha\beta} \nabla_\beta \xi^i_\alpha .$$

This becomes by the equations of structure :

$$N^i = \frac{1}{n-1} \gamma^{\alpha\beta} (-a^\rho_{\alpha\beta})\, \xi^i_\rho - \frac{1}{n-1} \gamma^{\alpha\beta} \gamma_{\alpha\beta} \xi^i$$

$$= - \xi^i - \frac{1}{n-1} \gamma^{\alpha\beta} a^\rho_{\alpha\beta}\, \xi^i_\rho .$$

The corresponding vector of the elementary euclidean theory of surfaces has the direction of the normal vector; here, N will be called as customary the "affine normal vector" of the surface S. N has not only a component -1 in normal direction, but in general also a tangential component $-t^i$, The latter will be seen to be rather significant. We shall find other expressions for t^i :

$$t^i = \frac{1}{n-1} \gamma^{\alpha\beta} a^\rho_{\alpha\beta}\, \xi^i_\rho = \frac{1}{2(n-1)} \gamma^{\alpha\beta} \gamma^{\rho\sigma} g_{jkl}\, \xi^j_\alpha\, \xi^k_\beta\, \xi^l_\rho\, \xi^i_\sigma$$

$$= \frac{1}{2(n-1)} (g^{lj} - \xi^l \xi^j)(g^{ik} - \xi^i \xi^k) g_{jkl}$$

$$= \frac{1}{2(n-1)} g^{lj} g^{ik} g_{jkl} = \frac{1}{2(n-1)} g^{ik} \frac{\partial \log g}{\partial \xi^k}$$

Here we used the following expression for g^{jk} :

$$g^{jk} = \xi^j \xi^k + \gamma^{\rho\sigma} \xi^j_\rho \xi^k_\sigma$$

Now, Cartan's vector

$$A^i = \frac{F}{2} g^{ik} \frac{\partial \log g}{\partial \xi^k} = (n-1)F.t^i$$

is essentially equal to the tangent component t^i of the affine normal of the indicatrix. A surface is called an affine sphere, if and only if $t^i = 0$. The spaces having $A^i = 0$, which CARTAN [1] called a natural generalization of the Riemannian spaces, are those Finsler spaces, the indicatrices of which are affine spheres. This was remarked by DEICKE [1] , who proved that the only closed affine spheres are central ellipsoids, generalizing a result of BLASCHKE [1] to $n \neq 3$ (See BOMPIANI p.36; there is a misprint in formula 8.13) [1].

I.4. Geometrical interpretations of the two fundamental tensors.

We have seen that the fundamental geometric objects of Finsler geometry (as far as they are of a punctal character) are closely related to the fundamental tensors $\gamma_{\alpha\beta}$ and $a_{\alpha\beta\gamma}$ of the

---

1) In fact, Deicke did not use the centro-affine geometry, but Blaschke's equi-affine geometry of BLASCHKE [1] . The definition of the affine normal vector according to the surface metric of this of this geometry fortunately leads to the same direction if the surface is an affine sphere. The following theorem holds (LAUGWITZ [12] , p.44) : The affine normal directions of centro-affine and of equi-affine geometry are identical if and only if

for $n \neq 3$: S is an affine sphere,
for $n = 3$: S is an arbitrary surface.

The latter result goes back to SALKOWSKI [1] , p.142.

affine geometry of the indicatrix S. To achieve a geometrical interpretation for the punctal objects of Finsler geometry, we shall have to furnish these two tensors γ and a with a more geometrical background. Both these tensors were hitherto deduced by rather formal applications of the well-known procedures for finding the equations of structure.

We shall discuss the deviations of our indicatrix S both from its tangent hyperplane and from its osculating central quadric.

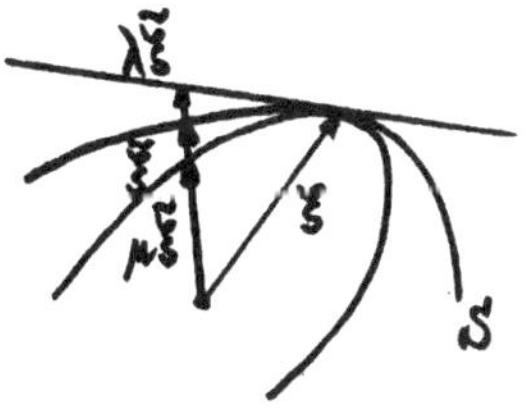

Figure 4

We consider a surface neighbourhood of a point $\xi = \xi(u)$. For any point $\tilde{\xi} = \xi(u+du)$, let $\lambda\tilde{\xi}$ and $\mu\tilde{\xi}$ be the points where the straight line with direction $\tilde{\xi}$ meets the tangent plane and the osculating quadric of ξ respectively. Then,

$$g_{ik}(\xi)\,\xi^i\,(\lambda\xi^k(u+du) - \xi^k(u)) = 0$$

or, by Taylor expansion,

$$g_{ik}(\xi)\,\xi^i\left[(\lambda-1)\,\xi^k + \lambda\,\xi^k_{\alpha}du^{\alpha} + \frac{\lambda}{2}\,\xi^k_{\alpha\beta}\,du^{\alpha}\,du^{\beta} + \dots\right] = 0$$

or, up to terms of at least third order in (du), which will be indicated by ((3)),

$$(\lambda - 1) = \frac{\lambda}{2} ds^2 + ((3)) = 0$$

or
$$ds^2 = 2(1 - \frac{1}{\lambda}) + ((3))$$

and, at last
$$\lambda - 1 = ds^2/2 + ((3)) \quad .$$

Our ds^2 is a centro-affine measure for the deviation of S from its tangent plane. A similar interpretation is due to O.MAYER [1] . The deviation of the surface from its osculating quadric will lead to a formula for the cubic fundamental form. The equation
$$1 = g_{ik}(\xi(u)) \mu \xi^i(u + du) \mu \xi^k(u + du)$$

yields by Taylor expansion
$$\frac{1}{\mu^2} = g_{ik}(\xi) \left[\xi^i(u) + \xi^i_\alpha du^\alpha + \frac{1}{2} \xi^i_{\alpha\beta} du^\alpha du^\beta + \ldots \right] \cdot$$
$$\cdot \left[\xi^k(u) + \xi^k_\rho du^\rho + \frac{1}{2} \xi^k_{\rho\sigma} du^\rho du^\sigma \right]$$

Since
$$\gamma_{\alpha\rho} = g_{ik} \xi^i_\alpha \xi^k_\rho = - g_{ik} \xi^i \xi^k_{\rho\alpha} ,$$

the second order term vanishes, and up to terms of fourth order ((4)), we receive
$$\frac{1}{\mu^2} = 1 + \frac{1}{3} g_{ik} \xi^i \xi^k_{\alpha\beta\gamma} du^\alpha du^\beta du^\gamma +$$
$$+ g_{ik} \xi^i_\alpha \xi^k_{\beta\gamma} du^\alpha du^\beta du^\gamma + ((4))$$

By

$$-\frac{\partial \gamma_{\alpha\beta}}{\partial u^\gamma} = \frac{\partial}{\partial u^\gamma}(g_{ik}\,\xi^i\,\xi^k_{\alpha\beta}) = g_{ik}\,\xi^i\,\xi^k_{\alpha\beta\gamma} + g_{ik}\,\xi^i_\gamma\,\xi^k_{\alpha\beta}$$

and because of the equations of structure, this yields :

$$\frac{1}{\mu^2} = 1 - \frac{2}{3}\,du^\alpha\,du^\beta\,du^\gamma\,a_{\alpha\beta\gamma} + ((4))$$

From

$$\sqrt{1-x} \approx 1 - \frac{x}{2} \ ,$$

$$\frac{1}{\mu} - 1 = -\frac{1}{3}\,du^\alpha\,du^\beta\,du^\gamma\,a_{\alpha\beta\gamma} + ((4))$$

or at least :

$$\mu - 1 = \frac{1}{3}\,du^\alpha\,du^\beta\,du^\gamma\,a_{\alpha\beta\gamma} + ((4))$$

The cubic fundamental form gives the deviation of a surface from its osculating quadric, expressed in a centro-affinely invariant way.

A similar result is

THEOREM. $a_{\alpha\beta\gamma} = 0$ if and only is S is (a piece of) a central quadric.

PROOF : If $a_{\alpha\beta\gamma} = 0$, then, by definition,

$$g_{ijk}\,\xi^i_\alpha\,\xi^j_\beta\,\xi^k_\gamma = 0 \ ,$$

and, because of the Euler relation, $g_{ijk}\,\xi^k_\alpha = 0$. Hence all components of g_{ijk} to the n linearly independent vectors ξ^i, ξ^i_α are vanishing, yielding $\dfrac{\partial g_{ij}}{\partial \xi^k} = g_{ijk} = 0$ and $g_{ij}(\xi) = g_{ij} =$ $=$ const., q.e.d.

(The converse, $a_{\alpha\beta\gamma} = 0$ for central quadrics, is trivial).

This theorem, though very easily proved, is nevertheless the foundation for many interesting characterizations of Riemannian spaces. We shall use this theorem in Part III.

I.5. Contributions to the differential geometry of the indicatrix.

The centro-affine differential geometry of the hypersurface S proves to be a subject of interest, even independent of its applications to the geometry of Finsler spaces. Now we shall indicate a few of the known results on this geometry, which is one of the less known geometries of Felix Klein's "Erlanger Programm". Some of the results are closely related to VAGNER's paper [1] .

Let $\xi^i(s)$ be a curve on the surface S , s being its arc length in the Riemannian metric $ds^2 = \gamma_{\alpha\beta}\, du^\alpha du^\beta$. We know that the normal curvature of this curve is equal to 1 . We are now interested in the tangential component of the curvature vector

$$\frac{d^2\xi^i}{ds^2} = \frac{d}{ds}\left(\xi^i_\alpha \frac{du^\alpha}{ds}\right) = \xi^i_{\alpha\beta}\frac{du^\alpha}{ds}\frac{du^\beta}{ds} + \xi^i_\alpha \frac{d^2u^\alpha}{ds^2},$$

which by means of the equations of structure may be written

$$\frac{d^2\xi^i}{ds^2} = \xi^i_\rho\left[\frac{d^2u^\rho}{ds^2} + \left(\begin{Bmatrix}\rho\\ \alpha\beta\end{Bmatrix} - a^\rho_{\alpha\beta}\right)\frac{du^\alpha}{ds}\frac{du^\beta}{ds}\right] - \xi^i$$

Its tangential part $k^i = k^\rho \xi^i_\rho$ is given by the expression in brackets, $[\quad]$. This surface vector will be called the intrinsic affine curvature of the curve. k^ρ is equal to the geodesic curvature,

$$k^\rho_g = \frac{d^2u}{ds^2} + \begin{Bmatrix}\rho\\ \alpha\beta\end{Bmatrix}\frac{du^\alpha}{ds}\frac{du^\beta}{ds},$$

if and only if S is a central quadric, $a^{\rho}_{\alpha\beta}$ being 0 only in this case.

The curves $k^{\rho}_{g} = 0$ are the Riemannian geodesics on S , the curves $k^{\rho} = 0$ are the autoparallel curves of the symmetric affine connection

$$A^{\rho}_{\alpha\beta} = \left\{ {\rho \atop \alpha\beta} \right\} - a^{\rho}_{\alpha\beta}$$

For such an autoparallel curve, we have, with an appropriate parameter t ,

$$\frac{d^2u^{\rho}}{dt^2} + A^{\rho}_{\alpha\beta} \frac{du^{\alpha}}{dt} \frac{du^{\beta}}{dt} = 0$$

or

$$\frac{d^2\xi^i}{dt^2} = \xi^i_{\rho} \left\{ \frac{d^2u^{\rho}}{dt^2} + A^{\rho}_{\alpha\beta} \frac{du^{\alpha}}{dt} \frac{du^{\beta}}{dt} \right\} - \gamma_{\alpha\beta} \frac{du^{\alpha}}{dt} \frac{du^{\beta}}{dt} \xi^i =$$

$$= -\left(\frac{ds}{dt}\right)^2 \xi^i$$

From this equation follows the

THEOREM. The autoparallel curves of $A^{\rho}_{\alpha\beta}$ are the curves, in which planes through 0 meet S , for short, they are the "plane sections" of S .

Hence, k^{ρ} is a measure for the deviation of a curve from such a plane section. By central projection from 0 to any hyperplane not meeting 0 the autoparallels are mapped onto the straight lines. $A^{\rho}_{\alpha\beta}$ is projectively euclidean, a fact remarked by VAGNER [1] . We may ask for which surfaces S the geodesics coincide with the plane sections of S . By a theorem of WEYL, the necessary and sufficient conditions will be

$$\left\{ {\rho \atop \alpha\beta} \right\} = A^{\rho}_{\alpha\beta} + \delta^{\rho}_{\alpha}\sigma_{\beta} + \delta^{\rho}_{\beta}\sigma_{\alpha}$$

or

$$a^{\rho}_{\alpha\beta} = \delta^{\rho}_{\alpha}\sigma_{\beta} + \delta^{\rho}_{\beta}\sigma_{\alpha}$$

which yields

$$a^{\rho}_{\rho\beta} \underset{\text{det}}{=} a_{\beta} = (n-1)\,\sigma_{\beta} + \sigma_{\beta} = n.\sigma_{\beta}$$

or

and, at last :

$$(*) \qquad n\, a_{\alpha\beta\gamma} = \gamma_{\gamma\beta}\, a_{\alpha} + \gamma_{\alpha\gamma}\, a_{\beta}$$

Now $a_{\alpha\beta\gamma}$ is a symmetrical tensor, hence

$$0 = n(a_{\alpha\beta\gamma} - a_{\alpha\gamma\beta}) = \gamma_{\alpha\gamma}\, a_{\beta} - \gamma_{\alpha\beta}\, a_{\gamma}$$

and

$$0 = \gamma^{\alpha\gamma}(\gamma_{\alpha\gamma}\, a_{\beta} - \gamma_{\alpha\beta}\, a_{\gamma}) = (n-1)a_{\beta} - a_{\beta} = (n-2)a_{\beta}$$

Since $n \geq 3$, we have $a_{\beta} = 0$ and from $(*)$ $a_{\alpha\beta\gamma} = 0$. Hence, S is a central quadric, this property being also sufficient.

THEOREM. The Riemannian geodesics are the plane sections of S if and only if S is a central quadric.

An interesting consequence of this theorem for $n = 3$ is a generalization of a famous theorem of H.BRUNN [1] , hitherto for convex surfaces only (cf.eg. DANZER, LAUGWITZ, and LENZ [1]).

THEOREM. If a surface S admits affine reflections with respect to every plane containing 0 and a tangent of S , the surface S is necessarily a piece of a central quadric.

PROOF : Let g be the geodesic passing through the point P of S having t as its tangent. Effecting the affine reflection with respect to the plane determined by O, P, and t, g has to remain fixed, since it is invariant under central affinities. Hence, g must be contained in this plane as well as in S , i.e., g has to be the plane section of S. The proposition follows by the preceding theorem.

This shows that Brunn's theorem is of a purely local character, not global, as might be assumed by the proofs of this theorem in the theory of convex bodies. It is valid for all surfaces without parabolic points.

The central plane sections generalize a property of the great circles of the euclidean sphere, being the autoparallel curves of the connection $A^{\rho}_{\alpha\beta}$. On the other hand, the Riemannian geodesics have another important property in common with the great circles. I.SANDOR [1] proved, that the torsion of these geodesics is 0. I shall give another proof for this theorem.

By cause of

$$g^{kj} = \gamma^{\alpha\beta}\,\xi^k_\alpha\,\xi^j_\beta + \xi^k\,\xi^j$$

the (spatial) covariant derivative of the tangent vector of a geodesic is

$$\frac{D\dot{\xi}^k}{ds} = \ddot{\xi}^k + \frac{1}{2}g^{ks}g_{jil}\,\dot{\xi}^i\,\dot{\xi}^l = -\,a^{\beta}_{\rho\sigma}\,\dot{u}^{\rho}\,\dot{u}^{\sigma}\,\xi^k_\beta +$$

$$-\,\gamma_{\alpha\beta}\,\dot{u}^{\alpha}\,\dot{u}^{\beta}\,\xi^k + \frac{1}{2}\,\xi^k_\alpha\,\xi^j_\beta\,\gamma^{\alpha\beta}g_{jil}\,\xi^i_\rho\,\dot{u}^{\rho}\,\xi^l_\sigma\,\dot{u}^{\sigma} = -\,\xi^k,$$

since the first and third parts add up to 0. The curvature of the geodesics (Absolute value of the second covariant derivative) is 1. The normal vector of the geodesic is $n^i = -\dot{\xi}^i$. The second of the Frenet formulas reads ($\dot{\xi}^i$ tangent, b^i binormal vector) :

$$\frac{Du^k}{ds} = -k\,\dot{\xi}^k + \tau\, b^k \ .$$

But, since here $\frac{Du^k}{ds} = -\dot{\xi}^k$ we obtain Sandor's result $\tau = 0$.

Some of the former results gave centro-affine characterizations of central quadrics among the set of all surfaces. All these characterizations will have immediate consequences for Minkowski and Finsler spaces. Since euclidean and Riemannian geometries are the only geometries having central quadrics as their indicatrices, our theorems on affine geometry correspond to characterizations of euclidean and Riemannian spaces. I shall treat this subject, the historical origin of which were results by Riemann, Helmholtz, and Lie, in the third part of my lectures.

II. Variational and geometrical methods for the deduction of connections in Finsler spaces.

The geometric objects, fundamental for all infinitesimal considerations, in differential geometry, are connections or parallel displacements. In Finsler geometry, many different connections have been used since the tensorial treatment of Finsler geometry commenced in 1925. We shall give two different methods, one variational and the other geometrical in their origins, for the deduction of the most important of these connections. The variational method will have as its source the equation of geodesics, which will be interpreted in the more general scheme of the geometry of paths. The geometrical method will use osculating Riemannian spaces, following an idea of O.VARGA [1] .

II.1. Connections in the geometry of paths.

A path space is a differentiable manifold the paths of which are the solutions $x^i(t)$ of a system of equations

$$\ddot{x}^i + 2\Gamma^i(x;\dot{x}) = 0$$

where

$$\Gamma^i(x;\lambda\dot{x}) = \lambda^2\,\Gamma^i(x;\dot{x})$$

This homogeneity condition is necessary and sufficient for the uniqueness of the path issuing from a point $\overset{\circ}{x}{}^i$ in a direction $\overset{\circ}{\xi}{}^i$.

The expressions

$$\Gamma^{i}_{kj} \underset{\text{def}}{=} \frac{\partial^2 \Gamma^i}{\partial \dot{x}^k \partial \dot{x}^j}$$

transform according to the law of transformations of a connection, since Γ^i has the transformation properties of the derivative of a vector ($\Gamma^i = -\frac{1}{2}\frac{d}{dt}(\dot{x}^i)$). Hence, by an idea of BERWALD [3] , [4],

$$(1) \qquad d\eta^i = -\Gamma^i_{kj}(x;x')\eta^k dx^j$$

is a connection, the coefficients Γ^i_{kj} being defined in the manifold of directions, since they depend only on the direction and not on x' itself :

$$\Gamma^i_{kj}(x;\lambda x') = \Gamma^i_{kj}(x;x')$$

We are led to different non-linear connections if we identify the still free directional vector x' to either η or dx. In these cases, the manifold is regarded as a set of points and not (as in Berwald's definition) as a fibre bundle of directions.

Since for

$$\Gamma^i_k(x;\, x') \underset{\text{def}}{=} \frac{\partial \Gamma^i}{\partial x'^k}$$

holds

$$\Gamma^i_k(x;\, x') = \Gamma^i_{kj}(x;\, x')x'^j = \Gamma^i_{jk}(x;\, x')x'^j$$

we may apply this to a definition of two different types of parallel displacements :

$$(2A) \qquad d\eta^i = -\Gamma^i_k(x;\, dx)\eta^k$$

(2B) $$d\eta^i = -\Gamma^i_k(x;\eta)dx^k .$$

These displacements are homogeneous of order 1 both for dx an for η . (2A) is in general non-linear in the displacement element dx, (2B) is non-linear in the displaced vector η.

II.2. Connections of the path-space of geodesics of a Finsler space.

If

$$\Gamma^i(x;\, x') = G^i(x;\, x')$$

where

$$G^i \underset{\text{def}}{=} \frac{1}{2}\left\{ {i \atop kj} \right\}_{(x,x')} x'^k x'^j \quad ,$$

the paths are the geodesics of the metric

$$ds^2 = g_{ik}(x,\, dx)dx^i dx^k \quad .$$

Since

$$\frac{\partial G^i}{\partial x'^k} = \left\{ {i \atop rk} \right\} x'^r + \frac{1}{2}\frac{\partial g^{il}}{\partial x'^k}\, g_{lm} \left\{ {m \atop rs} \right\} x'^r x'^s$$

and

$$0 = \frac{\partial \delta^i_m}{\partial x'^k} = \frac{\partial g^{il} g_{lm}}{\partial x'^k} = g^{il}\frac{\partial g_{lm}}{\partial x'^k} + \frac{\partial g^{il}}{\partial x'^k}\, g_{lm}$$

we have at last

$$\frac{\partial G^i}{\partial x'^k} = \left\{ {i \atop rk} \right\} x'^r - \frac{1}{2} g^{il}\frac{\partial g_{lm}}{\partial x'^k}\left\{ {m \atop rs} \right\} x'^r x'^s$$

The arguments of the Christoffel symbols as well as of the g_{lm} are $(x;x')$.

By specializing (2A) and (2B) to Finsler spaces we obtain two well-known parallel displacements, introduced by other methods in papers of RUND [2] and BARTHEL [1] :

Rund
$$\begin{cases} d\eta^i = -P^i_{kj}(x;dx)\eta^k dx^j \\ P^i_{kj}(x;dx) = \left\{ {i \atop kj} \right\}_{(x;dx)} - \frac{1}{2} g^{il}(x;dx) \dfrac{\partial g_{lm}}{\partial x'^k} \left\{ {m \atop jr} \right\} \dfrac{dx^r}{ds} \end{cases}$$

Barthel
$$\begin{cases} d\eta^i = -B^i_{kj}(x;\eta)\eta^k dx^j \\ B^i_{kj}(x;\eta) = \left\{ {i \atop kj} \right\}_{(x;\eta)} - \frac{1}{2} g^{il}(x;\eta) \dfrac{\partial g_{lm}(x;\eta)}{\partial \eta^j} \left\{ {m \atop kr} \right\} \eta^r. \end{cases}$$

The second derivatives $\dfrac{\partial^2 G^i}{\partial x'^r \partial x'^s}$ are the coefficients ob Berwald's connection.

General references for II.1, II.2 :
BARTHEL [1] , BERWALD [3] , [4] , BOMPIANI [1] , KNEBELMAN [1] , LAUGWITZ [4] , RUND [1] , [2] .

II.3. Applications to an inverse problem in the calculus of variations.

The Barthel parallelism has the important property that the length of a vector remains unchanged (BARTHEL [1]). The analytical expression for this invariance property is the equation

$$F^2(x+dx;\eta^i - \Gamma^i_k(x;\eta)dx^k) - F^2(x;\eta) = 0$$

or

$$\text{(3)}\qquad \frac{\partial F^2(x;\eta)}{\partial x^j} - \frac{\partial F^2(x;\eta)}{\partial \eta^k}\Gamma^k_j(x;\eta) = 0 \ .$$

The invariance of $F(x;\eta)$ under the parallelism $d\eta^i = -\Gamma^i_k(x;\eta)dx^k$ is a necessary condition that the solutions of $\ddot{x}^i + 2\Gamma^i(x;\dot{x}) = 0$ be the geodesics of the Finsler metric F. Now we shall show that this condition is even sufficient. Let $F(x;x')$ and $\Gamma^i(x;x')$ be connected by the equations (3). We shall have to prove

$$\Gamma^i(x;x') = \frac{1}{2}\left\{ {i \atop kj} \right\}_{(x;x')} x'^k x'^j$$

for the Christoffel symbols of F.

The equations (3) may be written :

$$\text{(4)}\qquad \frac{\partial g_{ik}}{\partial x^j}\eta^i\eta^k - 2g_{ir}\eta^i\Gamma^r_j = 0 \ .$$

We shall calculate the expression

$$D_r \underset{\text{def}}{=} \frac{\partial}{\partial \eta^r}\left\{ \frac{\partial g_{ik}}{\partial x^j}\eta^i\eta^k\eta^j \right\}$$

in two different ways.

Firstly, we have

$$D_r = 2\frac{\partial g_{ir}}{\partial x^j}\eta^i\eta^j + \frac{\partial g_{ij}}{\partial x^r}\eta^i\eta^j$$

since the expression obtained by differentiation of $\dfrac{\partial g\ldots}{\partial x\ldots}$ vanishes.

On the other hand, because of (4), we receive

$$D_r = 2\frac{\partial}{\partial \eta^r}\left\{g_{ir}\eta^i\Gamma^r_j\eta^j\right\}$$

$$= 4\frac{\partial}{\partial \eta^r}\left\{g_{il}\eta^i\Gamma^l\right\} = 4g_{rl}\Gamma^l + 4g_{il}\eta^i\Gamma^l_r .$$

Inserting (4) this gives

$$D_r = 4g_{rl}\Gamma^l + 2\frac{\partial g_{ij}}{\partial x^r}\eta^i\eta^j .$$

By equating the two expressions for D_r, we obtain

$$4g_{rl}\Gamma^l = 2\frac{\partial g_{ir}}{\partial x^j}\eta^i\eta^j - \frac{\partial g_{ij}}{\partial x^r}\eta^i\eta^j$$

or, by multiplication with g^{kr} and dividing by 2

$$2\Gamma^k = \left\{{k \atop ij}\right\}_{(x;\eta)}\eta^i\eta^j .$$

This gives the

THEOREM. A necessary and sufficient condition that the solutions of

$$\ddot{x}^i + 2\Gamma^i(x;\dot{x}) = 0$$

be the geodesics of the Finsler metric F is, that $F(x;\eta)$ be an invariant of the parallel displacement

$$d\eta^i = -\Gamma^i_k(x;\eta)dx^k .$$

An equivalent condition is, that F be an invariant of the holonomy group of this displacement. By this condition, all the Finsler metrics belonging to a system of paths may be constructed in a geometrical manner. We shall give another application of this condition in part II of these lectures.

II.4. Osculating Riemannian spaces and covariant differentiation along a curve in Finsler space.

A more geometrical method of deriving connections in Finsler space proceeds from the consideration of osculating Riemannian spaces. Let ξ (x) be a vector field in a Finsler space of metrical structure $ds = F(x;dx)$ where $\xi(x) \neq 0$. The osculating Riemannian metric approximating F in the second order in a neighbourhood of the field vectors ξ has as its metrical tensor

$$g_{ik}(x) = g_{ik}(x;\ \xi(x)).$$

Its Christoffel symbols are

$$(5) \qquad \Gamma^{i}_{kj}(x) = \left\{ {i \atop kj} \right\}_{(x;\xi(x))} + \frac{1}{2} g^{ir}(x;\xi(x)) \cdot \left\{ \frac{\partial g_{rk}}{\partial \xi^l} \frac{\partial \xi^l}{\partial x^j} - \frac{\partial g_{kj}}{\partial \xi^l} \frac{\partial \xi^l}{\partial x^r} + \frac{\partial g_{jr}}{\partial \xi^l} \frac{\partial \xi^l}{\partial x^k} \right\}.$$

The auxiliary field ξ will now be chosen is an invariant manner, in different ways, each of which will be closely related geometrically to the vector η , the point (x), and the displacement dx.

At first, we shall consider the parallel displacement of η along a curve $z^i(s)$. A natural choice for the auxiliary vector field ξ will be

$$\xi^i(z(s)) = \frac{dz^i}{ds} .$$

Since

$$\frac{\partial \xi^i}{\partial x^j} dz^j = \frac{d^2 z^i}{ds^2} ds ,$$

a covariant differentiation along the curve z(s) will be

$$(6) \qquad D\eta^i = d\eta^i + \left\{ {i \atop kj} \right\}_{(z;dz)} \eta^k dz^j + \\ + \frac{1}{2} g^{ir}(z;dz) \frac{\partial g_{rk}}{\partial x'^j} \frac{d^2 z^j}{ds^2} \eta^k ds \quad ,$$

the other terms re vanishing by Euler's relation. Without any further calcula ons, $D\eta^i$ is seen to be a vector, what follows immediately fro the corresponding property of the covariant differential in Riemannian spaces.

(6) is the formula of covariant differentiation along a curve in Finsler space, which was first indicated by TAYLOR [1] and SYNGE [1] in 1925. The differential geometry of a curve z(s) in Finsler space (e.g. its Frenet formulas) may be treated simply by regarding the corresponding geometric quantities of the osculating Riemannian space.

By the way, the covariant differentiation is independent of the special choice of the auxiliary field ξ, provided that $\xi(z(s)) = \frac{dz}{ds}$ along the curve z(s).

II.5. Cartan's euclidean connection.

The idea of using osculating Riemannian spaces goes back to the thesis of A.NAZIM [1] . O.VARGA [1] was first to use these spaces for the theory of connections. We shall now prove his theorem on the geometrical exposition of Cartan's euclidean connection.

If $\underset{0}{x}, \underset{0}{x}'$ is a fixed line-element and if the variations dx, dx' are given, a very natural and simple way of determining an auxiliary field $\xi(x)$ is the following :

$$\frac{\partial \xi^i(x_0)}{\partial x^j} = - \frac{\partial G^i(x_0)}{\partial x'^j} = - G^i_j(x_0, x'_0)$$

$$\frac{\partial \xi}{\partial x^j} dx^j = dx' \ ; \quad \xi(x_0) = x'_0$$

If $x' = \frac{dx}{ds}$, this is nothing else than the field used for Taylor-Synge's covariant differentiation. In general, the auxiliary field is chosen as covariant stationary at x_0 , because of

$$\frac{\partial \xi^i(x_0)}{\partial x^j} = - G^i_j(x_0).$$

By combining these conditions with equations (5) of the preceding section we obtain

$$D\eta^i = d\eta^i + \left\{ {i \atop kj} \right\}_{(x,x')} \eta^k dx^j + C^i_{kl}\eta^k dx'^l +$$
$$- g^{ir} C_{kjl} \frac{\partial G^l}{\partial x'^r} \eta^k dx^j +$$
$$+ C^i_{jl} \frac{\partial G^l}{\partial x'^k} \eta^k dx^j .$$

This is in fact the "euclidean connection" obtained by CARTAN [1] . This seems to be a rather short way of determining this connection, giving moreover a geometrical insight into the significance of this analytical method of dealing with Finsler geometry.

II.6. Rund's and Barthel's connections.

If the x' of the preceding section is not independent of dx, η , but chosen in a special way, other parallelisms are obtained. We first identify $x' = \xi(x)$ to dx/ds. The Riemannian space will then be chosen as the space osculating along the geodesic line issuing from x in the direction dx. This gives z = x, dz = dx and

$$\frac{d^2z^i}{ds^2} = - \left\{ {i \atop kj} \right\}_{(z,dz)} \frac{dz^k}{ds} \frac{dz^j}{ds}$$

Formula (6) the yields Rund's parallelism (RUND [1] , [2] , [3]).

Another choice for the auxiliary field is $\xi(\underset{0}{x}) = \eta$. We obtain

$$(7) \qquad D_B \eta^i = d\eta^i + \left\{ {i \atop kj} \right\}_{(x;\eta)} \eta^k dx^j + \frac{1}{2} g^{ir} \frac{\partial g_{jr}}{\partial \eta^s} \frac{\partial \xi^s}{\partial x^k} \eta^k dx^j$$

We have to make further assumptions on $\frac{\partial \xi}{\partial x^k} \eta^k$, which will be : $\xi(x+\eta\, ds)$ is postulated to be parallel according to (7) to $\xi(x)$, $D_B \xi = 0$ for $dx = \eta\, ds$. This postulate gives BARTHEL's parallelism [1] .

Even without making this special assumption, all the parallelisms given by (7) may be shown to leave lengths invariant :

$$d(g_{ik}(x;\eta)\eta^i\eta^k) = \frac{\partial g_{ik}(x;\eta)}{\partial x^j}\eta^i\eta^k dx^j + g_{ik} d\eta^i d\eta^k + g_{ik}\eta^i d\eta^k$$

$$= - g_{ik} \cdot \frac{1}{2} g^{ir} \frac{\partial g_{jr}}{\partial \eta^s} \frac{\partial \xi^s}{\partial x^l} \eta^l dx^j \eta^k + - g_{ik}\eta^i \cdot \frac{1}{2} g^{kr} \ldots$$

$$= - \frac{\partial g_{jr}}{\partial \eta^s} \frac{\partial \xi^s}{\partial x^l} \eta^l dx^j \eta^r = 0 .$$

General references for II.4, II.5, and II.6 :

BARTHEL [1] , BOMPIANI [1] , CARTAN [1] , LAUGWITZ [3] , [4] , NAZIM [1] , RUND [1] , [2] , SYNGE [1] , TAYLOR [1] , VARGA [1] .

II.7. A table of the different connections in Finsler spaces.

The results of this chapter may be combined in a schedule.

The still white fields of this table will be considered now.

At first, I mention that Taylor-Synge's and Cartan's connections cannot be derived by the variational method; this is seen by considering Minkowski spaces. The equations of geodesics are the same for all Minkowski spaces, $\ddot{x}^i = 0$, whereas Taylor-Synge's and Cartan's connections are different for the various Minkowski spaces.

On the other hand, Berwald's connection may be deduced by the method of osculating Riemannian spaces, only in a rather formal manner. E.g., Berwald's Γ may be obtained from Rund's coefficients, which were already deduced by using osculating Riemannian spaces :

$$\Gamma^{i}_{jk}(x,x') = \frac{\partial P^{i}_{jl}(x,x')x'^{l}}{\partial x'^{k}}$$

The method of osculating Riemannian spaces permits to find, without any new calculation, connection coefficients for Rund's and Barthel's processes which have useful properties. One may take simply the Christoffel symbols of the respective osculating Riemannian space. At present, we have not yet determined the auxiliary field $\xi(x)$ completely. It has been sufficient to fix the value of ξ in the point $(\underset{0}{x})$ under consideration, and the value of a certain directional differential. We postulate moreover, that ξ be a covariant constant field with respect to the parallelism on hand, in the osculating Riemannian space. This yields in the case of Rund's parallelism :

$$\frac{\partial \xi^{i}(x_{0})}{\partial x^{k}} = - P^{i}_{hk}(x;\xi)\xi^{h}$$

For $\xi = dx/ds$, we obtain the following coefficients for Rund's parallelism

$$P^{*i}_{hk} = \left\{ {i \atop hk} \right\}_{(x;dx)} - \frac{1}{2} g^{ir} \left\{ \frac{\partial g_{rh}}{\partial x^{m}} P^{m}_{lk} - \frac{\partial g_{hk}}{\partial x^{m}} P^{m}_{lr} + \frac{\partial g_{rk}}{\partial x^{m}} P^{m}_{lh} \right\} \frac{dx}{ds}$$

They have been obtained in a more complicated manner by RUND [3] . These coefficients are identical to Cartan's Γ^{*}, as has been remarked by DEICKE in his review (Zentralblatt) on Rund's paper. Of course, the Γ^{*} may be obtained in the same way as the P^{*}. The equivalent construction yields coefficients for Barthel's connection. For a more detailed description, confer LAUGWITZ [4].

Parallelism	Deduction from the equation of geodesics $\ddot{x}^i + 2G^i(x;\dot{x}) = 0$	Deduction from Levi-Civita's parallelism in the Riemannian space oscillating at $\xi($
TAYLOR e SYNGE, along a curve $x(s)$ $d\eta^i = -\left\{ {i \atop kj} \right\}_{(x;dx)} \eta^k dx^j - \frac{1}{2} g^{ir} g_{rk\ell} \ddot{x}^\ell \eta^k ds$	impossible	$\xi(x(s)) = \frac{dx}{ds}$
BERWALD, in the bundle of directions $d\eta^i = -\Gamma^i_{kj}(x;x')\eta^k dx^j$	$\Gamma^i_{kj} = \frac{\partial^2 G^i}{\partial x'^k \partial x'^j}$	
CARTAN, euclidean connection in the bundle of directions $d\eta^i = -\Gamma^i_{kj}(x;x')\eta^k dx^j - C^i_{kj}(x;x')\eta^k dx'^j$ $\Gamma^i_{kj} = \left\{ {i \atop kj} \right\} + g^{ir} \frac{\partial G^\ell}{\partial x^r} C_{kj\ell} - C^i_{j\ell} \frac{\partial G^\ell}{\partial x'^k}$ $C^i_{kj} = \frac{1}{2} g^{ir} g_{rkj}$	impossible	VARGA (in an elem. of supp. $\underset{\circ}{x}$, $\xi(\underset{\circ}{x}) = \underset{\circ}{x}'$ $\frac{\partial \xi(\underset{\circ}{x})}{\partial x^j} dx^j = dx'$ $\frac{\partial \xi(\underset{\circ}{x})}{\partial x} = -\frac{\partial G}{\partial x'}$
RUND, in point space $d\eta^i = \left[\left\{ {i \atop kj} \right\}_{(x;dx)} - \frac{1}{2} g^{im} g_{km\ell} \left\{ {\ell \atop rj} \right\} \dot{x}^r \right] \eta^k dx^j$	$d\eta^i = -\frac{\partial G^i(x;dx)}{\partial x'^k} \eta^k$	$\xi(\underset{\circ}{x}) = \frac{dx}{ds}$ $\frac{\partial \xi(\underset{\circ}{x})}{\partial x^j} \frac{dx^j}{ds} = -\left\{ {i \atop kj} \right\}_{(x;dx)} \frac{dx^k}{ds} \frac{dx^j}{ds}$
BARTHEL, in point space, isometric $d\eta^i = -\left[\left\{ {i \atop kj} \right\}_{(x;\eta)} - \frac{1}{2} g^{im} g_{jm\ell} \left\{ {\ell \atop rk} \right\} \eta^r \right] \eta^k dx^j$	$d\eta^i = -\frac{\partial G^i(x;\eta)}{\partial \eta^j} dx^j$	$\xi(\underset{\circ}{x}) = \eta$ $\frac{\partial \xi(\underset{\circ}{x})}{\partial x^j} \eta^j = -\left\{ {i \atop kj} \right\}_{(x;\eta)} \eta^k \eta^j$

D.Laugwitz

III. Geometrical characterizations of euclidean and Riemannian spaces.

Some of the theorems of the preceding sections gave necessary and sufficient conditions that a Minkowski or Finsler space be euclidean or Riemannian. The indicatrices are in these cases central quadrics. Our central-affine access to Finsler geometry shows, that every characterization of central quadrics amongst the set of all surfaces in vector spaces will correspond to a characterization of euclidean and Riemannian spaces. The significance of such characterizations will be evident from the fact, that, e.g., lots of theorems had been proved for Finsler spaces with $A_i = 0$, until DEICKE [1] showed that the Finsler spaces with a definite metric and $A_i = 0$ are Riemannian.

I shall now give a report on some characterizations of Riemannian geometry, including the famous theorems of Helmholtz-Lie and Weyl. Both theorems will be proved by new and, as I think, adequate methods, permitting shorter proofs than before.

III.1. The methods of central affine geometry and its applications.

As a base for many characterizations of Riemannian geometry serves the theorem, that an indicatrix with vanishing cubic fundamental form is necessarily a central hyperquadric. We already proved that the indicatrix is a central hyperquadric if and only if its geodesics are the central plane sections. For three-dimensional spaces, another characterization has already been proved : S admits affine reflections with respect to every plane determined by O, by a point P of S, and by a tangent of S in P. I have listed some similar theorems in a paper dedicated to E.Bompiani on the occasion of his scientific anniversary in 1960 (LAUGWITZ [14]).

Property A. For every point P of the hypersurface S and every hyperplane H through P and O there exists an affine reflection which leaves H pointwise invariant and maps S into S.

Property B. For every P and every tangent t of S through P there exists an affine reflection with direction t, mapping S into S .

(An affine reflection is said to have direction t, if t is mapped onto itself, changing the direction of the straight lines with direction t to its opposite).

THEOREM. Each of the two properties A and B is necessary and sufficient for S being a hyperquadric.

PROOF. We shall first consider property B. Let t be described by (du). The affine reflection with direction t maps (du) into a vector (-a.du), where $a > 0$. Since this map is a central affinity,

$$a_{\alpha\beta\gamma}\, du^{\alpha}\, du^{\beta}\, du^{\gamma} = a_{\alpha\beta\gamma}\,(-a du^{\alpha})(-a du^{\beta})(-a du^{\gamma})$$

hence $a_{\alpha\beta\gamma} = 0$, and S is a central hyperquadric.

Now, property A is easily reduced to property B. An affine reflection with respect to the hyperplane H has as its direction t the tangent of S orthogonal to H with respect to the Riemannian metric in P.

Another local characterization of hyperquadrics has

been known for a long time in the case of convex surfaces and ellipsoids. It is concerned with properties of "shade lines". Let a surface which is supposed to be not transparent be lighted by (e.g.) parallel beams of light. The line separating the dark part from the lighted part of the surface will be called a shade line.

THEOREM. If all of the central plane sections of a surface in the affine space of three dimensions are shade lines, the surface is a piece of a central quadric.

PROOF. Let $\xi(t)$ be a fixed central plane section, e^i the vector of the direction of the light beams. Then

$$g_{ik}(\xi)\,\xi^i e^k = 0$$

and by differentiation

$$(*)\qquad g_{ik}(\xi)\,\dot{\xi}^i e^k = 0$$

both equations being valid on the curve $\xi(t)$. Since this is a plane curve, $\ddot{\xi}$ is a linear combination of ξ and $\dot{\xi}$, hence

$$g_{ik}(\xi)\,\ddot{\xi}^i e^k = 0$$

Differentiation of the equation $(*)$ yields

$$g_{ikj}(\xi)\,\dot{\xi}^i e^k \dot{\xi}^j = 0$$

or :

The validity of $g_{ikj}(\xi)\,\eta^i\eta^k\zeta^j = 0$ is a consequence of $g_{ik}(\xi)\,\eta^i\zeta^k = 0$. In the same manner follows the equation

$$g_{ikj}(\xi)\,\eta^i\zeta^k\zeta^j = 0$$

Hence

$$l_k(\xi) \underset{def}{=} g_{ikj}(\xi)\eta^i\zeta^j = 0$$

since $l_k\eta^k = l_k\zeta^k = l_k\xi^k = 0$ for three linearly independent vectors. Hence, for fixed k, the bilinear form with coefficients $g_{ikj} = g_{jki}$ vanishes if $g_{ij}\eta^i\zeta^j = 0$. Hence

$$g_{ikj} = m_k g_{ij}$$

with certain coefficients $m_k(\xi)$. By a special choice of coordinates, $g_{ij}(\xi) = 0$ for $i \neq j$, $g_{jj} \neq 0$. For $i \neq j$

$$0 = m_j g_{ij} = g_{ijj} = m_i g_{jj}$$

Hence, $m_i = 0$, which implies $g_{ikj} = 0$ and g_{ij} = const., q.e.d.

III.2. <u>The theorem of Helmholtz-Lie</u>.

Now we shall give a very simple proof for the characterization of Riemannian spaces named after HELMHOLTZ [1] and LIE [1].

THEOREM. Let S be an indicatrix surface in n-dimensional vector space ($n \neq 2$). Let, for every two points P, P' of S and tangent directions t, t' in them, exist a linear homogeneous mapping T of the space having the following properties. T maps a surface neighbourhood of P onto a surface neighbourhood of P', such that TP = P', Tt = t'. Then, S is necessarily a central quadric of one of the following affine types :

$$\xi_1^2 + \varepsilon(\xi_2^2 + \ldots + \xi_n^2) = 1$$

where $\varepsilon = \pm 1$ or 0.

PROOF. No tangent hyperplane can meet O, since a tangent straight line meeting O might by a homogeneous linear map be transformed into any other tangent, such that all tangents of the surface would meet O, which is impossible for $n \geq 3$. (For n=2, indicatrix curves consisting of one or several straight lines through O do not lead to a metric and may be cancelled!). Let us now assume that the osculating central quadrics have maximal rank, the surface metric $\gamma_{\alpha\beta}$ then being regular. Let du be the surface vector of t in P. By our assumptions, there exists a homogeneous affinity mapping P to P' = P and du to du' = -a.du, where a > 0. The cubic fundamental form has to remain unchanged, hence it vanishes, which proves that S is a central quadric. Since the first fundamental form must have the same sign for all directions du because of the affine invariance of this sign, only the listed types with $\varepsilon = \pm 1$ may occur. The latter reasoning holds good even for a non-regular metric and gives $\varepsilon = 0$.

This theorem uses only very simple theorems of the central affine geometry of hypersurfaces. Former proofs of the Helmholtz-Lie theorem seem to be far more complicated. Even a proof by REIDEMEISTER [1], being rather ahead of his time in using methods of (equiaffine) geometry, is valid only for closed indicatrices and uses deeper theorems.

General references for III.1 and III.2 :
BLASCHKE [1] , LAUGWITZ [5] , [6] , [12] , [13] , [14] , LENZ [1] (Lenz's paper contains some theorems similar to those of III.1, proved by projective methods).

III.3. The method of Löwner's ellipsoid.

Another method, the importance of which goes back to BUSEMANN ([1] and earlier papers listed there), is based on the so-called Löwner's ellipsoid. Its advantage is, that it needs no differentiability properties of the surface S. A disadvantage in comparision to the method of III.1 and III.2 is, that this method can be applied only to closed surface S, or to Finsler metrics related to regular problems of the calculus of variations.

From now on, S will be supposed to be a closed, continuous but not necessarily convex surface in vector space. The unique ellipsoid with center O, containing S, and having minimal volume, is called Löwner's ellipsoid. Proofs of the existence and uniqueness of this ellipsoid, which are rather simple, are found in BUSEMANN [1] and DANZER, LAUGWITZ, and LENZ [1] , where some applications of this ellipsoid are listed. For our purposes, any other ellipsoid would serve as well, provided that it be uniquely determined by S in a central affinely invariant manner. Löwner's ellipsoid seems to be especially easy to deal with.

We shall now give another proof of Helmholtz-Lie's theorem for the case of a definite metric. (LAUGWITZ [5] , [11]).

We shall begin with a more general theorem on bounded groups of homogeneous linear transformations. A group G is called bounded, if an open bounded set S has a bounded image GS.

THEOREM. Any bounded linear-homogeneous group G leaves a definite quadratic form invariant.

PROOF. Take E as the Löwner's ellipsoid of the bounded point set

GS of the above definition. I shall prove that E is mapped by any $T \in G$ onto itself. Since TSG = SG because of TG = G, T is volume-preserving. Hence, the ellipsoid TE contains TSG = SG and has the same volume as E, hence TE = E from the uniqueness of the minimal ellipsoid. The quadratic form Q which defines E by Q=1 is invariant under G, which proves the theorem.

This theorem is due (furnished with a more complicated proof) to Auerbach. It has important consequences in the theory of representations of groups, which are outside the realm of the present lectures. We are interested in the following proof of

HELMHOLTZ' THEOREM. Let G be a group of linear homogeneous transformations of a vector space having the following properties :

A) G is bounded,

B) (free mobility) For any two vectors $\xi \neq 0$, $\eta \neq 0$ there exists a $T \in G$ and $a > 0$, such that $T\xi = a\eta$.

Then G leaves a quadratic form invariant, and every invariant depends only on this quadratic form.

PROOF. Only the last part needs a proof. It will be sufficient to show : There exists a $T \in G$, $T\xi = \eta$, if and only if $Q(\xi) = Q(\eta)$. By B), there exists $T \in G$ and $a > 0$, such that $T\xi = a\eta$, hence $Q(\xi) = Q(T\xi) = Q(a\eta) = a^2Q(\eta)$. If, on the other hand, $Q(\xi) = Q(\eta)$, then $a^2 = 1$, hence $a = 1$.

The application to Minkowski geometry follows directly: If F is a Minkowski metric invariant under a group G of free mobility, then $F = f(Q)$ for a certain definite quadratic form Q, and from homogeneity properties $F^2 = Q$. - A direct geometrical proof may be given, showing that the indicatrix of F coincides with its Löwner's ellipsoid.

III.4. Weyl's problem of space.

The characterization of Riemannian spaces by the theorem of Helmholtz-Lie should - according to H.WEYL [1] , [2] - be replaced from physical reasons by another characterization, avoiding the concept of homogeneous linear transformations, since these correspond to rigid motions of the space and since the concept of rigidity is rejected by Weyl from the standpoint of general relativity. WEYL [1] conjectured another characterization of Riemannian spaces and proved a similar theorem in [2] . I shall now prove the conjecture of WEYL [1] by the method of Löwner's ellipsoid. (LAUGWITZ [9] , FREUDENTHAL [1]).

Weyl observed, that a fundamental theorem of Riemannian geometry is the (unique) existence of a length-preserving symmetric affine connection. Of course, the Riemannian spaces share this property with other Finsler spaces, e.g. the Minkowski spaces. Obviously, since the parallel displacement permits linear homogeneous mappings of the different tangent spaces of an arcwise connected manifold, which are length-preserving, the indicatrices in the different points are necessarily affinely equivalent. Now WEYL [1] conjectured :

THEOREM. If K is closed indicatrix in an n-dimensional vector space, and if all Finsler spaces having indicatrices affinely equivalent to K possess a length-preserving symmetric affine connection, then K is a central ellipsoid.

PROOF. Let E be the Löwner ellipsoid of K. Since, if we assume the existence of an affine connection, parallel displacement maps the indicatrices of different points onto each other, the same holds for their uniquely determined Löwner ellipsoids. The parallel displacement leaves moreover the Riemannian metric invariant, the indicatrices of which are the Löwner ellipsoids. Now choose this Riemannian metric in such a way, that its holonomy group is of free mobility. This property is shared by all Riemannian spaces of non-vanishing constant curvature. Then, by the theorems of the preceding section, K has to coincide with E, what was to be proved.

It would be important to generalize this theorem to indicatrices which are not necessarily bounded; in this case, only group-theoretical proofs (which are rather involved) are known (see FREUDENTHAL [1]).

R E F E R E N C E S

BARTHEL, W. : [1] Über eine Parallelverschiebung mit Längeninvarianz in lokal-Minkowskischen Räumen, Arch. d.Math. 4, 346-365 (1953).

BERWALD, L. : [1] Über affine Geometrie XXX. Die oskulierenden Flächen zweiter Ordnung in der affinen Flächentheorie, Math.Z. 10, 160-172 (1921).

[2] Die Grundgleichungen der Hyperflächen im euklidischen Raum gegenüber den inhaltstreuen Affinitäten, Monatsh.f.Math. 32, 89-106 (1922).

[3] Über Parallelübertragung in Räumen mit allgemeiner Massbestimmung, Jahresb, deutsch.Math. Ver. 34, 213-220 (1926).

[4] Untersuchung der Krümmung allgemeiner metrischer Raüme auf Grund des in ihnen herrschenden Parallelismus, Math.Z.25, 40-73 (1926).

BLASCHKE, W. : [1] Vorlesungen über Differentialgeometrie II. Affine Differentialgeometrie, Berlin (1923).

BOMPIANI, E. : [1] Sulle connessioni affini non posizionali, Arch.d.Math.3, 183-186 (1952).

BRUNN, H. : [1] Über Kurven ohne Wendepunkte, Habilitationsschrift München (1889).

BUSEMANN, H. : [1] The geometry of geodesics, New York (1955).

CARTAN, E. : [1] Les espaces de Finsler, Paris (1934).

DANZER, L., D.LAUGWITZ und H.LENZ : [1] Über das Löwnersche Ellipsoid, Arch.d.Math.8, 214-219 (1957).

DEICKE, A. : [1] Über die Finsler-Räume mit $A_i = 0$, Arch.d.Math. 4, 45-51 (1953).

DELENS, P. : [1] La métrique angulaire des espaces de Finsler et la géométrie différentielle projective, Paris (1934).

FREUDENTHAL, H.: [1] Zu den Weyl-Cartanschen Raumproblemen, Arch. d.Math. 11, 107-115 (1960).

HELMHOLTZ, H.v.: [1] Über die Tatsachen, die der Geometrie zugrunde liegen, Nachr.Ges.Wiss.Göttingen 193-221 (1868).

KAWAGUCHI, A. :[1] On the theory of non-linear connections II-Theory of Minkowski space and of non-linear connections in a Finsler space, Tensor 6, 165-190 (1956).

KAWAGUCHI, A. and D.LAUGWITZ :[1] Remarks on the theory of Minkowski spaces, Tensor 7, 190-199 (1957).

KNEBELMAN, M.S.:[1] Collineations and motions in generalized space, Amer.J.Math.51, 527-564 (1929).

LANDSBERG, G. :[1] Krümmungstheorie und Variationsrechnung, Jahresb, DMV. 16, 547-551 (1907).

LAUGWITZ, D. :[1] Differentialgeometrie, Teubner-Verlag Stuttgart (1960).

[2] Grundlagen für die Geometrie der unendlichdimensionalen Finslerräume, Annali mat.pura appl. 41, 21-41 (1955).

[3] Zur geometrischen Begründung der Parallelverschiebung in Finslerschen Räumen, Arch.d.Math. 6, 448-453 (1955).

[4] Vektorübertragungen in der Finslerschen Geometrie und der Wegegeometrie, Indagationes math. 18, 21-29 (1956).

[5] Über die Invarianz quadratischer Formen bei linearen Gruppen und das Räumproblem, Nachr. Ges.Wiss.Göttingen 21-25 (1956).

[6] Über die Rolle der quadratischen Metrik in der Physik, Zeitschr.f.Naturforschung, 9a, 827-832 (1954).

[7] Zur Differentialgeometrie der Hyperflächen in Vektorräumen und zur affingeometrischen Deutung der Theorie der Finslerräume, Math.Z.67, 63-74 (1957).

[8] Eine Beziehung zwischen Minkowskischer und affiner Differentialgeometrie, Publ.math.Debrecen, 5, 72-76 (1957).

[9] Über eine Vermutung von H.Weyl zum Räumproblem, Arch.d.Math.9, 128-133 (1958).

[10] Geometrische Behandlung eines inversen Problems der Variationsrechnung, Annales Univ.Sarav. Sciences 5, 235-244 (1956).

LAUGWITZ, D. :[11] Die Geometrien von H.Minkowski, Der Mathematikunterricht (1958), 27-42 (Heft4).

[12] Beiträge zur affinen Flächentheorie mit Anwendungen auf die allgemeinmetrische Differentialgeometrie, Abhandl.Bayer.Akad.Wiss. München, Math.nat.Kl.Neue Folge Heft 93, München 1959, 59 S.

[13] Über einen Typ von Geometrien, welche die Riemannsche verallgemeinern, Publ.math.Debrecen 7, 72-77 (1960).

[14] Einige differentialgeometrische Charakterisierungen der Quadriken. Annali di mat.pura ed appl. (To appear).

LENZ, H. : [1] Einige Anwendungen der projektiven Geometrie auf Fragen der Flächentheorie, Math.Nachr.18, 346-359 (1958).

LIE, S. : [1] Theorie der Transformationsgruppen III, Abt. V. Leipzig (1893).

LORCH, E.R. : [1] A curvature study of convex bodies in Banach spaces, Annali di mat.pura ed appl. 34, 105-112 (1955).

LORCH, E.R. and D.LAUGWITZ : [1] Riemann metrics associated with convex bodies in normed spaces, Amer.J.Math. 78, 889-894 (1956).

MAYER, O. : [1] Géométrie centro-affine différentielle des surfaces, Ann.Scient.Univ.Jassy 21, 1-77 (1935).

NAZIM, A. : [1] Über Finslersche Räume, Thesis München (1936).

REIDEMEISTER,K.: [1] Das Lie-Helmholtzsche Raumproblem und ein Satz von Maschke, Abh.Math.Sem.Hamburg 4, 172-173 (1925).

RIEMANN, B. : [1] Über die Hypothesen, die der Geometrie zugrunde liegen, Lecture Göttingen(1854). Reprint Berlin(1919).

RUND, H. : [1] The differential Geometry of Finsler spaces, Berlin (1959).

[2] Über die Parallelverschiebung in Finslerschen Räumen, Math.Z. 54, 115-128 (1951).

RUND, H. : [3] On the analytical properties of curvature tensors in Finsler spaces, Math.Ann. 127, 82-104 (1954).

SALKOWSKI, E. : [1] Affine Differentialgeometrie, Berlin-Leipzig (1934).

SANDOR, I. : [1] A Minkovski-féle tér mértékfelületére vonatkozó Darboux képletek, 2nd Hungarian Math. Congress, Budapest (1960), supplement to the abstracts of lectures, 48-50.

SYNGE, J.L. : [1] A generalisation of the Riemannian line-element, Trans.Amer.math.Soc. 27, 61-67 (1925).

TAYLOR, J.H. : [1] A generalisation of Levi-Civitas parallelism and the Frenet formulas. Trans.Amer.math. Soc. 27, 246-264 (1925).

VAGNER, V.V. : [1] The geometry of Finsler as the theory of a field of local hypersurfaces in X_n, Trudy sem.vektor.tenzor analizu 7, 65-166 (1949) (in Russian).

VARGA, O. : [1] Zur Herleitung des invarianten Differentials in Finslerschen Räumen, Monatsh.Math.Phys. 50, 165-175 (1941).

[2] Die Krummung der Eichfläche des Minkowskischen Raumes und die geometrische Deutung des einen Krümmungstensors des Finslerschen Raumes, Abh.math.Sem.Univ.Hamburg. 20, 41-51 (1955).

WEYL, H. : [1] "Kommentar" to Riemann's lecture, Berlin (1919).

[2] Mathematische Analyse des Raumproblems, Berlin (1923).

CENTRO INTERNAZIONALE MATEMATICO ESTIVO

(C. I. M. E.)

V. V. WAGNER

GEOMETRIA DEL CALCOLO DELLE VARIAZIONI

INDICE

INTRODUZIONE

Non ho l'intenzione di fare una rassegna storica delle teorie geometriche sorte dalle applicazioni di geometria differenziale al calcolo delle variazioni, le quali possiamo unire sotto il nome generale di geometria del calcolo delle variazioni. Voglio solamente attirare la loro attenzione ai tratti caratteristici che distinguono la teoria geometrica del calcolo delle variazioni, che sarà da me presentata in questo corso, dalle altre teorie ben conosciute.

Anzitutto bisogna notare che gran parte delle teorie geometriche collegate col calcolo delle variazioni rappresentano piuttosto teorie geometriche di invarianti differenziali di problemi del calcolo delle variazioni che uno studio geometrico del calcolo delle variazioni stesso. Riconoscendo che la teoria di equivalenza di problemi del calcolo delle variazioni è importantissima e perciò è importantissima la teoria degli invarianti differenziali tuttavia non dobbiamo trascurare la possibilità di applicare direttamente i metodi geometrici al calcolo delle variazioni stesso.

Come dimostrò Caratheodory all'inizio del nostro secolo per interpretazioni geometriche delle condizioni di estremo nel calcolo delle variazioni è importantissima la nozione di indicatrice, la quale fu introdotta da lui pel caso del problema semplice del calcolo delle variazioni. Però la gran parte dei geometri avendo sviluppato le teorie geometriche del calcolo delle variazioni tendeva a conservare una somiglianza formale con la geometria riemanniana, non avendo prestato la dovuta attenzione alla nozione di indicatrice.

Come esempio di teoria geometrica dal calcolo delle variazioni possiamo considerare la geometria finsleriana di Cartan. Allo stesso modo che la metrica riemanniana definisce localmente una metrica euclidea nello spazio tangente associato ad un punto dello spazio riemanniano, nella geometria finsleriana di Cartan la metrica finsleriana definisce una metrica euclidea nello spazio tangente associato ad un elemento lineare dello spazio finsleriano. Inoltre questa riduzione della geometria finsleriana alla geometria euclidea è artificiale e non risponde alle esigenze del calcolo delle variazioni. Un metodo analogo fu applicato da Cartan nella teoria geometrica del problema del calcolo delle variazioni per un integrale multiplo e dai geometri giapponesi nelle teorie geometriche dei problemi del calcolo delle variazioni di ordine superiore. Ma in tutte le teorie menzionate le condizioni di estremo in sostanza non sono considerate e pertanto queste teorie non hanno grande importanza per il calcolo delle variazioni stesso benchè siano interessanti dal punto di vista geometrico.

Abbiamo anche un altro metodo per costruire la geometria finsleriana, più adatto dal punto di vista del calcolo delle variazioni, in cui la metrica finsleriana è considerata come definita dalle metriche minkowskiane negli spazi tangenti associati ai punti dello spazio finsleriano. In questo caso le uniche sfere degli spazi minkowskiani tangenti sono le indicatrici del problema corrispondente del calcolo delle variazioni. Sembra che tale costruzione della geometria finsleriana si trovi per la prima volta nell'articolo di Winternitz pubblicato nel 1930 ([35]).

Scegliendo la nozione di indicatrice come una nozione fondamentale della geometria finsleriana vediamo che il trasporto parallelo di vettori nello spazio finsleriano può essere determinato mediante rappresentazioni puntuali delle indicatrici lungo curve dello spazio, giacchè il modulo del vettore trasportato parallelamente è invariante.
Con questo è naturale di considerare le indicatrici come spazi $\mathfrak{X}_{n-1}$ associati ai punti dello spazio $\mathfrak{X}_n$ dove la metrica finsleriana è definita. In tal modo veniamo alla connessione lineare generale nello spazio fibrato il cui spazio base è lo spazio finsleriano considerato e le cui fibre sono le indicatrici.

V. V. Wagner

Nel 1942, quando non erano note applicazioni della teoria degli spazi fibrati alla geometria differenziale, sviluppai una teoria formale generale per la geometria finsleriana, che nominai la "teoria delle varietà composte".

La varietà composta $\mathfrak{X}_{n+(\mathfrak{s})}$ fu definita da me nello spirito della teoria classica delle connessioni come uno spazio $\mathfrak{X}_n$ ai punti del quale sono associati spazi $\mathfrak{X}_{\mathfrak{s}}$ locali. Una connessione (più precisamente una connessione parziale) lineare fu introdotta in $\mathfrak{X}_{n+(m)}$ mediante un sistema di Pfaff

$$d\eta^i + \Gamma^i_\alpha(\xi^\beta, \eta^j)\, d\xi^\alpha = 0 \qquad (1) \qquad \begin{array}{l}(\alpha, \beta = 1, \ldots\ldots, n) \\ (i, j = 1, \ldots\ldots, s)\end{array}$$

ove ξ^α sono coordinate nello spazio base $\mathfrak{X}_n$ e η^i sono coordinate negli spazi locali $\mathfrak{X}_{\mathfrak{s}}$ ([27] , [29] , [41] , [44]).

E' evidente che dal punto di vista moderno la varietà composta $\mathfrak{X}_{n+(m)}$ è niente altro che uno spazio fibrato differenziale generale e che la connessione parziale (1) è una connessione parziale nello spazio fibrato definita da un "allestimento" delle fibre ([9]).

Con l'aiuto della connessione (1) poi fu introdotta un'operazione di derivazione assoluta nello spazio base, di oggetti differenziali qualunque non obbligatoriamente geometrici i quali erano dati negli spazi locali associati. Per il passaggio dalla teoria di derivazione assoluta nella varietà composta all'operazione corrispondente nella teoria moderna degli spazi fibrati è necessario solamente di cambiare la terminologia conservando tutto l'apparato formale.

L'applicazione della teoria generale delle varietà composte ha comportato naturalmente una generalizzazione della geometria finsleriana. Considerando le indicatrici dello spazio finsleriano come gli spazi $\mathfrak{X}_{n-1}$ associati ai punti dello spazio base $\mathfrak{X}_n$, noi le determiniamo mediante un sistema di equazioni parametriche. Questa circostanza risulta importante. E' facile immaginare come la geometria differenziale elementare diventerebbe complicata se determinassimo curve e superficie mediante equazioni implicite. Però determiniamo l'indicatrice nella geometri finsleriana, di solito con l'aiuto della funzione metrica, da un'equazione implicita. La determinazione delle indicatrici da equazioni parametriche è più comoda per lo sviluppo della geometria finsleriana. Nella geometria finsleriana abbiamo sempre una non conformità fra il punto di vista locale nello studio dello spazio finsleriano stesso e il punto di vista globale nello studio delle indicatrici. Ci preme far notare che praticamente molti geometri, per esempio Berwald ([2]), studiando gli spazi finsleriani particolari non sempre usano il metodo puramente globale nello studio delle indicatrici e considerano come indicatrici di uno spazio finsleriano ipersuperficie non chiuse. Per questa ragione è razionale di generalizzare la nozione di spazio finsleriano non mantenendo la condizione che ogni curva differenziabile debba essere misurabile, cioè che debba essere definita per essa la lunghezza degli archi, ma conservando la condizione che le indicatrici siano ipersuperficie, in generale non più chiuse.

Questa generalizzazione della geometria finsleriana corrisponde al problema ordinario del calcolo delle variazioni ove la funzione di linea di solito non è definita per tutte le curve differenziabili. Per la determinazione di una connessione intrinseca e lo sviluppo della teoria degli invarianti differenziali nella geometria finsleriana è superflua la supposizione che la metrica finsleriana sia convessa, cioè che siano convesse le indicatrici. E' importante solamente l'ipotesi che la metrica finsleriana sia regolare, la quale ipotesi si conserva anche nel caso della geometria finsleriana generalizzata.

L'applicazione del metodo della teoria delle varietà composte non è limitata alla geometria finsleriana ma può estendersi a qualunque problema del calcolo delle variazioni.

Come è detto, la nozione di indicatrice fu usata da Caratheodory solo nel caso del problema semplice del calcolo delle variazioni. E' vero che Vessiot ([24]) usava la nozione di indicatrice anche nel caso del problema di Lagrange ma egli non sviluppò una teoria geometrica di questo problema del calcolo delle variazioni che avrebbe incluso la teoria degli invarianti differenziali. Perciò in modo naturale sorge il problema di sviluppare la teoria geometrica completa del problema di Lagrange del calcolo delle variazioni.

Di solito la definizione del "problema di Lagrange" in forma geometrica è la seguente:

Data una funzione numerica di linea orientata nello spazio $\mathfrak{X}_n$, rappresentata da un integrale

$$(2) \qquad \int_{t_1}^{t_2} L(\xi^\alpha, \dot{\xi}^\alpha)dt \quad ,$$

ove L è una funzione positivamente omogenea di grado 1, rispetto alle variabili $\dot{\xi}^\alpha$, si cerca un estremo di questa funzione sull'insieme delle curve orientate che congiungono due punti dati e soddisfano ad un sistema di equazioni differenziali

$$(3) \qquad \overset{z}{f}(\xi^\alpha, \dot{\xi}^\alpha) = 0 \qquad (z = \ell + 1, \ldots, n),$$

ove $\overset{z}{f}$ sono funzioni positivamente omogenee rispetto alle variabili $\dot{\xi}^\alpha$.

Le curve soddisfacenti al sistema (3) di equazioni differenziali si dicono le curve ammissibili del problema di Lagrange.

Indicando con X^α le coordinate di punto negli spazi centro-affini tangenti che sono associati ai punti dello spazio $\mathfrak{X}_n$, vediamo che dal sistema di equazioni

$$\overset{z}{f}(\xi^\alpha, X^\alpha) = 0 \tag{4}$$

rimane determinato, nel caso generale, in ogni spazio tangente, un semicono ℓ-dimensionale, generato da semirette uscenti dal centro dello spazio tangente.

Le curve ammissibili del problema di Lagrange considerato coincideranno con le curve orientate le cui semitangenti sono generatrici dei semiconi corrispondenti.

E' naturale considerare la possibilità di ridurre il problema di Lagrange dove si cerca un estremo condizionato ad un problema dove la funzione di linea è definita solo sull'insieme delle curve ammissibili, il quale perciò può essere considerato come un problema del calcolo delle variazioni dove si cerca un estremo non condizionato ([30] , [31] , [34] , [40]).

Per questo scopo determiniamo il campo (4) dei semiconi negli spazi tangenti mediante un sistema di equazioni parametriche

$$X^\alpha = L^\alpha(\xi^\beta, \chi^\lambda) \tag{5}$$

ove L^α sono funzioni positivamente omogenee di grado 1 rispetto ai parametri χ^λ.

Le curve ammissibili devono soddisfare le equazioni

$$\dot{\xi}^\alpha = L^\alpha(\xi^\beta, \chi^\lambda) \qquad (\lambda = 1, \ldots\ldots, \ell), \tag{6}$$

e quindi per ogni curva ammissibile

$$\xi^\alpha = \xi^\alpha(t) \tag{7}$$

possiamo esprimere le coordinate curvilinee χ^λ determinanti il vettore tangente della curva (7) come funzioni del parametro t

$$\chi^\lambda = \chi^\lambda (t) \quad . \tag{8}$$

Una funzione di linea definita solo sull'insieme delle curve soddisfacenti al sistema (6) di equazioni differenziali può essere definita dall'integrale

$$\int_{t_1}^{t_2} L(\xi^\alpha, \chi^\lambda) dt \quad , \tag{9}$$

dove L è una funzione positivamente omogenea di grado 1 rispetto alle variabili χ^λ.

Il problema del calcolo delle variazioni dove si cerca un estremo di una funzione di linea di forma (9) si dice problema di Lagrange intrinseco.

Per ridurre un problema di Lagrange ordinario dato dalle (2), (3) al problema di Lagrange intrinseco dobbiamo prendere la funzione $L(\xi^\alpha, L^\alpha(\xi^\beta, \chi^\lambda))$, dove le funzioni $L^\alpha(\xi^\beta, \chi^\lambda)$ sono prese dal sistema di equazioni del campo (3) dei semiconi nella forma parametrica (5), come la funzione $L(\xi^\alpha, \chi^\lambda)$ nell'integrale (9).

Se la funzione di linea (9) è positivamente definita, il che equivale alla condizione che la funzione L sia positivamente definita $L > 0$, possiamo determinare il problema di Lagrange intrinseco mediante il campo delle superficie $(\ell-1)$-dimensionali che sono le direttrici dei semiconi (5) e sono date dall'equazione

$$L(\xi^\alpha, \chi^\lambda) = 1 \quad . \tag{10}$$

Queste superficie $(\ell-1)$-dimensionali si dicono le indicatrici del problema di Lagrange.

Il campo delle indicatrici può essere determinato da un sistema di equazioni parametriche

$$X^{\alpha} = 1^{\alpha}(\xi^{\beta}, \eta^{i}) \quad . \qquad (i = 1, \dots, \ell-1). \tag{11}$$

La teoria degli invarianti differenziali del problema di Lagrange intrinseco può essere considerata come la teoria degli invarianti differenziali del campo delle sue indicatrici.

In tal modo veniamo alla teoria del campo di superficie $(\ell-1)$-dimensionali date negli spazi tangenti dello spazio $\mathfrak{X}_n$. Considerando queste superficie come gli spazi $\mathfrak{X}_{\ell-1}$ associati ai punti dello spazio base $\mathfrak{X}_n$ avremo una varietà composta o ciò, che è la stessa cosa, uno spazio fibrato. Ora abbiamo il problema fondamentale di trovare una connessione lineare nello spazio fibrato $\mathfrak{X}_{n+(\ell-1)}$ che sia associata in modo intrinseco al campo di superficie $(\ell-1)$-dimensionali considerato. La soluzione di questo problema esige dapprima uno studio di una superficie del campo associata ad un punto qualunque dello spazio $\mathfrak{X}_n$ con l'aiuto della teoria generale delle superficie nello spazio centro-affine. Si può dire che le complicazioni della teoria geometrica del problema di Lagrange in primo luogo dipendono dalla complessità della teoria centro-affine delle superficie di dimensione arbitraria.

Si può dimostrare che una connessione lineare intrinseca nello spazio fibrato si trova senza pena se un iperpiano tangente sia associato intrinsecamente ad ogni punto della superficie $(\ell-1)$-dimensionale del campo considerato in tal modo che la iperstriscia ottenuta sia regolare ([46]).

Indipendente da questa teoria generale una teoria geometrica degli invarianti differenziali del problema di Lagrange intrinseco era costruita nel caso particolare $\ell = n - 1$ ([34]).

Quanto alla teoria geometrica del problema di Lagrange stesso è comodo ottenere la condizione di Eulero per l'estremo riducendo il problema di Lagrange al problema di Mayer nello spazio avente una dimensione di più e poi considerando il problema di Mayer come la teoria del campo di semiconi negli spazi tangenti. Con questo si vede che la condizione di Eulero è collegata con una proprietà di curve ammissibili del problema di Mayer la quale si può considerare indipendentemente del problema di estremo ([47]). Questo metodo geometrico di ottenere la regola ben conosciuta dei moltiplicatori indeterminati di Eulero per il problema di Lagrange, in sostanza rappresenta una geometrizzazione del metodo dovuto a Radon ([16]).

Per ottenere condizioni sufficienti d'estremo nel problema di Lagrange è comodo applicare il metodo di Caratheodory il quale consiste in una trasformazione del problema del calcolo delle variazioni tale che tutte le curve estremanti si conservino ([5]).

Dal punto di vista geometrico la trasformazione di Caratheodory del problema di Lagrange positivamente definito equivale a trasformazioni delle sue indicatrici mediante trasformazioni proiettive degli spazi tangenti, nel modo seguente :

$$\bar{X}^{\alpha} = \frac{X^{\alpha}}{\varepsilon(1 - \varphi_{\beta} X^{\beta})} \tag{12}$$

ove il vettore covariante φ_{β} è un gradiente $\varphi_{\beta} = \partial_{\beta} \varphi(\xi^{\alpha})$ e l'indicatrice si trova dallo stesso lato del centro dello spazio rispetto all'iperpiano

$$1 - \varphi_{\beta} X^{\beta} = 0 \quad , \tag{13}$$

se $\varepsilon = 1$ e nel lato opposto se $\varepsilon = -1$.

E' interessante studiare la teoria degli invarianti differenziali del problema di Lagrange data a meno di una trasformazione di Caratheodory arbitraria. Questa teoria esige l'applicazione della teoria delle superficie nello spazio centro-proiettivo. In tal modo la geometria centro-proiettiva trova un'applicazione al calcolo delle variazioni.

La teoria degli invarianti differenziali della metrica finsleriana definita a meno di una trasformazione di Caratheodory arbitraria fu sviluppata da me nel 1945 ([28]) e quella del problema di Lagrange definita a meno d'una trasformazione di Caratheodory arbitraria da Cabanov in tempo recente ma solamente per il caso particolare $\ell = n-1$ ([52]) .

In relazione con la teoria geometrica del problema di Mayer del calcolo delle variazioni è interessante la teoria degli invarianti differenziali del campo di semiconi negli spazi tangenti dello spazio $\mathfrak{X}_n$.

Giacchè questa teoria non differisce dalla teoria degli invarianti differenziali del campo di coni essa può applicarsi anche alla teoria degli invarianti differenziali di una equazione differenziale alle derivate parziali di primo ordine.

Sviluppando la teoria del campo di semiconi ℓ-dimensionali è comodo considerare i semiconi non come insiemi di punti ma come insiemi di semirette generatrici, cioè come spazi $\mathfrak{X}_{\ell-1}$ associati ai punti dello spazio base $\mathfrak{X}_n$. In tal modo veniamo allo spazio fibrato $\mathfrak{X}_{n+(\ell-1)}$ ove dobbiamo determinare una connessione lineare intrinsecamente collegata con il campo di semiconi. Questo problema è risoluto per il caso d'un campo di semi-iperconi regolari. Per il caso particolare $n = 3$ la teoria degli invarianti differenziali del campo di semi-iperconi fu sviluppata da me nel 1948 e per il caso generale da Ciascecinicov nel 1957 ([59] , [60]).

Passando a teorie geometriche di problemi del calcolo delle variazioni con integrali multipli cominceremo dal problema più generale, cioè dal problema di Lagrange per integrali multipli.

Di solito la definizione del problema di Lagrange per integrali multipli in forma geometrica è la seguente :

Data una funzione numerica di superficie orientata m-dimensionale nello spazio $\mathfrak{X}_n$ rappresentata dall'integrale

$$\int \cdots\cdots \int L(\xi^\alpha, \xi^\alpha_a)\, dt^1 \cdots\cdots dt^m \tag{14}$$

ove L rispetto alle variabili ξ^α_a è una funzione positivamente omogenea di grado 1 della matrice $\|\xi^\alpha_a\|$, si cerca un estremo di questa funzione sull'insieme delle superficie orientate m-dimensionali che hanno una frontiera comune fissa e soddisfano ad un sistema di equazioni differenziali alle derivate parziali

$$\overset{z}{f}(\xi^\alpha, \xi^\alpha_a) = 0 \quad , \qquad (z = \ell+1, \ldots\ldots, m(n-m)+1), \tag{15}$$

ove le $\overset{z}{f}$ rispetto alle variabili ξ^α_a sono funzioni positivamente omogenee della matrice $\|\xi^\alpha_a\|$.

Le superficie soddisfacenti al sistema (15) di equazioni differenziali alle derivate parziali si dicono le <u>superficie ammissibili</u> del problema di Lagrange.

Considerando lo spazio $\binom{n}{m}$-dimensionale $\mathcal{M}_{\binom{n}{m}}$ di tutti gli m-vettori contravarianti n-dimensionali $X^{\alpha_1 \cdots\cdots \alpha_m}$ possiamo interpretare le condizioni di semplicità d'un m-vettore $X^{\alpha_1 \cdots\cdots \alpha_m}$

(16) $$X^{[\alpha_1 \ldots \alpha_m} X^{\beta_1] \ldots \beta_m} = 0$$

come le equazioni d'un cono (m(n-m)+1)-dimensionale dello spazio $\mathcal{M}_{\binom{n}{m}}$, che chiameremo <u>cono di Grassmann</u>.

Possiamo rappresentare ogni m-vettore semplice $X^{\alpha_1 \ldots \ldots \alpha_m}$ come un prodotto alternato di m vettori X^{α}_{a} (a=1, ..., m)

(17) $$X^{\alpha_1 \ldots \alpha_m} = m! \; X^{[\alpha_1}_{1} \ldots X^{\alpha_m]}_{m}$$

dati a meno di una combinazione lineare arbitraria, i cui coefficienti soddisfano alla sola condizione che il loro determinante sia uguale a 1 .

Allora possiamo considerare gli mn numeri X^{α}_{a} come coordinate soprannumerarie di un punto del cono di Grassmann, date a meno di una trasformazione

(18) $$X^{\alpha}_{a'} = C^{a}_{a'} \; X^{\alpha}_{a} \qquad , \qquad \mathrm{Det} \left| C^{a}_{a'} \right| = 1 \; .$$

In questo caso il sistema delle equazioni

(19) $$\overset{z}{f} \, (\xi^{\alpha}, \; X^{\alpha}_{a} \;) = 0$$

determina un semicono ℓ-dimensionale giacente sul cono di Grassmann in ogni spazio $\mathcal{M}_{\binom{n}{m}}$ associato ad un punto dello spazio $\mathcal{X}_n$.

Le superficie ammissibili coincideranno con le superficie orientate m-dimensionali la cui m-direzione tangente in ogni punto corrisponde ad una semiretta generatrice del semicono (19) nello spazio $\mathcal{M}_{\binom{n}{m}}$ associato a questo punto della superficie ([32] , [33] , [37]) .

Analogamente al problema di Lagrange per integrali semplici, il problema di Lagrange per integrali multipli può essere ridotto ad un problema ove si cerca un estremo non condizionato.

Con questo scopo determineremo il campo (19) di semiconi negli spazi $\mathcal{M}_{\binom{n}{m}}$ mediante un sistema di equazioni parametriche

$$X^{\alpha_1 \dots \alpha_m} = L^{\alpha_1 \dots \alpha_m}(\xi^\beta, \chi^\lambda) \tag{20}$$

ove $L^{\alpha_1 \dots \alpha_m}$ rispetto alle variabili χ^λ sono funzioni positivamente omogenee di grado 1 .

Le superficie ammissibili devono soddisfare le equazioni

$$m!\, \xi_1^{[\alpha_1} \dots \xi_m^{\alpha_m]} = L^{\alpha_1 \dots \alpha_m}(\xi^\beta, \chi^\lambda) \tag{21}$$

donde per ogni superficie ammissibile

$$\xi^\alpha = \xi^\alpha(t^a) \tag{22}$$

possiamo esprimere le coordinate curvilinee χ^λ determinanti l'm-vettore tangente della superficie (22) come funzioni dei parametri t^a

$$\chi^\lambda = \chi^\lambda(t^a). \tag{23}$$

Una funzione di superficie orientata definita solo sull'insieme delle superficie soddisfacenti al sistema (19) di equazioni differenziali può essere definita dall'integrale

$$\int \dots \int L(\xi^\alpha, \chi^\lambda)\, dt^1 \dots\dots dt^m \tag{24}$$

ove L è una funzione positivamente omogenea di grado 1 rispetto alle variabili x^{λ}.

Il problema del calcolo delle variazioni dove si cerca un estremo d'una funzione di superficie della forma (24) si dice problema di Lagrange intrinseco per gli integrali multipli.

Se la funzione di superficie (24) è positivamente definita, il che equivale alla condizione che la funzione L sia positivamente definita, $L > 0$, possiamo determinare il problema di Lagrange intrinseco mediante il campo di superficie $(\ell-1)$-dimensionali che sono le direttrici dei semiconi (20) e sono date dall'equazione

$$L(\xi^{\alpha}, x^{\lambda}) = 1. \tag{25}$$

Queste superficie $(\ell-1)$-dimensionali giacenti sui coni di Grassmann si dicono le <u>indicatrici</u> del problema di Lagrange per gli integrali multipli.

Il campo delle indicatrici può essere determinato da un sistema di equazioni parametriche

$$X^{\alpha_1 \ldots \alpha_m} = 1^{\alpha_1 \ldots \alpha_m}(\xi^{\alpha}, \eta^{i}) \tag{26}$$

oppure in termini delle coordinate soprannumerarie X^{α}_{a} sul cono di Grassmann

$$X^{\alpha}_{a} = 1^{\alpha}(\xi^{\beta}, \eta^{i}). \tag{27}$$

Per quanto io sappia la necessità della condizione di Eulero nella teoria del problema di Lagrange per l'integrale multiplo, la quale è espressa dalla regola dei moltiplicatori indeterminati, finora non è dimostrata. Nel 1945 Barker dimostrò la necessità della condizione di Eulero solamente nel caso particolare $n=4$ con l'ausilio di un metodo che non può essere applicato nel caso generale ([1]).

La dimostrazione della necessità della condizione di Eulero nel problema di Lagrange per l'integrale multiplo è difficile a causa di difficoltà analitiche di costruzione di una famiglia di superficie ammissibili colla frontiera comune alla quale appartenga una superficie ammissibile data. Poco probabile che metodi geometrici possano aiutare qui in modo essenziale.

Però le cose stanno diversamente per condizioni sufficienti dell'estremo. Il metodo di Caratheodory ([4]) applicato da lui al problema del calcolo delle variazioni per l'integrale multiplo senza condizioni complementari può essere applicato anche nel caso del problema di Lagrange per l'integrale multiplo senza modificazioni sostanziali. Al problema di Lagrange per lo integrale multiplo può essere applicata anche la teoria dei campi geodetici dovuta a Lepage nel caso del problema ordinario per l'integrale multiplo (14).

Per ciò che concerne la teoria geometrica degli invarianti nei problemi del calcolo delle variazioni per integrali multipli qui abbiamo risultati solamente per il problema ordinario senza condizioni complementari.

Applicando i risultati dovuti a Kawaguchi ([10], [11], [12]) possiamo trovare una connessione lineare intrinseca nello spazio fibrato associato al campo di indicatrici se i numeri n e m sono primi tra loro ([38]).

Nel caso $m = n - 1$ possiamo introdurre la connessione lineare intrinseca nello spazio fibrato $\mathcal{X}_{n+(n-1)}$ associato al campo di indicatrici, la quale corrisponde alla connessione nello spazio di elementi iperpiani introdotta da Cartan ([6], [45])

La geometria di Kawaguchi è una teoria di invarianti differenziali del problema del calcolo delle variazioni di ordine superiore per l'integrale ordinario, la quale è costruita come una generalizzazione della geometria finsleriana di Cartan. Dal punto di vista del calcolo delle variazioni essa ha gli stessi difetti della geometria finsleriana di Cartan.

In altro modo la teoria geometrica di problemi d'ordine superiore del calcolo delle variazioni può essere sviluppata sulla base della nozione di indicatrice. Per questo dobbiamo considerare gli spazi tangenti d'ordine superiore associati ai punti dello spazio $\mathfrak{X}_n$.

Tutti i differenziali fino all'ordine ν incluso delle coordinate di punto dello spazio $\mathfrak{X}_n$ sono considerati come coordinate di punto nello spazio tangente di ordine ν associato al punto corrispondente. E' anche sviluppata la teoria puramente algebrica degli spazi tangenti d'ordine superiore.

Considerando un problema di ordine superiore del calcolo delle variazioni che consiste nella ricerca dell'estremo di una funzione positivamente definita di linea orientata in uno spazio $\mathfrak{X}_n$, data dall'integrale

$$\int_{t_1}^{t_2} L(\xi^\alpha, \xi^{(u)\alpha})dt , \qquad (u = 1, \ldots, \nu) \tag{28}$$

ove L è una funzione soddisfacente a certe condizioni rispetto alle variabili $\xi^{(u)\alpha}$ che sono generalizzazioni di quelle di omogeneità positiva di grado 1, introdurremo come indicatrici le superficie negli spazi tangenti d'ordine ν definiti dall'equazione

$$L(\xi^\alpha, X^{(u)\alpha}) = 1 . \tag{29}$$

La teoria degli invarianti differenziali del problema d'ordine superiore del calcolo delle variazioni può essere considerata come la teoria degli invarianti differenziali del campo di ipersuperficie (29) .

In tal modo possiamo anche ottenere interpretazioni geometriche di condizioni diverse dell'estremo, che mancano nella teoria di Kawaguchi.

La teoria suddetta finora non è sviluppata ma certi risultati preliminari sono ottenuti da Losic ([55]).

Poi è naturale considerare la teoria geometrica del problema di Lagrange d'ordine superiore la quale finora non era studiata.

Kawaguchi considerava anche la teoria geometrica del problema d'ordine superiore per gli integrali multipli usando lo spazio di elementi superficiali d'ordine superiore. E' evidente che questa teoria può essere anche sviluppata sulla base della nozione di indicatrice.

Riassumendo possiamo dire che la teoria geometrica di ogni problema del calcolo delle variazioni può essere considerata come un caso particolare della teoria generale del campo di superficie ad s dimensioni le quali sono date in spazi di oggetti differenziali geometrici di una certa specie associati ai punti dello spazio $\mathfrak{X}_n$.

Dal punto di vista della teoria degli spazi fibrati questa teoria del campo di superficie può essere considerata come una teoria di una sola superficie ad $n + s$ dimensioni nello spazio fibrato corrispondente di oggetti differenziali geometrici, la quale taglia ogni fibra secondo una superficie ad s dimensioni.

Questa superficie ad $n + s$ dimensioni può essere considerata come uno spazio fibrato generale $\mathfrak{X}_{n+(s)}$ con lo stesso spazio base $\mathfrak{X}_n$.

Lo studio della superficie comincia con lo studio delle sue sezioni con le fibre le quali sono superficie negli spazi di Klein.

Perciò questo studio sarà fondato sulla teoria generale delle superficie nello spazio di Klein di specie corrispondente.

Usando oggetti differenziali geometrici ottenuti in questo modo poi dobbiamo cercare una connessione lineare intrinseca nello spazio fibrato $\mathfrak{X}_{n+(s)}$, cioè un allestimento intrinseco di fibre.

Se questa connessione lineare intrinseca nello spazio fibrato $\mathfrak{X}_{n+(s)}$ è trovata possiamo usare l'operazione di derivazione covariante corrispondente per ottenere nuovi oggetti differenziali nelle fibre dagli oggetti differenziali già ottenuti durante lo studio delle superficie sezioni negli spazi di Klein.

V. V. Wagner

I - NOZIONI PRELIMINARI

§ 1. Relazioni binarie.

Come è noto una relazione binaria fra elementi degli insiemi A e B è un sottoinsieme ρ del prodotto cartesiano $A \times B$.

Se A = B la relazione binaria ρ si chiama una relazione binaria omogenea.

Oltre le operazioni ben note della teoria degli insiemi abbiamo nella teoria delle relazioni binarie tre nuove operazioni fondamentali, cioè l'inversione, la moltiplicazione, la moltiplicazione diretta.

La relazione binaria $\overset{-1}{\rho} \subset B \times A$ definita dalla formula[1]

$$(1.1) \qquad (b,a) \in \overset{-1}{\rho} \longleftrightarrow (a,b) \in \rho$$

dicesi la relazione binaria inversa della relazione binaria $\rho \subset A \times B$.

La relazione binaria $\sigma \circ \rho \subset A \times C$ definita dalla formula

$$(1.2) \qquad (a,c) \in \sigma \circ \rho \longleftrightarrow \bigvee_{b} (a,b) \in \rho \wedge (b,c) \in \sigma$$

dicesi il prodotto delle relazioni binarie $\rho \subset A \times B$ e $\sigma \subset B \times C$.

La relazione binaria $\rho_1 \square \rho_2 \subset (A_1 \times A_2) \times (B_1 \times B_2)$ definita dalla formula

---

1)Usiamo alcuni simboli della logica matematica : $\wedge$ -la congiunzione, $\rightarrow$ -l'implicazione, $\leftrightarrow$ l'equivalenza, $\bigvee_{b}$ -il quantificatore esistenziale rispetto alla variabile b .

(1.3) $$((a_1, a_2), (b_1, b_2)) \in \rho_1 \square \rho_2 \longleftrightarrow (a_1, b_1) \in \rho_1 \wedge (a_2, b_2) \in \rho_2$$

dicesi il prodotto diretto delle relazioni binarie $\rho_1 \subset A_1 \times B_1$ a $\rho_2 \subset A_2 \times B_2$.

La relazione binaria

(1,4) $$\rho_3 \circ \rho_2^{-1} \circ \rho_1$$

dicesi il prodotto ternario delle relazioni binarie ρ_1, ρ_2, ρ_3 fra elementi degli insiemi A e B .

Sia P un insieme di relazioni binarie fra elementi degli insiemi A e B . Indicheremo con P^{-1} l'insieme di tutte le relazioni binarie inverse delle relazioni binarie di P .

Siano P e Σ rispettivamente un insieme di relazioni binarie fra elementi degli insiemi A e B e un insieme di relazioni binarie fra elementi degli insiemi B e C . Indicheremo con $\Sigma \circ P$ l'insieme di tutti i prodotti $\sigma \circ \rho$ ove $\rho \in P$ e $\sigma \in \Sigma$.

Analogamente siano P_1 e P_2 rispettivamente un insieme di relazioni binarie fra elementi degli insiemi A_1 e B_1 e un insieme di relazioni binarie fra elementi degli insiemi A_2 e B_2 ;indicheremo con $P_1 \square P_2$ l'insieme di tutti i prodotti diretti $\rho_1 \square \rho_2$ ove $\rho_1 \in P_1$ e $\rho_2 \in P_2$.

Il sottoinsieme $\rho\langle a\rangle$ dell'insieme B definito dalla formula

(1.5) $$b \in \rho\langle a\rangle \longleftrightarrow (a, b) \in \rho$$

dicesi la sezione della relazione binaria $\rho \subset A \times B$ nell'elemento a .

L'unione

(1.6) $$\rho(\mathfrak{a}) = \bigcup_{a \in \mathfrak{a}} \rho\langle a\rangle$$

delle sezioni della relazione binaria ρ in tutti gli elementi di un sottoinsieme $\mathfrak{a} \subset A$ dicesi la sezione della relazione binaria ρ nel sottoinsieme $\mathfrak{a}$.

La sezione della relazione binaria inversa ρ^{-1} nell'insieme B coincide con la prima proiezione di ρ e la sezione della relazione binaria ρ stessa nell'insieme A coincide con la seconda proiezione di ρ

(1.7) $$\mathrm{pr}_1\, \rho = \rho^{-1}(B) \quad , \quad \mathrm{pr}_2\, \rho = \rho(A) \quad .$$

Nella teoria delle relazioni binarie sono importantissime le formule seguenti :

(1.8) $$\sigma \circ \rho(\mathfrak{a}) = \sigma(\rho(\mathfrak{a})) \; ,$$

la quale dà un'espressione della sezione del prodotto di relazioni binarie in un sottoinsieme $\mathfrak{a}$;

(1.9) $$\rho_1 \square \rho_2(\rho) = \rho_2 \circ \rho \circ \rho_1^{-1}$$

la quale dà un'espressione della sezione del prodotto diretto di relazioni binarie in un sottoinsieme $\rho \subset A_1 \times A_2$.

Un sottoinsieme $\mathfrak{a}$ dicesi <u>stabile</u> rispetto ad una relazione binaria omogenea $\rho \subset A \times A$ se la sezione della ρ in $\mathfrak{a}$ è contenuta in $\mathfrak{a}$ stesso

(1.10) $$\rho(\mathfrak{a}) \subset \mathfrak{a} \; .$$

Un sottoinsieme $\mathfrak{A}$ dicesi <u>co-stabile</u> rispetto a ρ se il suo complementare $\mathfrak{A}'$ è stabile rispetto a ρ. Si può dimostrare che la co-stabilità d'un sotto-insieme $\mathfrak{A}$ rispetto ad una relazione binaria ρ equivale alla sua stabilità rispetto alla relazione binaria inversa ρ^{-1}.

Un sottoinsieme $\mathfrak{A}$ dicesi <u>bi-stabile</u> rispetto a ρ se nello stesso tempo $\mathfrak{A}$ è stabile e co-stabile rispetto alla ρ.

Un sottoinsieme $\mathfrak{A}$ dicesi <u>universale</u> rispetto ad una relazione binaria omogenea $\rho \subset A \times A$ se ciascuna coppia (a_1, a_2) di elementi di $\mathfrak{A}$ appartiene a ρ, cioè

(1.11) $$\mathfrak{A} \times \mathfrak{A} \subset \rho \quad .$$

Usando il teorema di Zorn possiamo dimostrare che ogni sottoinsieme universale rispetto a ρ è contenuto in un sottoinsieme universale massimale rispetto a ρ.

§ 2. Rappresentazioni parziali.

Una relazione binaria $\varphi \subset A \times B$ si dice una rappresentazione parziale dell'insieme A nell'insieme B se tutte le sezioni non vuote della φ consistono di un solo elemento.

Secondo questa definizione la relazione binaria vuota è una rappresentazione parziale. E' evidente che una rappresentazione parziale non vuota φ è una rappresentazione, nel senso ordinario, del sottoinsieme $pr_1\varphi$ sul sottoinsieme $pr_2\varphi$. Indicando con $\varphi(a)$ l'immagine di un elemento $a \in pr_1$ rispetto a questa rappresentazione abbiamo $\varphi\langle a\rangle = \{\varphi(a)\}$. Indicheremo con $\mathfrak{F}(A \times B)$ l'insieme di tutte le rappresentazioni parziali dell'insieme A nell'insieme B.

Se una rappresentazione parziale $\varphi \subset A \times B$ è una rappresentazione cioè $pr_1\varphi = A$ allora il prodotto $\varphi^{-1} \circ \varphi$ è una relazione di equivalenza fra gli elementi dell'insieme A, la quale si dice il nucleo della rappresentazione φ.

Una coppia (φ_1, φ_2) di rappresentazioni parziali di $\mathfrak{F}(A \times B)$ si dice $\mathfrak{F}$-unibile se l'unione $\varphi_1 \cup \varphi_2$ di queste rappresentazioni parziali è ancora una rappresentazione parziale.

L'insieme di tutte le coppie $\mathfrak{F}$-unibili di rappresentazioni parziali di $\mathfrak{F}(A \times B)$ è una relazione binaria fra rappresentazioni parziali la quale si dice relazione di $\mathfrak{F}$-unibilità.

Un insieme $\Phi \subset \mathfrak{F}(A \times B)$ di rappresentazioni parziali si dice $\mathfrak{F}$-unibile se l'unione $\cup\Phi$ è ancora una rappresentazione parziale. Si può dimostrare che la $\mathfrak{F}$-unibilità di un insieme $\Phi \subset \mathfrak{F}(A \times B)$ di rappresentazioni parziali equivale al fatto che Φ sia un sottoinsieme universale dell'insieme $\mathfrak{F}(A \times B)$ rispetto alla relazione di $\mathfrak{F}$-unibilità.

Una rappresentazione parziale φ si dice invertibile se la relazione binaria inversa $\overset{-1}{\varphi}$ è anche una rappresentazione parziale.

La rappresentazione parziale vuota è manifestamente una rappresentazione parziale invertibile. Una rappresentazione parziale invertibile non vuota è una rappresentazione biunivoca nel senso ordinario del sottoinsieme $pr_1\varphi$ sul sottoinsieme $pr_2\varphi$.

Indicheremo con $\mathfrak{J}(A \times B)$ l'insieme di tutte le rappresentazioni parziali invertibili dell'insieme A nell'insieme B .

Una rappresentazione parziale di un insieme A in sè stesso si dice una trasformazione parziale dell'insieme A .

La trasformazione parziale invertibile $\Delta_{\mathfrak{a}}$ definita dalla formula

$$\Delta_{\mathfrak{a}} = \bigcup_{a \in \mathfrak{a}} \{(a, a)\} \tag{2.1}$$

si dice la trasformazione parziale identica definita sul sottoinsieme $\mathfrak{a}$.

Un insieme di trasformazioni parziali invertibili $\Gamma \subset \mathfrak{J}(A \times A)$ di un insieme A si dice gruppo generalizzato di trasformazioni parziali invertibili se Γ è stabile rispetto all'operazione di moltiplicazione

$$\overset{2}{\Gamma} \subset \Gamma \tag{2.2}$$

e rispetto all'operazione di inversione

$$\overset{-1}{\Gamma} \subset \Gamma \tag{2.3}$$.

Notiamo che se una rappresentazione invertibile γ appartiene ad un gruppo generalizzato Γ allora ambedue le trasformazioni parziali identiche $\Delta_{pr_1\gamma}$ e $\Delta_{pr_2\gamma}$ appartengono pure a Γ .

Un gruppo generalizzato $\Gamma \subset \mathfrak{F}(A \times A)$ si dice effettivo se $\bigcup_{\gamma \in \Gamma} pr_1 \gamma = A$.

Ogni gruppo generalizzato di trasformazioni parziali invertibili è un semigruppo astratto speciale il quale si dice un gruppo generalizzato astratto.

Un gruppo generalizzato di trasformazioni parziali invertibili è un gruppo astratto solo nel caso che esso contenga l'unica trasformazione parziale identica, cioè quando tutte le trasformazioni parziali che appartengono al gruppo generalizzato sono definite sullo stesso sottoinsieme.

E' evidente che il prodotto ternario di rappresentazioni parziali invertibili di un insieme A in un insieme B è anche una rappresentazione parziale invertibile di questi insiemi.

Un insieme $K \subset \mathfrak{F}(A \times B)$ di rappresentazioni parziali invertibili si dice t-stabile, se esso è stabile rispetto all'operazione di moltiplicazione ternaria. L'insieme di tutti i sottoinsiemi t-stabili dell'insieme $\mathfrak{F}(A \times B)$ definisce in questo un'operazione di chiusura f_t la quale si dice l'operazione di t-chiusura.

Se $K \subset \mathfrak{F}(A \times B)$ è un insieme t-stabile di rappresentazioni parziali invertibili allora gl'insiemi $\overset{-1}{K} \circ K$ e $K \circ \overset{-1}{K}$ di trasformazioni parziali invertibili degli insiemi A e B sono gruppi generalizzati.

Una coppia $(\varkappa_1, \varkappa_2)$ di rappresentazioni parziali invertibili di $\mathfrak{F}(A \times B)$ si dice compatibile se l'unione $\varkappa_1 \cup \varkappa_2$ è anche una trasformazione parziale invertibile. E' evidente che la compatibilità di una coppia $(\varkappa_1, \varkappa_2)$ di rappresentazioni parziali invertibili equivale alla $\mathfrak{F}$-unibilità delle coppie $(\varkappa_1, \varkappa_2)$ e $(\overset{-1}{\varkappa}_1, \overset{-1}{\varkappa}_2)$.

L'insieme di tutte le coppie compatibili di trasformazioni parziali invertibili dell'insieme $\mathfrak{F}(A \times B)$ è una relazione binaria che si dice la relazione di compatibilità .

Un insieme $K \subset \mathfrak{G}(A \times B)$ di rappresentazioni parziali invertibili si dice compatibile se l'unione $\bigcup K$ è pure una rappresentazione parziale invertibile . Si può dimostrare che la compatibilità di un insieme $K \subset \mathfrak{G}(A \times B)$ di rappresentazioni parziali invertibili equivale al fatto che K sia un sottoinsieme universale dell'insieme $\mathfrak{G}(A \times B)$ rispetto alla relazione di compatibilità. Un insieme $K \subset \mathfrak{G}(A \times B)$ di rappresentazioni parziali invertibili si dice u-stabile, se esso è stabile rispetto all'operazione di unione di sottoinsiemi compatibili, cioè se l'unione di ogni sottoinsieme compatibile di K appartiene a K stesso.

L'insieme di tutti i sottoinsiemi u-stabili dell'insieme $\mathfrak{G}(A \times B)$ definisce in questo un'operazione di chiusura f_u la quale si dice operazione di u-chiusura.

Il prodotto $f_u \circ f_t$ delle operazioni di t-chiusura e di u-chiusura è ancora un'operazione di chiusura in $\mathfrak{G}(A \times B)$ la quale si dice l'operazione di c-chiusura. I sottoinsiemi c-chiusi coincidono con i sottoinsiemi i quali sono t-chiusi e u-chiusi nello stesso tempo.

Sia $\Gamma \subset \mathfrak{G}(B \times B)$ un gruppo generalizzato di trasformazioni parziali dell'insieme B . Una coppia $(\varkappa_1, \varkappa_2)$ di rappresentazioni parziali invertibili di $\mathfrak{G}(A \times B)$ si dice Γ-coniugata se $\varkappa_2 \circ \varkappa_1^{-1} \in \Gamma$. L'insieme di tutte le coppie Γ-coniugate è una realzione binaria fra rappresentazioni parziali invertibili di $\mathfrak{G}(A \times B)$ la quale si dice relazione di Γ-coniugio . Un sottoinsieme universale rispetto alla relazione di Γ-coniugio si dice un insieme di rappresentazioni parziali invertibili Γ-coniugato.

E' facile vedere che affinchè un insieme K di rappresentazioni parziali invertibili sia Γ-coniugato è necessario e sufficiente che

$$K \circ K^{-1} \subset \Gamma \quad (2.4)$$

In virtù della proprietà generale di sottoinsiemi universali rispetto ad una relazione binaria, ogni insieme Γ-coniugato di rappresentazioni parziali invertibili è contenuto in un insieme Γ-coniugato massimale.

Si può dimostrare che ogni insieme Γ-coniugato massimale di rappresentazioni parziali invertibili è t-stabile e, se Γ è u-stabile, anche u-stabile.

§ 3. Quasi-rappresentazioni parziali.

Una relazione binaria $\rho \subset A \times B$ si dice quasi-rappresentazione parziale dell'insieme A nell'insieme B se tutte le sue sezioni non vuote definiscono una partizione dell'insieme $pr_1 \rho$ cioè se per una qualunque coppia di elementi (a_1, a_2)

(3.1) $$\{\rho\langle a_1\rangle \cap \rho\langle a_2\rangle \neq \emptyset\} \rightarrow \{\rho\langle a_1\rangle = \rho\langle a_2\rangle\}.$$

Si può dimostrare che affinchè una relazione binaria ρ sia una quasi-rappresentazione parziale è necessario e sufficiente che essa soddisfi alla condizione

(3.2) $$\rho \circ \rho^{-1} \circ \rho \subset \rho \quad .$$

Poichè per ogni relazione binaria ρ

(3.3) $$\rho \subset \rho \circ \rho^{-1} \circ \rho$$

la condizione (3.2) è equivalente alla condizione

(3.4) $$\rho \circ \rho^{-1} \circ \rho = \rho \quad .$$

E' evidente che ogni rappresentazione parziale è anche una quasi-rappresentazione parziale.

Da (3.2) si deduce che se ρ è non vuota i prodotti $\rho^{-1} \circ \rho$ e $\rho \circ \rho^{-1}$ sono relazioni di equivalenza negli insiemi $pr_1 \rho$ e $pr_2 \rho$.

Indicando con $\frac{pr_1 \rho}{\rho^{-1} \circ \rho}$ e $\frac{pr_2 \rho}{\rho \circ \rho^{-1}}$ gli insiemi fattoriali corrispondenti si può dimostrare che ogni sezione $\rho(\mathfrak{a})$ della quasi-rappresentazione parziale ρ in un sottoinsieme, che è un elemento dell'insieme fattoriale $\frac{pr_1 \rho}{\overset{-1}{\rho} \circ \rho}$, è un elemento dell'insieme fattoriale $\frac{pr_2 \rho}{\rho \circ \overset{-1}{\rho}}$.

Con questo risulta che la rappresentazione di $\frac{pr_1 \rho}{\overset{-1}{\rho} \circ \rho}$ in $\frac{pr_2 \rho}{\rho \circ \overset{-1}{\rho}}$ è una rappresentazione invertibile di $\frac{pr_1 \rho}{\overset{-1}{\rho} \circ \rho}$ su $\frac{pr_2 \rho}{\rho \circ \overset{-1}{\rho}}$.

Questa rappresentazione invertibile si dice la <u>rappresentazione fattoriale</u> per la quasi-rappresentazione parziale ρ. La rappresentazione inversa di questa rappresentazione fattoriale coincide con la rappresentazione fattoriale della quasi-rappresentazione parziale inversa $\overset{-1}{\rho}$.

Una quasi-rappresentazione parziale $\rho \subset A \times B$ si dice una <u>quasi-rappresentazione</u> se $pr_1 \rho = A$.

Le quasi-rappresentazioni sono state introdotte da Riguet ([17])[2]. Esse sono importanti per diverse questioni di geometria differenziale; in particolare avremo bisogno di loro nella teoria geometrica della condizione di Eulero nel problema di Lagrange del calcolo delle variazioni.

---

2) Riguet le chiama "relations difonctionnelles".

§ 4. Spazi aritmetici.

Come è noto lo spazio aritmetico ad n dimensioni è l'n-esima potenza cartesiana R^n dell'insieme R di tutti i numeri reali. Giacchè l'operazione di moltiplicazione cartesiana d'insiemi non è associativa, il prodotto cartesiano $R^n \times R^s$ non è lo spazio aritmetico ad $n+s$ dimensioni. Indicheremo con $\sigma_{(n,s)}$ la rappresentazione invertibile del prodotto cartesiano $R^n \times R^s$ sullo spazio aritmetico R^{n+s} ad $n+s$ dimensioni definita dalla formula:

$$\sigma_{(n,s)} \left((\xi^1, \ldots, \xi^n), (\eta^1, \ldots, \eta^s)\right) = (\xi^1, \ldots, \xi^n, \eta^1, \ldots, \eta^s) . \tag{4.1}$$

Indicheremo poi con $p_{(n+s,n)}$ la proiezione di R^{n+s} sul prodotto R^n dei primi n fattori e con $p^*_{(n+s,s)}$ la proiezione dell' R^{n+s} sul prodotto R^s degli ultimi s fattori

$$\begin{aligned} p_{(n+s,n)} \left((\xi^1, \ldots, \xi^n, \eta^1, \ldots, \eta^s)\right) &= (\xi^1, \ldots, \xi^n) \\ p^*_{(n+s,s)} \left((\xi^1, \ldots, \xi^n, \eta^1, \ldots, \eta^s)\right) &= (\eta^1, \ldots, \eta^s). \end{aligned} \tag{4.2}$$

Una trasformazione parziale invertibile f dello spazio aritmetico R^n si dice un <u>diffeomorfismo</u> se $pr_1 f$ è un sottoinsieme aperto di R^n e f è definita da un sistema di equazioni

$$\bar{\xi}^\alpha = f^\alpha (\xi^\beta) \qquad (\alpha, \beta = 1, \ldots, n) \tag{4.3}$$

ove f^α sono funzioni dotate di derivate continue, il cui jacobiano è differente da zero. In conformità con questa definizione generale chiameremo diffeomorfismo anche la trasformazione vuota.

Se un diffeomorfismo non vuoto f è definito da funzioni differenziabili

ϑ volte e con le derivate continue fino a tale ordine incluso allora f è detto un <u>diffeomorfismo di classe ϑ</u>. Se un diffeomorfismo non vuoto è definito da funzioni indefinitamente differenziabili e con derivate di ogni ordine continue allora f è detto un diffeomorfismo di classe C^{∞}. Considereremo il diffeomorfismo vuoto come diffeomorfismo di classe C^{∞}. In seguito considereremo in generale solamente diffeomorfismi di classe C^{∞}.

Indichiamo con $\mathfrak{X}_n$ l'insieme di tutti gli isomorfismi di classe C^{∞} in R^n.

E' facile vedere che $\mathfrak{X}_n$ è un gruppo generalizzato di trasformazioni parziali invertibili il quale è u-stabile.

In seguito considereremo solamente gruppi generalizzati di diffeomorfismi cioè sottogruppi generalizzati del gruppo generalizzato $\mathfrak{X}_n$.

Un gruppo generalizzato di diffeomorfismi $\mathfrak{G}$ in R^n si dice <u>localizzato</u>, se ogni diffeomorfismo di $\mathfrak{X}_n$ che è contenuto in un diffeomorfismo di $\mathfrak{G}$ appartiene pure a $\mathfrak{G}$.

Indicheremo con $\mathfrak{I}_n$ il sottogruppo generalizzato di $\mathfrak{X}_n$ che consiste di tutte le trasformazioni parziali identiche di $\mathfrak{X}_n$.

Dunque $\mathfrak{I}_n$ è l'insieme di tuttele trasformazioni parziali identiche definite sui sottoinsiemi aperti di R^n.

Si può dimostare che i gruppi generalizzati localizzati $\mathfrak{G}$ di diffeomorfismi in R^n sono caratterizzati dalla condizione

$$\mathfrak{G} \circ \mathfrak{I}_n \subset \mathfrak{G} . \tag{4.4}$$

Un gruppo generalizzato di diffeomorfismi $\mathfrak{G}$ si dice <u>perfetto</u> se esso è localizzato e u-stabile.

Un esempio semplicissimo di gruppo generalizzato perfetto di diffeomorfismi in R^n differente da $\mathfrak{X}_n$ stesso è il gruppo $\vec{\mathfrak{X}}_n$ di tutti i diffeomorfismi orientati cioè di diffeomorfismi definiti da un sistema di funzioni f^{α} con jacobiano positivo

(4.5) $$\mathrm{Det}\,|f^{\alpha}_{\beta}| > 0 \quad .$$

Una classe importante di gruppi generalizzati perfetti è quella formata dai gruppi generalizzati di trasformazioni parziali di Lie, definiti da un sistema di equazioni differenziali (Lie stesso insieme con i suoi discepoli erroneamente diceva "gruppi di trasformazioni").

Come un esempio di un gruppo generalizzato di Lie possiamo considerare il gruppo generalizzato $\mathcal{W}_n$ di tutti i diffeomorfismi che conservano i volumi in R^n, il quale è definito da un'equazione differenziale

(4.6) $$\mathrm{Det}\,|f^{\alpha}_{\beta}| = 1 \quad .$$

Però si trovano anche gruppi generalizzati non perfetti importanti.

Come un esempio importante dal punto di vista di applicazioni ulteriori considereremo il cosidetto "gruppo generalizzato di diffeomorfismi semiconici".

Un sottoinsieme di uno spazio aritmetico si dice semiconico se esso è l'unione di semirette uscenti dal punto zero .

Se ρ è una relazione binaria fra punti di R^n allora ρ è un sottoinsieme del prodotto cartesiano $R^n \times R^n$. L'immagine $\sigma_{(n,n)} \circ \rho \circ \overset{-1}{\sigma}_{(n,n)}$ di ρ rispetto alla rappresentazione $\sigma_{(n,n)} \square \sigma_{(n,n)}$ si dice il grafico aritmetico di ρ .

Un diffeomorfismo f nello spazio R^n si dice semiconico se il suo grafico aritmetico è semiconico. E' facile vedere che se f è non vuoto questo equivale a dire che $pr_1 f$ è un sottoinsieme semiconico di R^n e che f è determinato da un sistema di funzioni positivamente omogenee di grado 1.

Indichiamo con $\mathfrak{H}_n$ l'insieme di tutti i diffeomorfismi semiconici nello spazio R^n. E' facile vedere che $\mathfrak{H}_n$ è un gruppo generalizzato u-stabile ma non localizzato.

Da ogni gruppo generalizzato $\mathfrak{G}$ di diffeomorfismi in R^n possiamo ottenere il gruppo generalizzato localizzato $\mathfrak{G} \circ \mathfrak{J}_n$, il quale si dice la <u>chiusura localizzata</u> del gruppo generalizzato $\mathfrak{G}$.

La chiusura localizzata di $\mathfrak{H}_n$ è il gruppo generalizzato di Lie definito dal sistema di equazioni differenziali

$$f^\alpha_\beta \xi^\beta = f^\alpha \tag{4.7}$$

le quali coincidono con le identità di Eulero per funzioni positivamente omogenee di grado 1.

Un diffeomorfismo f in R^{n+s} si dice <u>n-proiettabile</u> se la sua immagine rispetto alla rappresentazione $p_{(n+s,n)} \square p_{(n+s,n)}$ è un diffeomorfismo in R^n, cioè se f è definito da un sistema di equazioni della forma seguente

$$\bar{\xi}^\alpha = f^\alpha(\xi^\beta) \quad , \quad \bar{\eta}^i = f^{n+i}(\xi^\beta, \eta^j) \tag{4.8}$$

$$(i, j = 1, \ldots, s)$$

ove le prime n equazioni definiscono il diffeomorfismo $p_{(n+s,n)} \circ f \circ \overset{-1}{p}_{(n+s,n)}$ in R^n.

L'insieme di tutti i diffeomorfismi n-proiettabili in R^{n+s} è un gruppo generalizzato che indicheremo con $\mathcal{R}_{n+(s)}$.

Il gruppo generalizzato $\mathcal{R}_{n+(s)}$ non è u-stabile. Indichiamo con $\mathfrak{X}_{n+(s)}$ il gruppo generalizzato che è la u-chiusura di $\mathcal{R}_{n+(s)}$.

E' evidente che $\mathfrak{X}_{n+(s)}$ è il gruppo generalizzato di Lie definito dal sistema di equazioni

$$f^{\alpha}_{n+i} = 0 \quad . \tag{4.9}$$

Siano $\mathcal{G}$ un gruppo di Lie astratto e Γ un gruppo di trasformazioni in R^s il quale è una rappresentazione di $\mathcal{G}$.

Le trasformazioni di Γ hanno la forma generale

$$\bar{\eta}^i = \varphi^i(g, \eta^j) \qquad \text{ove} \quad g \in \mathcal{G} \tag{4.10}$$

Il Γ-prolungamento generale di un gruppo generalizzato di diffeomorfismi $\mathfrak{g}$ in R^n è il gruppo generalizzato di diffeomorfismi n-proiettabili in R^{n+s} della forma seguente

$$\bar{\xi}^{\alpha} = f^{\alpha}(\xi^{\beta}) \quad , \quad \bar{\eta}^i = \varphi^i(g(\xi^{\beta}), \eta^j), \tag{4.11}$$

dove il primo sistema di equazioni definisce un diffeomorfismo f di $\mathfrak{g}$ e $g(\xi^{\beta})$ è una funzione differenziabile la quale è definita su $pr_1 f$ e prende valori nel gruppo di Lie $\mathcal{G}$.

Un sottogruppo $\mathfrak{g}_p$ del Γ-prolungamento generale di un gruppo generalizzato $\mathfrak{g}$ si dice un Γ-prolungamento parziale di $\mathfrak{g}$ se le due seguenti condizioni sono soddisfatte : 1) l'insieme delle n-proiezioni di tutti i diffeomorfismi di $\mathfrak{J}_P$ coincide con $\mathfrak{J}$ stesso, 2) non esiste un sottogruppo $\bar{\Gamma}$ proprio di Γ tale che $\mathfrak{J}_P$ sia un sottogruppo del $\bar{\Gamma}$-prolungamento generale di $\mathfrak{J}$.

§ 5. Spazi dotati di coordinate .

Un atlante n-dimensionale su un insieme A avente la potenza del continuo è un insieme $\mathfrak{X}_n$-coniugato di rappresentazioni parziali invertibili $K \subset \mathfrak{F}(A \times R^n)$ il quale soddisfa la condizione che l'insieme delle prime proiezioni delle rappresentazioni di K è un ricoprimento dell'insieme A , cioè :

$$(5.1) \qquad \bigcup_{\varkappa \in K} pr_1 \varkappa = A \qquad .$$

Le rappresentazioni parziali invertibili di K si dicono sistemi di coordinate(o coordinati)nell'insieme A .

Un sottoinsieme $\overline{K}$ dell'atlante coordinato K si dice una sua base se K è contenuto nella u-chiusura di $\overline{K}$, cioè se $K \subset f_u(\overline{K})$.

E' evidente che una base di un atlante è anche un atlante.

Come è noto se un atlante n-dimensionale K è dato su un insieme A allora possiamo definire in A in modo intrinseco, una topologia tale che tutti i sistemi coordinati siano omeomorfismi parziali aperti dello spazio topologico A in R^n , cioè omeomorfismi di sottoinsiemi aperti di A su sottoinsiemi aperti di R^n .

Un atlante coordinato si dice separabile se questa topologia intrinseca è separabile.

Uno spazio dotato di coordinate ad n-dimensioni è una coppia (A, K) ove A è un insieme avente la potenza del continuo e K è un suo atlante n-dimensionale separabile. L'insieme A si dice l'insieme puntuale dello spazio dotato di coordinate e gli elementi di questo insieme si dicono punti dello spazio.

Diffeomorfismi della forma $\varkappa_2 \circ \varkappa_1^{-1}$, ove $\varkappa_1, \varkappa_2$ sono sistemi coordinati, si dicono diffeomorfismi coordinati.

Nel seguito indicheremo uno spazio dotato di coordinate a n dimensioni con una sola lettera per esempio C_n, ove l'indice n indica la dimensione dello spazio.

Allora con $\Pi_1(C_n)$ indicheremo l'insieme puntuale che è il primo elemento della coppia C_n e con $\Pi_2(C_n)$ indicheremo l'atlante che è il secondo elemento della coppia C_n .

Sia $\mathcal{G}$ un gruppo generalizzato di diffeomorfismi in R^n .

Uno spazio dotato di coordinate ad n dimensioni si dice un $\mathcal{G}$ -spazio se il suo atlante è un insieme $\mathcal{G}$ -coniugato massimale di rappresentazioni parziali invertibili in R^n .

Abbiamo due classi importantissime di $\mathcal{G}$ -spazi : quello di spazi semplici di Klein, ove $\mathcal{G}$ è un gruppo di trasformazioni di Lie, e quello di spazi di Veblen-Whitehead, ove $\mathcal{G}$ è un gruppo generalizzato di trasformazioni parziali di Lie.

Elenchiamo alcuni fra i $\mathcal{G}$-spazi più importanti dal punto di vista di applicazioni ulteriori.

Un $\mathcal{X}_n$-spazio si dice uno spazio generale dotato di coordinate ed è indicato con X_n .

Un $\vec{\mathcal{X}}_n$-spazio si dice uno spazio orientato generale ed è indicato con $\vec{X}_n$.

Un $\vec{\mathcal{V}}_n$-spazio si dice uno spazio "equivoluminale" orientato ed è indicato con $\vec{V}_n$.

Un $\mathcal{H}_n$-spazio si dice uno spazio semiconico ed è indicato con H_n .

Un $\mathcal{X}_{n+(s)}$-spazio si dice uno spazio "fogliettato" generale ed è indicato con $X_{n+(s)}$.

Tra gli spazi semplici di Klein ci occorrono quelli corrispondenti a gruppi di trasformazioni di Lie in R^n : il gruppo centro-affine $\mathcal{A}_n$, il gruppo-centro-affine orientato $\vec{\mathcal{A}}_n$, il gruppo centro-equiaffine orientato $\vec{\mathcal{U}}_n$.

Un $\mathcal{A}_n$-spazio si dice uno spazio centro-affine ed è indicato con A_n .

Un $\vec{\mathcal{A}}_n$-spazio si dice uno spazio centro-affine orientato ed è indicato con $\vec{A}_n$.

Un $\vec{\mathfrak{U}}_n$-spazio si dice uno spazio centro-equiaffine orientato ed è indicato con $\vec{U}_n$.

Oltre a ciò un $\mathcal{L}$-spazio, ove $\mathcal{L}$ è un sottogruppo del gruppo centro-affine $\mathfrak{A}_n$, lo chiameremo uno spazio lineare e lo indicheremo con L_n . Ogni sottoinsieme aperto non vuoto φ dell'insieme puntuale $\Pi_1(C_n)$ di uno spazio dotato di coordinate C_n si può considerare come l'insieme puntuale di uno spazio dotato di coordinate ad n dimensioni di cui l'atlante è l'insieme $\Pi_2(C_n) \circ \{\Delta_\varphi\}$ delle restrizioni di tutti i sistemi coordinati dello spazio C_n rispetto al sottoinsieme φ . Questo spazio dotato di coordinate si dice la restrizione dello spazio C_n rispetto al sottoinsieme aperto φ .

Un sottoinsieme aperto non vuoto φ dell'insieme puntuale dello spazio dotato di coordinate C_n si dice una regione normale di C_n se la restrizione $\varkappa \circ \Delta_\varphi$ di ogni sistema coordinato di C_n è ancora un sistema coordinato. La restrizione di uno spazio dotato di coordinate rispetto a una regione normale si dice una restrizione normale.

E' facile vedere se uno spazio dotato di coordinate C_n è un $\mathfrak{G}$-spazio ove $\mathfrak{G}$ è un gruppo generalizzato localizzato di diffeomorfismi, poichè ogni sottoinsieme puntuale aperto non vuoto di C_n sarà una regione normale. In particolare ogni spazio $\mathfrak{X}_n$ possiede questa proprietà.

Uno spazio dotato di coordinate C_n si dice subordinato ad uno spazio dotato di coordinate D_n se gli insiemi puntuali di questi spazi coincidono $\Pi_1(C_n) = \Pi_1(D_n)$ e l'atlante del primo spazio è contenuto nell'atlante del secondo spazio : $\Pi_2(C_n) \subset \Pi_2(D_n)$.

Se $\mathfrak{G}$ è un gruppo generalizzato u-stabile di diffeomorfismi allora ogni spazio dotato di coordinate C_n il cui atlante è $\mathfrak{G}$-coniugato è subordinato ad un unico $\mathfrak{G}$-spazio, precisamente allo spazio $(\Pi_1(C_n), f_n(\mathfrak{G} \circ \Pi_2(C_n))$.

Sia : $\mathcal{K} \in \mathcal{Q}(A \times R^n)$ e $\mathcal{X} \in \mathcal{Q}(B \times R^s)$. La rappresentazione $\mathcal{K} \boxtimes \mathcal{X}$ del prodotto cartesiano $A \times B$ in R^{n+s} definita dalla formula

$$\mathcal{K} \boxtimes \mathcal{X} = \sigma_{(n,s)} \circ (\mathcal{K} \square \mathcal{X}) , \tag{5.2}$$

ove $\mathcal{K} \square \mathcal{X}$ è il prodotto diretto delle rappresentazioni $\mathcal{K}$ e $\mathcal{X}$ si dice il <u>prodotto diretto aritmetizzato</u> di queste rappresentazioni.

Il prodotto di spazi dotati di coordinate C_n e D_s è definito come lo spazio dotato di coordinate ad $n+s$ dimensioni

$$C_n \boxtimes D_s = (\Pi_1(C_n) \times \Pi_1(D_s), \ \Pi_2(C_n) \boxtimes \Pi_2(D_s)) . \tag{5.3}$$

Una rappresentazione puntuale θ di uno spazio dotato di coordinate C_{n+s} su uno spazio dotato di coordinate B_n si dice una <u>proiezione coordinata</u> se esiste un sottoatlante $K \subset \Pi_2(C_{n+s})$ dello spazio C_{n+s} tale che l'immagine di ogni sistema coordinato $\mathcal{K} \in K$ mediante la rappresentazione $\theta \square P_{(n+s,n)}$ sia un sistema coordinato dello spazio B_n ; cioè per ogni $\mathcal{K} \in K$ avremo $P_{(n+s,n)} \circ \mathcal{K} \circ \theta^{-1} \in \Pi_2(B_n)$.

Indichiamo con K_θ il sottoatlante massimo dello spazio C_{n+s} che gode di questa proprietà e chiamiamo K_θ <u>l'atlante ammissibile</u> della proiezione coordinata θ , per lo spazio C_n .

L'insieme X_θ dei sistemi coordinati dello spazio B_n i quali sono le immagini dei sistemi coordinati di K_θ mediante la rappresentazione $\theta \square P_{(n+s,n)}$ è evidentemente un sottoatlante dello spazio B_n . Questo sottoatlante X_θ si chiama l'<u>atlante ammissibile</u> della proiezione coordinata θ per lo spazio B_n .

Una proiezione coordinata θ dello spazio C_{n+s} sullo spazio B_n si dice un morfismo coordinato locale se K_θ è una base dell'atlante $\Pi_2(C_{n+s})$ e X_θ è una base dell'atlante $\Pi_2(B_n)$.

Un morfismo coordinato locale si dice un morfismo coordinato se $K_\theta = \Pi_2(C_{n+s})$ e $X_\theta = \Pi_2(B_n)$.

E' evidente che un morfismo coordinato può essere una rappresentazione invertibile solamente nel caso che ambedue gli spazi siano di dimensioni eguali. Tale morfismo coordinato si dice un isomorfismo degli spazi. Un isomorfismo di uno spazio su se stesso si dice un automorfismo.

E' facile vedere che per una proiezione coordinata θ di uno spazio dotato di coordinate C_{n+s} su uno spazio dotato di coordinate B_n l'atlante ammissibile dello spazio C_{n+s} è $\mathcal{R}_{n+(s)}$ -coniugato.

Una rappresentazione parziale puntuale θ di uno spazio dotato di coordinate C_{n+s} in uno spazio dotato di coordinate B_n si dice una proiezione coordinata parziale se $pr_1\theta$ e $pr_2\theta$ sono regioni normali di questi spazi e θ è una proiezione coordinata della restrizione di C_{n+s} rispetto a $pr_1\theta$ sulla restrizione di B_n rispetto a $pr_2\theta$.

Analogamente sono definite le nozioni di morfismo coordinato locale parziale , morfismo coordinato parziale, isomorfismo parziale e automorfismo parziale.

Una rappresentazione parziale puntuale di uno spazio dotato di coordinate C_{n+s} in uno spazio dotato di coordinate B_n si dice regolare se essa è una proiezione coordinata parziale dello spazio $\mathfrak{X}_{n+s}$ nello spazio $\mathfrak{X}_n$ ai quali sono subordinati rispettivamente gli spazi C_{n+s} e B_n .

Una fibrazione è una terna (C_{n+s}, B_n, θ) ove C_{n+s} e B_n sono spazi dotati di coordinati e θ è un morfismo coordinato locale dello spazio C_{n+s} sullo spazio B_n . Lo spazio C_{n+s} si dice spazio fibrato , lo spazio B_n si dice spazio base e il morfismo coordinato locale θ si dice la proiezione della fibrazione data.

Indicheremo con Θ la rappresentazione dell'atlante ammissibile K_θ dello spazio fibrato C_{n+s} sull'atlante ammissibile X_θ dello spazio base B_n determinata dalla rappresentazione $\theta \square \mathfrak{p}_{(n+s,n)}$

(5.4) $$\Theta(\varkappa) = \mathfrak{p}_{(n+s,n)} \circ \varkappa \circ \theta^{-1} \quad .$$

Se la rappresentazione Θ è invertibile allora la fibrazione (C_{n+s}, B_n, θ) si dice olonoma .

Il sottoinsieme puntuale $\theta^{-1}\langle p\rangle$ dello spazio fibrato C_{n+s} , ove p è un punto dello spazio base B_n , si dice la fibra di C_{n+s} sopra il punto p .

Indicheremo con φ_p la rappresentazione dell'atlante $\Pi_2(C_{n+s})$ dello spazio C_{n+s} in $\mathfrak{A}(\theta^{-1}\langle p\rangle \times R^s)$ definita dalla formula

(5.5) $$\varphi_p(\varkappa) = \mathfrak{p}^{*}_{(n+s,s)} \circ \varkappa \circ \Delta_{\theta^{-1}\langle p\rangle} \quad .$$

Allora $pr_2 \varphi_p$ sarà un atlante s-dimensionale sulla fibra $\theta^{-1}\langle p\rangle$. Lo spazio dotato di coordinate $C_s(p) = (\theta^{-1}\langle p\rangle, pr_2 \varphi_p)$ ad s dimensioni si dice la fibra sopra il punto p dello spazio base .

Una fibrazione si dice una fibrazione generale se lo spazio fibrato è uno spazio fibrato generale $\mathfrak{X}_{n+(s)}$ e lo spazio base è uno spazio dotato di coordinate generale $\mathfrak{X}_n$.

Una fibrazione si dice una fibrazione semplice di Klein se tutte le fibre sono spazi semplici di Klein con il gruppo coordinato comune, ai cui atlanti è aggiunto il sistema coordinato vuoto.

E' evidente che se lo spazio base di una fibrazione semplice di Klein è un $\mathcal{G}$-spazio, lo spazio fibrato sarà un $\overline{\mathcal{G}}$-spazio ove il gruppo generalizzato $\overline{\mathcal{G}}$ è un Γ-prolungamento del gruppo generalizzato $\mathcal{G}$. Se $\overline{\mathcal{G}}$ è il Γ-prolungamento generale di $\mathcal{G}$ questa fibrazione semplice di Klein si dice generale. Come è noto, per ogni superficie S ad s dimensioni in uno spazio dotato di coordinate C_n si ottiene intrinsecamente un atlante s-dimensionale K tale che (S, K) sia uno spazio $\mathcal{X}_s$. Questo spazio $\mathcal{X}_s$ è detto <u>immerso</u> nello spazio C_n in modo intrinseco. In casi speciali si può determinare intrinsecamente un atlante speciale più ristretto e in tal modo ottenere invece di $\mathcal{X}_n$ uno spazio dotato di coordinate più particolare.

Chiameremo una superficie ad $n+\ell$ dimensioni nello spazio fibrato C_{n+s} di una fibrazione (C_{n+s}, B_n, θ) una <u>superficie ℓ-secante</u> se la proiezione verticale determina una proiezione coordinata dello spazio $\mathcal{X}_{n+\ell}$, intrinsecamente immerso nello C_{n+s}, corrispondente alla superficie sullo spazio base B_n.

Se $\ell=0$ e la proiezione coordinata corrispondente è invertibile una superficie 0-secante si dice una superficie secante.

Si può dimostrare che ogni superficie ℓ-secante S può essere considerata come l'insieme puntuale dello spazio fibrato di una fibrazione $(D_{n+\ell}, B_n, \overline{\theta})$ il quale ha lo stesso spazio base B_n e la cui proiezione $\overline{\theta}$ è una restrizione della θ, cioè $\overline{\theta} = \theta \circ \Delta_S$.

La superficie ℓ-secante S taglia ogni fibra dello spazio fibrato C_{n+s} secondo una superficie ad ℓ dimensioni la quale è l'insieme puntuale della fibra corrispondente dello spazio fibrato $D_{n+\ell}$.

Se lo spazio base è uno spazio $\mathcal{X}_n$ allora ad una superficie ℓ-secante corrisponde una fibrazione generale $(\mathfrak{X}_{n+(s)}, \mathfrak{X}_n, \theta)$.

Se una superficie ℓ-secante S è data in uno spazio fibrato C_{n+s} , ad ogni punto dello spazio base B_n corrisponde una superficie ad ℓ dimensioni nella fibra sopra questo punto. Dunque la superficie ℓ-secante S può essere considerata come un campo, definito sullo spazio base B_n, di superficie ad ℓ dimensioni nelle fibre.

Considerando una superficie ℓ-secante come l'insieme puntuale dello spazio fibrato corrispondente, la determineremo mediante un sistema di equazioni parametriche

$$(5.6) \qquad \eta^i = \eta^i(\xi^\alpha, \xi^u) , \qquad (u = 1, \dots, \ell; \ \alpha = 1, \dots, n; \ i = 1, \dots, s)$$

ove ξ^α -sono coordinate di punto dello spazio base B_n e η^i , ξ^u sono coordinate di punto nelle fibre C_s e D_ℓ rispettivamente.

Consideriamo in particolare una superficie ℓ-secante nello spazio fibrato di una fibrazione semplice di Klein.

Questa superficie si dice una superficie ℓ-secante <u>isomorficamente</u> se le sue sezioni con le fibre sono superficie ad ℓ-dimensioni le quali si corrispondono tutte l'una all'altra in certi isomorfismi delle fibre di Klein.

Considerando la superficie ℓ-secante isomorficamente come un campo di superficie ad ℓ dimensioni, chiameremo un tale campo, un <u>campo di superficie isomorfe.</u>

Una superficie ℓ-secante si dice una superficie ℓ-secante <u>costante</u> se nello spazio fibrato si può scegliere un sottoatlante tale che in tutti i sistemi coordinati nelle fibre di Klein, indotti dai sistemi coordinati di questo sottoatlante , tutte le sezioni delle fibre abbiano come loro immagini la stessa superficie ad ℓ dimensioni nello spazio aritmetico R^s .

Una superficie ℓ-secante costante rispetto ai sistemi coordinati del sottoatlante scelto può essere determinata dal sistema di equazioni parametriche ove mancano le coordinate ξ^{α} della base

(5.7) $$\eta^{i} = \eta^{i}(\xi^{n}) .$$

Considerando la superficie ℓ-secante costante come un campo di superficie ad ℓ dimensioni chiameremo un tale campo, un campo di superficie costanti.

Una fibrazione (C_{n+s}, B_n, θ) si dice un prolungamento dello spazio base B_n se questo è identificato con una superficie secante. In questo caso in ogni fibra è dato un punto, il quale si dice punto base .

Nelle applicazioni avremo sovente prolungamenti lineari di uno spazio $\mathcal{X}_n$. Le fibre saranno spazi lineari isomorfi i cui centri saranno i punti base.

§ 6. Oggetti coordinati. [3]

Sia C_n uno spazio dotato di coordinate. Una coppia $(p, \varkappa)$, ove p è un punto di C_n e $\varkappa$ un sistema coordinato di C_n nel quale il punto p è rappresentato, cioè $p \in pr_1 \varkappa$, si dice un sistema coordinato centralizzato dello spazio C_n. Il punto p si chiama il centro del sistema coordinato centralizzato $(p, \varkappa)$.

In senso lato un oggetto coordinato nello spazio C_n definito su un sottoinsieme puntuale φ è una funzione qualunque $\Omega(p, \varkappa)$ del sistema coordinato centralizzato $(p, \varkappa)$ definita sull'insieme di tutti i sistemi coordinati centralizzati i cui centri appartengono a φ.

Un oggetto coordinato si dice un oggetto coordinato puro se esso può essere considerato come una funzione del solo sistema coordinato cioè se per ogni sistema coordinato $\varkappa$ e punti $p_1, p_2 \in \varphi \cap pr_1 \varkappa$ si ha $\Omega(p_1, \varkappa) = \Omega(p_2, \varkappa)$.

Dimostreremo che ogni oggetto coordinato Ω in C_n può essere ridotto a un oggetto coordinato puro Ω_f i cui valori sono funzioni definite su sottoinsiemi di R^n, le quali prendono valori nell'insieme $pr_2 \Omega$ di valori dell'oggetto coordinato Ω stesso.

Per questo facciamo corrispondere a ogni sistema coordinato $\varkappa$, ove $pr_1 \varkappa \cap \varphi \neq \emptyset$, una funzione $f_\varkappa$ definita sul sottoinsieme $\varkappa(\varphi)$ di R^n dalle equazioni

$$(6.1) \qquad f_\varkappa(\xi^\alpha) = \Omega(\overset{-1}{\varkappa}((\xi^1, \ldots, \xi^n)), \varkappa) \quad ,$$

allora definiamo l'oggetto coordinato puro Ω_f con le equazioni

---

3) Diremo "oggetti coordinati" invece di "oggetti". Sulla teoria generale v. [15], [21].

(6.2) $$\Omega_f(\varkappa) = f_\varkappa(\xi^\alpha) \quad .$$

Questo oggetto coordinato puro Ω_f si dice <u>oggetto coordinato funzionale</u> corrispondente all'oggetto coordinato Ω .

Siano Ω_1 e Ω_2 due oggetti coordinati definiti su lo stesso sottoinsieme puntuale $\mathcal{Y}$. Ω_2 si dice un <u>comitante locale</u> di Ω_1 se esiste una funzione f tale che per tutti i sistemi coordinati centralizzati $(p, \varkappa)$, ove $p \in \mathcal{Y}$, è

(6.3) $$\Omega_2(p, \varkappa) = f(p, \Omega_1(p, \varkappa)) \quad .$$

Se questa funzione non dipende da p , cioè se esiste una rappresentazione φ di $pr_2\Omega_1$ su $pr_2\Omega_2$ tale che

(6.4) $$\Omega_2 = \varphi \circ \Omega_1 \quad ,$$

allora Ω_2 si dice un <u>comitante</u> di Ω_1 .

Un oggetto coordinato Ω si dice <u>aritmetico</u> se esso prende valori in uno spazio aritmetico R^N . In questo caso Ω può essere considerato come il sistema di N oggetti coordinati numerici $\Omega^{\mathcal{A}}$ $(\mathcal{A} = 1, \dots, N)$ che si dicono i <u>componenti</u> dell'oggetto coordinato aritmetico Ω .

N funzioni $\Omega^{\mathcal{A}}(\xi^\alpha)$ le quali siano valori dell'oggetto coordinato funzionale corrispondente in un sistema coordinato $\varkappa$ si dico <u>componenti funzionali</u> dell'oggetto coordinato aritmetico Ω in questo sistema coordinato.

Avendo due oggetti coordinati aritmetici $\Omega_1^{\mathcal{A}_1}$ $(\mathcal{A}_1 = 1, \dots, N_1)$ e $\Omega_2^{\mathcal{A}_2}$ $(\mathcal{A}_2 = 1, \dots, N_2)$ possiamo costruire un nuovo oggetto coordinato aritmetico con $N_1 + N_2$ componenti $\Omega_1^{\mathcal{A}_1}$, $\Omega_2^{\mathcal{A}_2}$, che si dice l'<u>unione</u> degli oggetti coordinati precedenti. In particolare ogni oggetto coordinato aritmetico può essere considerato come l'unione dei suoi componenti.

Inversamente prendendo un sistema ordinato di certi componenti di un oggetto coordinato aritmetico dato $\Omega^{\mathcal{A}}$ otterremo un nuovo oggetto coordinato aritmetico il quale si dice una parte dell'oggetto coordinato precedente.

Finora abbiamo considerato solamente oggetti coordinati i quali sono funzioni di un solo argomento, che è un sistema coordinato centralizzato. Come una generalizzazione naturale possiamo considerare anche oggetti coordinati che sono funzioni di parecchi sistemi coordinati centralizzati in spazi diversi o coincidenti. Tali oggetti coordinati possono essere ridotti a oggetti coordinati ordinari nello spazio prodotto degli spazi corrispondenti.

Dal punto di vista di applicazioni ulteriori un interesse speciale hanno i cosidetti oggetti coordinati "connettenti" e in particolare oggetti coordinati intermedi.[4)]

Sia D_s uno spazio dotato di coordinate immerso in uno spazio dotato di coordinate C_n : Cioè $\Pi_1(D_s) \subset \Pi_1(C_n)$. Un oggetto coordinato di due argomenti $(p, \varkappa)$; (q, χ) ove $p \in \Pi_1(C_n), \varkappa \in \Pi_2(C_n)$, $q \in \Pi_1(D_s), \chi \in \Pi_2(D_s)$ si dice "connettente" se esso è definito su un insieme di coppie $(p, \varkappa), (p, \chi)$ di sistemi coordinati centralizzati con centro comune $p \in \mathcal{G}$ ove $\mathcal{G}$ è un sottoinsieme puntuale dello spazio D_s .

Questo oggetto coordinato può essere considerato come una funzione $\Omega(p, \varkappa, \chi)$ della terna $(p, \varkappa, \chi)$ la quale si chiama una coppia di sistemi coordinati centralizzati col centro in comune.

Se gli spazi C_n e D_s coincidono allora un oggetto coordinato connettente $\Omega(p, \varkappa_1, \varkappa_2)$ si dice intermedio.

Un oggetto coordinato intermedio aritmetico $A^{\alpha}_{\alpha_1 \dots \alpha_n}(p, \varkappa_1, \varkappa_2)(n=1, \dots, v)$

---

4) Per il caso speciale di tensori di connessione e in particolare intermedie v. [20], [22].

definito dalle equazioni

$$A^{\alpha}_{\alpha_1 \ldots\ldots \alpha_n}(p, \mathcal{X}_1, \mathcal{X}_2) = f^{\alpha}_{\alpha_1 \ldots\ldots \alpha_n}(\xi^{\beta}_p) \qquad (6.5) \qquad (n = 1, \ldots . \nu)$$

ove le funzioni f^{α} determinano il diffeomorfismo coordinato $\mathcal{X}_2 \circ \mathcal{X}_1^{-1}$ e ξ^{β}_p sono le coordinate del punto p rispetto al sistema coordinato $\mathcal{X}_1$, si dice l'oggetto coordinato intermedio unitario d'ordine ν.

Questa denominazione di $A^{\alpha}_{\alpha_1 \ldots\ldots \alpha_n}$ è dovuta al fatto che per ogni sistema coordinato $\mathcal{X}$, ove $p \in pr_1 \mathcal{X}$, è

$$A^{\alpha}_{\alpha_1 \ldots\ldots \alpha_n}(p, \mathcal{X}, \mathcal{X}) = \delta^{\alpha}_{\alpha_1 \ldots\ldots \alpha_n} \qquad (6.6) \qquad (n = 1, \ldots, \nu)$$

ove $\delta^{\alpha}_{\alpha_1 \ldots\ldots \alpha_n}$ per $n=1$ coincidono con i simboli di Kronecker e per $n > 1$ sono eguali a zero.

Due sistemi coordinati centralizzati $(p_1, \mathcal{X}_1)$ e $(p_2, \mathcal{X}_2)$ si dicono aventi la <u>vicinanza differenziale d'ordine ν</u> se i loro centri coincidono $p_1 = p_2 = p$ e se sussistono le relazioni seguenti

$$A^{\alpha}_{\alpha_1 \ldots\ldots \alpha_n}(p, \mathcal{X}_1, \mathcal{X}_2) = \delta^{\alpha}_{\alpha_1 \ldots\ldots \alpha_n} \qquad (6.7) \qquad (n = 1, \ldots, \nu)$$

Due sistemi coordinati centralizzati si dicono aventi la vicinanza differenziale esattamente d'ordine ν, se essi hanno la vicinanza differenziale di ordine ν ma non quella d'ordine $\nu + 1$.

Un oggetto coordinato Ω si dice <u>locale</u>, se per tutti i sistemi coordinati $\mathcal{X}_1, \mathcal{X}_2$ e $p \in pr_1 \mathcal{X}_1$

$$\mathcal{X}_1 \subset \mathcal{X}_2 \longrightarrow \Omega(p, \mathcal{X}_1) = \Omega(p, \mathcal{X}_2) \qquad (6.8)$$

E' facile vedere che se due oggetti coordinati locali, definiti sullo stesso sottoinsieme puntuale di uno spazio, coincidono sull'insieme di tutti i sistemi coordinati centralizzati $(p, \varkappa)$, ove $\varkappa$ appartiene a una base dell'atlante $\Pi_2(C_n)$ e $p \in \mathcal{P} \cap pr_1 \varkappa$, allora tali oggetti coordinati sono uguali. Dunque un oggetto coordinato locale può essere definito unicamente se i suoi valori sono dati dati solamente per un insieme di sistemi coordinati centralizzati in modo tale che la condizione (6.8) sia soddisfatta.

Un caso particolare importantissimo di oggetto coordinato locale è quello di <u>oggetto coordinato differenziale</u>, o più brevemente <u>oggetto differenziale</u>, di classe ν.

Un oggetto coordinato si dice un oggetto differenziale di classe ν se i suoi valori per ogni coppia di sistemi coordinati centralizzati, aventi la vicinanza differenziale dell'ordine ν, coincidono.

Un oggetto differenziale si dice un oggetto differenziale esattamente di classe ν, se esso è di classe ν ma non di classe $\nu + 1$.

Considerando l'oggetto coordinato intermedio unitario d'ordine ν come un oggetto coordinato nello spazio prodotto, abbiamo che questo è un oggetto differenziale esattamente di classe ν.

Fissando un punto p_0 e un sistema coordinato $\varkappa_0$ ove $p_0 \in pr_1 \varkappa_0$ possiamo ottenere dall'oggetto coordinato intermedio unitario d'ordine ν, due oggetti differenziali di classe ν definiti nel punto p_0:

(6.9) $$\mathcal{X}^{\alpha}_{\alpha_1 \ldots \alpha_n}(p_0, \varkappa) = A^{\alpha}_{\alpha_1 \ldots \ldots \alpha_n}(p_0, \varkappa_0, \varkappa)$$

il quale si dice un oggetto differenziale <u>universale</u> d'ordine ν e

(6.10) $$\mathcal{Y}^{\alpha}_{\alpha_1 \ldots \ldots \alpha_n}(p_0, \varkappa) = A^{\alpha}_{\alpha_1} \ldots \alpha_n(p_0, \varkappa, \varkappa_0)$$

che si dice un oggetto differenziale universale inverso d'ordine ν .

Si può dimostrare che ogni oggetto differenziale di classe ν può essere rappresentato come un comitante locale di un oggetto differenziale universale d'ordine ν , oppure come un comitante di un oggetto differenziale universale inverso d'ordine ν .

Un oggetto differenziale Ω di classe ν si dice localmente geometrico se esiste una funzione $\mathfrak{F}$ tale che per ogni coppia $(p, \varkappa_1)$ e $(p, \varkappa_2)$ di sistemi coordinati centralizzati con il centro comune risulti :

$$\Omega(p, \varkappa_2) = \mathfrak{F}(p, \Omega(p, \varkappa_1), A^{\alpha}_{\alpha_1 \ldots \alpha_n}(p, \varkappa_1, \varkappa_2)) \quad (n = 1, \ldots, \nu). \tag{6.11}$$

Se la funzione $\mathfrak{F}$ non dipende dal punto p , cioè se

$$\Omega(p, \varkappa_2) = \mathfrak{F}(\Omega(p, \varkappa_1), A^{\alpha}_{\alpha_1 \ldots \alpha_n}(p, \varkappa_1, \varkappa_2)) \qquad (n = 1, \ldots, \nu) \tag{6.12}$$

allora Ω si chiama un oggetto differenziale geometrico.

L'insieme delle trasformazioni dell'insieme $pr_2 \Omega$ di valori di un oggetto differenziale localmente geometrico determinati dal sistema di equazioni

$$\overline{\Omega} = \mathfrak{F}(p_o, \Omega, A^{\alpha}_{\alpha_1 \ldots \alpha_n}) \quad , \tag{6.13}$$

ove $A^{\alpha}_{\alpha_1 \ldots \alpha_n}$ sono parametri che prendono valori sull'insieme di valori dell'oggetto differenziale intermedio unitario d'ordine ν , è un gruppo transitivo di trasformazioni, che si chiama il gruppo dell'oggetto differenziale localmente geometrico nel punto p_o .

L'oggetto differenziale universale d'ordine ν ha il gruppo semplicemente transitivo, che si dice il gruppo differenziale d'ordine ν nel punto p_o .

Si può dimostrare che il gruppo di un oggetto differenziale localmente geometrico in un punto p_0 è una rappresentazione del gruppo differenziale d'ordine v corrispondente a questo punto.

Consideriamo un insieme $\mathcal{O}_p$ di oggetti differenziali geometrici con N componenti definiti in un punto dato p, soddisfacente alle condizioni seguenti :

1) Per ogni sistema coordinato $\mathcal{X}$, in cui si può rappresentare il punto p, l'insieme dei punti aritmetici che sono valori di oggetti differenziali geometrici di $\mathcal{O}_p$ corrispondenti a $\mathcal{X}$, è lo stesso sottoinsieme aperto $\mathcal{W}$ di R^n.

2) L'insieme di funzioni che determinano trasformazioni di componenti di oggetti differenziali geometrici di $\mathcal{O}_p$ è $\mathcal{F}$-unibile[5] e la sua unione è una funzione differenziabile $\mathcal{F}$.

E' facile vedere che tale insieme $\mathcal{O}_p$ di oggetti differenziabili geometrici può essere considerato in modo intrinseco come l'insieme puntuale di uno spazio semplice di Klein.

Infatti, ad ogni sistema coordinato $\mathcal{X} \in K_p$, ove K_p è l'insieme di tutti i sistemi coordinati in cui è rappresentato il punto p, corrisponde una rappresentazione $h(\mathcal{X})$ dell'insieme $\mathcal{O}_p$ sul sottoinsieme aperto $\mathcal{W}$ dello spazio aritmetico R^n. E' facile vedere che questa rappresentazione è invertibile poichè se in un sistema coordinato $\mathcal{X}$ i componenti di due oggetti differenziali geometrici diversi di $\mathcal{O}_p$ fossero eguali, le funzioni che determinano le trasformazioni dei loro componenti non sarebbero $\mathcal{F}$-unibili.

Poi è evidente che per ogni coppia $h(\mathcal{X}_1)$, $h(\mathcal{X}_2)$, con $\mathcal{X}_1, \mathcal{X}_2 \in K_p$, di rappresentazioni invertibili di $\mathcal{O}_p$ su $\mathcal{W}$, la trasformazione

---

5) La nozione di funzione $\mathcal{F}$-unibile è identica a quella di rappresentazioni parziali $\mathcal{F}$-unibili v. § 2 .

$h(\varkappa_2) \circ \overline{h(\varkappa_1)}^{-1}$ è il diffeomorfismo

$$\overline{\Omega}^{A} = \mathcal{F}^{A}(\Omega^{B}, A(p, \varkappa_1, \varkappa_2)) . \tag{6.14}$$

L'insieme $h(K_p) \circ \overline{h(K_p)}^{-1}$ di tutti questi diffeomorfismi è un gruppo che è una rappresentazione del gruppo differenziale di ordine v corrispondente al punto p . Questo gruppo di trasformazioni, nel caso generale, può essere non transitivo.

Dunque $(\mathcal{O}_p, h(K_p))$ è uno spazio semplice di Klein il quale si dice uno spazio di oggetti differenziali geometrici nel punto p dello spazio C_n .

Consideriamo ora un insieme $\mathcal{O}$ di oggetti differenziali geometrici ad N componenti dati nei punti di uno spazio coordinizzato C_n .

$\mathcal{O}$ gode della proprietà che ogni suo sottoinsieme $\mathcal{O}_p$ costituito da tutti gli oggetti differenziali geometrici di $\mathcal{O}$ definiti nel punto p , definisce, nel senso precedente, uno spazio semplice di Klein e che tutti questi spazi hanno lo stesso gruppo coordinato .

Allora è facile vedere che questo insieme di oggetti differenziali geometrici $\mathcal{O}$ può essere considerato come l'insieme puntuale di uno spazio fibrato D_{n+N} di una fibrazione di Klein (D_{n+N}, C_n, θ) ove la proiezione θ fa corrispondere ad ogni oggetto differenziale geometrico di $\mathcal{O}$ il punto dello spazio C_n in cui esso è definito. Le fibre sono gli spazi semplici di Klein isomorfi a oggetti differenziali geometrici definiti in punti corrispondenti dello spazio base C_n .

Chiameremo tali fibrazioni, fibrazioni <u>differenziali.</u>

E' evidente che ogni fibrazione differenziale è olonoma.

Infatti, ad ogni sistema coordinato dello spazio base C_n corrisponde un sistema coordinato unico in ogni fibra e perciò corrisponde un sistema coordinato unico nello spazio fibrato stesso. Questo dimostra che la proiezione

Θ di sistemi coordinati è una rappresentazione invertibile.

Nelle applicazioni avremo da fare di solito con oggetti differenziali geometrici i quali appartengono ad uno spazio fibrato di fibrazione differenziale data. Dunque nelle applicazioni la teoria degli oggetti differenziali geometrici in sostanza si riduce alla teoria di fibrazioni differenziali con lo spazio base dato.

Sia φ una rappresentazione differenziabile arbitraria di un segmento numerico in uno spazio dotato di coordinate C_n. I valori $\varphi^{(n)\alpha}(t_o)$ $(n = 1, \ldots, \nu)$ delle derivate fino all'ordine ν incluso delle funzioni φ^{α} che determinano la rappresentazione φ in termini coordinati, sono componenti di un oggetto differenziale geometrico di classe ν il quale si chiama un vettore d'ordine ν, nel punto $p(t_o)$ dello spazio C_n. Se la rappresentazione definisce in C_n una curva parametrizzata, questo vettore d'ordine ν si dice allora il vettore tangente d'ordine ν nel punto $p(t_o)$ della curva.

Lo spazio $T_{\nu n}(p)$, a νn dimensioni, di tutti i vettori d'ordine ν nel punto p dello spazio C_n si dice lo spazio tangente d'ordine ν associato a questo punto p. Questo spazio $T_{\nu n}(p)$ è la fibra sopra il punto p dello spazio fibrato $T_{n+(\nu n)}$ di una fibrazione $(T_{n+(\nu n)}, C_n, \theta)$ la quale si dice la fibrazione tangenziale d'ordine ν.

Di solito i punti dello spazio base sono identificati coi vettori nulli delle fibre e dunque la fibrazione tangenziale d'ordine ν è considerata come un prolungamento dello spazio C_n, il quale si dice il prolungamento tangenziale d'ordine ν. Consideriamo ora oggetti coordinati nello spazio fibrato C_{n+s} di una fibrazione arbitraria (C_{n+s}, B_n, θ).

Ad ogni sistema coordinato centralizzato $(p, \varkappa)$, ove $\varkappa \in pr_1 \Theta$, corrispondono il sistema coordinato centralizzato $(\theta(p), \Theta(\varkappa))$ dello spazio base e il sistema coordinato centralizzato $(p, \varphi_p(\varkappa))$ della fibra, che chiameremo rispettivamente la proiezione basica e la proiezione fibrale del sistema coordinato centralizzato $(p, \varkappa)$.

Un oggetto coordinato locale Ω si dice riducibile se sull'insieme dei sistemi coordinati centralizzati di $pr_1 \theta$ esso può essere considerato come una funzione della proiezione basica $(\theta(p), \theta(\varkappa))$ e della proiezione fibrale $(p, \varphi_p(\varkappa))$.

Dunque un oggetto coordinato reducibile in un punto p di uno spazio C_{n+s} fibrato può essere considerato come un oggetto coordinato di due argomenti negli spazi B_n e $C_s(p)$.

Considerando l'oggetto differenziale intermedio unitario d'ordine ν in un punto dello spazio fibrato C_{n+s} vediamo che i suoi componenti di specie $A^{\alpha}_{n+i_1 \dots n+i_m}$ sono sempre nulli. I componenti di specie

$A^{\alpha}_{\alpha_1 \dots \alpha_\nu}$ possono essere considerati come componenti dell'oggetto differenziale intermedio unitario d'ordine ν dello spazio base B_n e le componenti di specie $A^{n+i}_{n+i_1 \dots n+i_\nu}$ come componenti dell'oggetto differenziale intermedio unitario d'ordine ν della fibra $C_s(p)$.

Chiameremo questi oggetti differenziali intermedi $A^{\alpha}_{\alpha_1 \dots \alpha_m}$ e $A^{n+i}_{n+i_1 \dots n+i_m}$ rispettivamente le componenti nella base e nella fibra dell'oggetto differenziale intermedio unitario dello spazio C_{n+s} e la loro unione la parte riducibile di questo ultimo.

Analogamente possiamo definire la parte basica fibrale e riducibile di un oggetto differenziale universale d'ordine ν e di un oggetto differenziale universale inverso d'ordine ν in uno spazio fibrato C_{n+s}.

E' facile dimostrare che ogni oggetto differenziale riducibile in uno spazio fibrato C_{n+s} può considerarsi come un comitante locale della parte riducibile di un oggetto differenziale universale d'ordine ν corrispondente, o come un comitante locale della parte riducibile di un oggetto differenziale universale inverso.

Per un oggetto differenziale geometrico locale riducibile di classe ϑ si ha

(6.15) $$\Omega(p, \varkappa_2) = \mathcal{F}(p, \Omega(p, \varkappa_1), A^{\alpha}_{\alpha_1 \ldots \alpha_n}(p, \varkappa_1, \varkappa_2), A^{n+i}_{n+i_1 \ldots n+i_n}(p, \varkappa_1, \varkappa_2))$$

cioè le equazioni di trasformazione delle componenti comprendono solo la parte riducibile dell'oggetto differenziale intermedio unitario d'ordine ϑ .

I casi particolari di oggetti differenziali riducibili in uno spazio fibrato C_{n+s} sono oggetti coordinati locali basici i quali sono funzioni solo della proiezione basica del sistema coordinato centralizzato e oggetti coordinati locali fibrali i quali sono funzioni solo della proiezione fibrale del sistema coordinato centralizzato.

§ 7. Derivata di Lie. [6]

Un campo vettoriale v^α definito in una regione normale $\mathcal{G}$ di uno spazio coordinizzato C_n si dice isomorfo se il sistema di equazioni differenziali

$$\dot{\xi}^\alpha = v^\alpha \tag{7.1}$$

definisce una famiglia ad un parametro t di isomorfismi parziali τ_t dello spazio C_n, ove $\tau_0 = \Delta_{\mathcal{G}}$.

Ogni isomorfismo parziale τ_t induce una trasformazione parziale T_t dell'insieme dei sistemi coordinati centralizzati

$$T_t((p, \varkappa)) = (\tau_t(p), \varkappa \circ \overset{-1}{\tau}_t) \tag{7.2}$$

Sia Ω un oggetto coordinato aritmetico definito nella regione $\mathcal{G}$. Il prodotto $\Omega \circ \overset{-1}{T}_t$ può essere considerato come una funzione di $p, \varkappa, t$.

Supponiamo che questa funzione abbia la derivata rispetto a t per $t = 0$. Allora possiamo definire l'oggetto coordinato aritmetico

$$\mathfrak{L}\,\Omega = -\left[\frac{d}{dt}\,(\Omega \circ \overset{-1}{T}_t)\right]_{t=0} \tag{7.3}$$

che si dice <u>la derivata di Lie</u> di Ω rispetto al vettore v^α.

Si può dimostrare che la derivata di Lie di un oggetto differenziale di classe v è un oggetto differenziale della stessa classe.

Sia Ω un oggetto differenziale geometrico e siano

---

6) Sulla teoria generale di derivata di Lie v. [20], [22], [36], [44].

$$(7.4)\qquad \Omega^{\mathcal{A}}(p, x_2) = \mathcal{F}^{\mathcal{A}}(\Omega^{\mathcal{B}}(p, x_1), A^{\alpha}_{\alpha_1 \dots \alpha_m}(p, x_1, x_2))$$

le equazioni di trasformazione dei suoi componenti ove $\mathcal{F}^{\mathcal{A}}$ sono funzioni differenziabili.[7)] Si ottengono le espressioni seguenti per i componenti $\mathcal{L}\,\Omega^{\mathcal{A}}$ della derivata di Lie ([44])

$$(7.5)\qquad \mathcal{L}\,\Omega^{\mathcal{A}} = v^{\alpha}\partial_{\alpha}\Omega^{\mathcal{A}} - \sum_{m=1}^{v} \partial_{\alpha_1 \dots \alpha_m} v^{\alpha} \overset{\circ}{\mathcal{F}}{}^{\mathcal{A}\alpha_1, \dots, \alpha_m}_{\alpha \dots\dots}(\Omega^{\mathcal{B}}) \quad ,$$

ove

$$(7.6)\qquad \overset{\circ}{\mathcal{F}}{}^{\mathcal{A}\alpha_1 \dots \alpha_m}_{\alpha \dots\dots}(\Omega^{\mathcal{B}}) = \mathcal{F}^{\mathcal{A}\alpha_1 \dots \alpha_m}_{\alpha}(\Omega^{\mathcal{B}}, \delta^{\beta}_{\beta_1 \dots \beta_m}), \quad \mathcal{F}^{\mathcal{A}\alpha_1 \dots \alpha_m}_{\alpha} = \frac{\partial \mathcal{F}^{\mathcal{A}}}{\partial A^{\alpha}_{\alpha_1 \dots \alpha_m}}$$

Poi si può dimostrare che i componenti della derivata di Lie per due sistemi coordinati centralizzati con centro comune stanno nella relazione seguente

$$(7.7)\qquad \mathcal{L}\,\Omega^{\mathcal{A}} = \mathcal{F}^{\mathcal{A}}_{\mathcal{B}}(\Omega^{\Gamma}(p, x_1), A^{\alpha}_{\alpha_1 \dots \alpha_m}(p, x_1, x_2))\,\mathcal{L}\,\Omega^{\mathcal{B}}(p, x_1) \quad .$$

Da (7.7) consegue che la derivata di Lie di un oggetto differenziale geometrico è un oggetto differenziale geometrico solamente nel caso che le funzioni $\mathcal{F}^{\mathcal{A}}$ siano lineari rispetto ai componenti $\Omega^{\mathcal{B}}$. Ciò nonostante

---

7) Una funzione numerica definita su un insieme qualunque dello spazio aritmetico si dice differenziabile se è possibile il suo prolungamento differenziabile su un insieme aperto. Nelle formule successive useremo le derivate di prolungamento differenziabile se le funzioni considerate non sono definite esse stesse su un insieme aperto.

l'unione $\Omega^{\mathfrak{A}}, \mathfrak{L}\Omega^{\mathfrak{A}}$ dell'oggetto $\Omega^{\mathfrak{A}}$ e della sua derivata di Lie sarà sempre un oggetto differenziale geometrico.

Affinchè un oggetto differenziale Ω sia stabile rispetto a tutte le trasformazioni $\mathcal{T}_t$, cioè che sia :

$$(7.8) \qquad \Omega \circ \overset{-1}{T_t} \subset \Omega$$

è necessario e sufficiente che i componenti della derivata di Lie di Ω siano eguali a zero

$$(7.9) \qquad \mathfrak{L}\Omega^{\mathfrak{A}}(p, \varkappa) = 0$$

per tutte le coppie $(p, \varkappa)$.

Secondo (7.7) abbiamo che per un oggetto differenziale geometrico la condizione (7.9) è equivalente alla condizione che per ogni punto $p \in \mathcal{G}$, $\mathfrak{L}\Omega^{\mathfrak{A}}(p, \varkappa)$ siano eguali a zero, per un certo sistema coordinato $\varkappa$. Se un oggetto differenziale Ω_2 è un comitante di un oggetto differenziale Ω_1

$$(7.10) \qquad \Omega_2^{\mathfrak{A}_2}(p, \varkappa) = f^{\mathfrak{A}_2}\left(\Omega_1^{\mathfrak{A}_1}(p, \varkappa)\right) ,$$

ove $f^{\mathfrak{A}_2}$ sono funzioni differenziabili, allora le sue derivate di Lie stanno nella relazione ([44])

$$(7.11) \qquad \mathfrak{L}\Omega_2^{\mathfrak{A}_2}(p, \varkappa) = f^{\mathfrak{A}_1}_{\mathfrak{A}_2}\left(\Omega_1^{\mathfrak{B}_1}(p, \varkappa)\right) \mathfrak{L}\Omega^{\mathfrak{A}_1}(p, \varkappa) .$$

Se un oggetto differenziale Ω_2 è un comitante locale di un oggetto differenziale Ω_1 e se questi oggetti sono definiti in una regione rappresentabile in un solo sistema coordinato $\varkappa_0$ allora abbiamo che i loro componenti stanno nella relazione

(7.12) $$\Omega_2^{A_2}(p,\varkappa) = f^{A_2}\left(\xi_0^{\alpha}(p), \Omega_1^{A_1}(p,\varkappa)\right) ,$$

dove le funzioni f^{A_2} dipendono dalla scelta del sistema coordinato $\varkappa_0$ e ξ_0^{α} sono funzioni numeriche del punto p uguali alle sue coordinate rispetto a $\varkappa_0$. Dunque l'oggetto differenziale $\Omega_2^{A_2}$ può essere considerato come un comitante dell'unione $\Omega_1^{A_1}$, ξ_0^{α} dell'oggetto differenziale $\Omega_1^{A_1}$ e degli n scalari ξ_0^{α}.

La formula precedente (7.12) dà ora l'espressione della derivata di Lie $\mathcal{L}\Omega_2^{A_2}$

(7.13) $$\mathcal{L}\Omega_2^{A_2}(p,\varkappa) =$$

$$= f_{A_1}^{A_2}\left(\xi_0^{\alpha}(p), \Omega_1^{B_1}(p,\varkappa)\right)\mathcal{L}\Omega^{A_1}(p,\varkappa) + f_{\alpha}^{A_2}\left(\xi_0^{\beta}(p), \Omega_1^{A_1}(p,\varkappa)\right)\mathcal{L}\,\xi_0^{\alpha}(p) .$$

Se $\Omega^{A}(\xi^{\beta})$ sono componenti funzionali di un oggetto differenziale di classe v e sono funzioni differenziabili allora le derivate parziali $\partial_{\alpha}\Omega^{A}$ sono componenti di un oggetto differenziale di classe $v+1$, che si dice la derivata dell'oggetto differenziale Ω^{A}. Con questo se Ω^{A} è un oggetto differenziale geometrico distinto dell'unione di scalari allora la sua derivata non sarà un oggetto differenziale geometrico, ma l'unione Ω^{A}, $\partial_{\alpha}\Omega^{A}$ sarà sempre un oggetto differenziale geometrico il quale si dice il prolungamento differenziale dell'oggetto differenziale geometrico Ω^{A}.

Si può dimostrare che le operazioni di derivazione di Lie e di derivazione ordinaria sono permutabili ([44])

$$\mathcal{L}\,\partial_\alpha \Omega^A = \partial_\alpha \mathcal{L}\,\Omega^A \quad . \tag{7.14}$$

Come è noto il commutatore di due campi vettoriali $\underset{1}{v}^\alpha$ e $\underset{2}{v}^\alpha$ è il campo vettoriale $\underset{12}{v}^\alpha = \underset{1}{v}^\beta \partial_\beta \underset{2}{v}^\alpha - \underset{2}{v}^\beta \partial_\beta \underset{1}{v}^\alpha$.

Si può dimostrare che per ogni oggetto differenziale Ω^A la differenza delle sue derivate seconde di Lie rispetto ai vettori $\underset{1}{v}^\alpha$ e $\underset{2}{v}^\alpha$ prese in ordine scambiato coincide con la sua derivata di Lie rispetto al commutatore di questi vettori $\underset{12}{v}^\alpha$:

$$\mathcal{L}_1 \mathcal{L}_2 \Omega^A - \mathcal{L}_2 \mathcal{L}_1 \Omega^A = \mathcal{L}_{12} \Omega^A \tag{7.15}$$

Notiamo che il commutatore dei campi vettoriali $\underset{1}{v}^\alpha$, $\underset{2}{v}^\alpha$ può essere rappresentato come la derivata di Lie del secondo vettore rispetto al primo vettore

$$\underset{12}{v}^\alpha = \mathcal{L}_1 \underset{2}{v}^\alpha \quad . \tag{7.16}$$

Ne consegue che l'identità di Jacobi può essere scritta nella forma seguente

$$\mathcal{L}_1 \underset{23}{v}^\alpha + \mathcal{L}_2 \underset{31}{v}^\alpha + \mathcal{L}_3 \underset{12}{v}^\alpha = 0 \tag{7.17}$$

ove $\underset{1}{v}^\alpha, \underset{2}{v}^\alpha, \underset{3}{v}^\alpha$ sono tre campi di vettori qualunque e $\mathcal{L}_1, \mathcal{L}_2, \mathcal{L}_3$ i simboli delle derivate di Lie rispetto a questi vettori .

II - CONNESSIONE PARZIALE IN UNO SPAZIO FIBRATO E DERIVAZIONE ASSOLUTA

§ 8 Connessione parziale lineare in uno spazio fibrato.

Una curva nello spazio fibrato C_{n+s} di una fibrazione (C_{n+s}, B_n, θ) si dice proiettabile se la sua proiezione sulla base è una curva curva nello spazio base B_n . E' facile vedere che questo equivale al fatto che in ogni suo punto la curva non è tangente alla fibra che passa per quel punto.

Sia c una curva nello spazio base. Ogni curva proiettabile nello spazio fibrato C_{n+s} è detta ricoprente la curva c se la sua proiezione sulla base coincide con c .

Una connessione parziale nello spazio fibrato di una fibrazione (C_{n+s}, B_n, θ) è un insieme $\mathcal{L}$ di curve orientate proiettabili nello spazio fibrato C_{n+s} il quale soddisfa alle condizioni seguenti [8]:

1) $\mathcal{L}$ è stabile rispetto all'operazione di "congiunzione" di curve .

2) $\mathcal{L}$ è stabile rispetto all'operazione di restrizione di curve.

3) Per ogni curva orientata c nello spazio base B_n la relazione binaria σ_c , che è l'insieme delle coppie $(\bar{p}_1, \bar{p}_2)$ di punti terminali di tutte le curve ricoprenti la curva c che appartengono a $\mathcal{L}$, se non è vuoto, è una rappresentazione parziale invertibile regolare della fibra $C_s(p_1)$ nella fibra $C_s(p_2)$ ove p_1, p_2 sono i punti terminali della curva c .

Le curve dell'insieme $\mathcal{L}$ si dicono le curve di connessione e le rappresentazioni parziali σ_c si dicono le rappresentazioni parziali di connessione.

---

8) Consideriamo curve orientate con un numero finito di punti angolosi.

E' facile vedere che se una curva c è ottenuta dal prodotto delle curve c_1 e c_2 allora

$$(8.1) \qquad \sigma_c = \sigma_{c_2} \circ \sigma_{c_1} .$$

Curve dello spazio base soddisfacenti alla condizione $\sigma_c \neq 0$ si dicono le curve ammissibili della connessione.

Una connessione parziale si dice basicamente completa se ogni curva orientata dello spazio base è ammissibile e si dice basicamente completa localmente se per ogni punto di una curva orientata qualunque dello spazio base esiste un arco a cui appartiene questo punto che è una curva ammissibile.

Una connessione parziale si dice simmetrica se $\mathcal{L}$ è stabile rispetto all'operazione di inversione di orientazione di curva. Nel caso di una connessione parziale simmetrica l'insieme delle curve basiche ammissibili sarà anche stabile rispetto all'operazione di inversione di orientazione di curva. Indicando con c^{-1} la curva ottenuta dall'inversione di orientazione della curva c abbiamo nel caso di una connessione parziale simmetrica

$$(8.2) \qquad \sigma_{c^{-1}} = \sigma_c^{-1} .$$

Una connessione parziale si dice isomorfa se tutte le rappresentazioni parziali di connessione sono isomorfismi parziali delle fibre corrispondenti.

Una connessione paziale si dice semplicemente una connessione se tutte le rappresentazioni parziali di connessione sono rappresentazioni.

Come è noto una superficie ad s-dimensioni in uno spazio ad $s+n$ dimensioni si dice "allestita" se in ogni suo punto nello spazio tangente associato è dato un piano centrale ad n dimensioni il quale ha solo un punto in comune con il piano tangente alla superficie.[9]

Uno spazio fibrato C_{n+s} è detto "allestito" se tutte le sue fibre sono superficie allestite.

Considerando i differenziali $d\xi^\alpha, d\eta^i$ delle coordinate ξ^α, η^i di punto nello spazio fibrato come coordinate di un punto nello spazio tangente corrispondente possiamo scrivere il sistema di equazioni dei piani "allestenti" come un sistema di Pfaff

$$d\eta^i + \Gamma_\alpha^i(\xi^\beta, \eta^j)\, d\xi^\alpha = 0 \quad . \tag{8.3}$$

Passando a notazioni più brevi e ponendo

$$\delta\eta^i = d\eta^i + \Gamma^i \quad , \quad \Gamma^i = \Gamma_\alpha^i\, d\xi^\alpha \quad , \tag{8.4}$$

trascriviamo il sistema di Pfaff (8.3) nella forma seguente

$$\delta\eta^i = 0 \quad . \tag{8.5}$$

I coefficienti delle forme di Pfaff (8.5) sono componenti funzionali di un oggetto differenziale geometrico il quale si dice l'oggetto di "allestimento".

Le equazioni di trasformazione dei componenti dell'oggetto di allestimento sono[10].

---

9) "Allestimento" è una versione italiana del termine inglese "rigging" dovuto a Schouten v. [22] .

10) Nella formula (8.6) usiamo che $A_\alpha^\beta(p, \varkappa_2, \varkappa_1)$ è uguale al minore del determinante $|A_\beta^\alpha(p, \varkappa_1, \varkappa_2)|$ diviso per questo determinante.

$$(8.6) \qquad \Gamma_{\alpha}^{\ i}(p, \varkappa_2) =$$

$$= \Gamma_{\beta}^{\ j}(p, \varkappa_1) A_{\alpha}^{\beta}(p, \varkappa_2, \varkappa_1) A_{n+j}^{n+i}(p, \varkappa_1, \varkappa_2) - A_{\beta}^{n+i}(p, \varkappa_1, \varkappa_2) A_{\alpha}^{\beta}(p, \varkappa_2, \varkappa_1) .$$

Ne segue che l'oggetto di allestimento non è riducibile.

Una curva si dice ammissibile per un campo di n-direzioni dato se in ogni punto la direzione tangente appartiene alla n-direzione corrispondente del campo. Le curve ammissibili per il campo delle n-direzioni allestenti lo spazio fibrato C_{n+s} si dicono le curve allestenti.

E' facile vedere che l'insieme di tutte le curve orientate allestenti in uno spazio fibrato C_{n+s} è una connessione parziale basicamente completa localmente e simmetrica.

Tali connessioni parziali determinate da un allestimento dello spazio fibrato si dicono connessioni parziali lineari. L'oggetto dell'allestimento si dice l'oggetto della connessione parziale lineare.

Se c è una curva orientata nello spazio base determinata dalle equazioni $\xi^{\alpha} = \varphi^{\alpha}(t)$ allora le curve di connessione ricoprenti questa curva c sono definite dal sistema di equazioni differenziali

$$(8.7) \qquad \dot{\eta}^i + \Gamma_{\alpha}^{\ i} \dot{\xi}^{\alpha} = 0 \quad .$$

Una superficie ad n dimensioni si dice ammissibile rispetto ad un campo di n-direzioni se in ogni punto la sua n-direzione tangente coincide con la n-direzione del campo. Il campo di n-direzioni si dice olonomo se per ogni punto dello spazio passa una superficie ad n dimensioni ammissibile per questo campo di n-direzioni. La olonomia di un campo di n-direzioni è caratterizzata dalla condizione che il cosidetto tensore di anolonomia di Schouten sia nullo ([19]).

Rappresentiamo l'operatore d di differenziazione totale nello spazio C_{n+s} nel modo seguente

(8.8) $$d = \partial + d\eta^i \partial_i$$

ove $\partial = d\xi^\alpha \partial_\alpha$ è un operatore differenziale dipendente dal sistema coordinato nello spazio fibrato C_{n+s} che chiameremo l'operatore di differenziazione basica nello spazio C_{n+s}.

Calcolando il differenziale esterno delle forme di Pfaff $\delta\eta^i$ otterremo

(8.9) $$[d\,\delta\eta^i] = \frac{1}{2} R^i + [\delta\eta^j \Gamma^i_j] \qquad (\Gamma^i_j - \partial_j \Gamma^i)$$

ove

(8.10) $$R^i = R_{\beta\alpha}^{\cdot\cdot\, i}[d\xi^\beta d\xi^\alpha] = 2([\partial\Gamma^i] - [\Gamma^j \Gamma^i_j])$$

sono forme differenziali bilineari emisimmetriche, le quali sono componenti di un vettore nella fibra che si dice il vettore di curvatura della connessione parziale lineare. I coefficienti $R_{\beta\alpha}^{\cdot\cdot\, i}$ di queste forme di Pfaff sono componenti di un tensore riducibile nello spazio fibrato C_{n+s}, che si dice il <u>tensore di curvatura</u> della connessione parziale. Si può dimostrare che il tensore di curvatura differisce dal tensore di olonomia di Schouten dell'allestimento per il fattore -2 .

Ne consegue da (8.9) che per l'integrabilità completa del sistema di Pfaff o, che è lo stesso, per l'olonomia dell'allestimento dello spazio fibrato è necessario e sufficiente che il tensore di curvatura sia nullo.

Allora la connessione parziale lineare si dice di curvatura nulla. In questo caso possiamo scegliere un sottoatlante coordinato nello spazio fibrato C_{n+s} in modo tale che tutti i componenti dell'oggetto di connessione parziale siano nulli

$$\Gamma_{\alpha}^{i} = 0 \quad . \tag{8.11}$$

Sia c una curva proiettabile dello spazio fibrato c_{n+s} determinata dalle equazioni $\xi^{\alpha} = \xi^{\alpha}(t)$, $\eta^{i} = \eta^{i}(t)$. Per ogni punto $p(t)$ della curva c passa una curva di connessione unica c_p la cui proiezione sulla base è contenuta nella proiezione della curva c.

Siano $\bar{p}(t+\Delta t)$ un punto della curva c vicino al punto $p(t)$ e $\bar{p}_o$ il punto della curva c_p che appartiene alla stessa fibra a cui appartiene il punto $\bar{p}$. Indicate con $\Delta\eta^{i}$ le differenze di coordinate nella fibra dei punti $\bar{p}$ e $\bar{p}_o$, consideriamo $\Delta\eta^{i}$ come funzioni di Δt,

E' facile vedere che

$$\lim_{\Delta t \to 0} \frac{\Delta\eta^{i}}{\Delta t} = \dot{\eta}^{i} + \Gamma_{\alpha}^{i}\dot{\xi}^{\alpha} = \frac{\delta\eta^{i}}{dt} \tag{8.12}$$

ove $\frac{\delta\eta^{i}}{dt}$ sono componenti di un vettore della fibra nel punto p della curva c. Questo vettore $\frac{\delta\eta^{i}}{dt}$ si dice il <u>vettore di deviazione assoluta</u> della curva c nel punto p. Se il vettore di deviazione assoluta è nullo in un punto p di una curva allora la curva ha contatto con la curva di connessione corrispondente al punto p. Se il vettore di deviazione assoluta è nullo identicamente allora la curva è una curva di connessione.

§ 9. Derivazione assoluta basica.

Supponiamo ora che in uno spazio fibrato C_{n+s} di fibrazione (C_{n+s}, B_n, θ) sia data una connessione parziale isomorfa.

Fissiamo nello spazio base B_n un sistema coordinato $\mathcal{X}$ e indichiamo con X l'insieme di tutti i sistemi coordinati di $\pi_2(B_n)$ contenuti in $\mathcal{X}$. L'insieme dei sistemi coordinati $\overset{-1}{\Theta}(X)$ è un atlante per la regione $\overset{-1}{\theta}(pr_1 \mathcal{X})$ dello spazio C_{n+s}.

Consideriamo nella regione $\overset{-1}{\theta}(pr_1 \mathcal{X})$ n congruenze di curve di connessione le quali ricoprano le linee coordinate del sistema coordinato fissato $\mathcal{X}$.

E' facile vedere che i componenti del vettore tangente alle curve della congruenza con indice α saranno $(\delta_\alpha^\beta, -\Gamma_\alpha^i)$. In virtù dell'isomorfia della connessione parziale data nello spazio C_{n+s} questi n campi vettoriali $(\delta_\alpha^\beta, -\Gamma_\alpha^i)$ sono isomorfi per lo spazio dotato di coordinate $(\overset{-1}{\theta}(pr_1 \mathcal{X}), \overset{-1}{\Theta}(X))$.

Sia ora Ω^A un oggetto differenziale riducibile nello spazio C_{n+s}, definito su un sottoinsieme aperto di $\overset{-1}{\theta}(pr_1 \mathcal{X})$. Considerando Ω^A come un oggetto differenziale dello spazio $(\overset{-1}{\theta}(pr_1 \mathcal{X}), \overset{-1}{\Theta}(X))$ indichiamo con $D_\alpha \Omega^A$ la sua derivata di Lie rispetto al campo vettoriale $(\delta_\alpha^\beta, -\Gamma_\alpha^i)$. Consideriamo di nuovo il sistema coordinato $\mathcal{X}$ come un sistema coordinato variabile allora $D_\alpha \Omega^A$ sono componenti di un oggetto differenziale nello spazio C_{n+s}, il quale è detto la <u>derivata covariante basica</u> dell'oggetto differenziale Ω^A.

Si può dimostrare che la derivata covariante basica $D_\alpha \Omega^A$ è anche un oggetto differenziale riducibile.

Il significato formale di derivazione covariante basica è che essa è una operazione differenziale basica per ottenere da oggetti differenziali riducibili altri oggetti differenziali riducibili. La derivazione ordinaria rispetto a coordinate basiche non gode di questa proprietà.

Sia $\Omega^{\mathcal{A}}$ un oggetto differenziale geometrico riducibile le cui equazioni di trasformazione di componenti siano

$$\Omega^{\mathcal{A}}(p, \varkappa_2) = \mathcal{F}^{\mathcal{A}}(\Omega^{\mathcal{B}}(p, \varkappa_1), A^{\alpha}_{\alpha_1 \ldots \alpha_n}(p, \varkappa_1, \varkappa_2), A^{n+i}_{n+i_1, \ldots, n+i_n}(p, \varkappa_1, \varkappa_2); \tag{9.1}$$

allora la formula generale (7.5) dà le espressioni delle componenti della derivata covariante basica $D_\alpha \Omega^{\mathcal{A}}$

$$D_\alpha \Omega^{\mathcal{A}} = \partial_\alpha \Omega^{\mathcal{A}} - \Gamma^i_\alpha \partial_i \Omega^{\mathcal{A}} + \sum_{n=1}^{v} \Gamma^i_{\alpha i_1 \ldots i_n} \overset{\circ}{\mathcal{F}}{}^{\mathcal{A}\, n+i}_{\quad n+i} {}_{, \ldots, n+i_n}(\Omega^{\mathcal{B}}) \tag{9.2}$$

ove

$$\Gamma^i_{\alpha i_1 \ldots i_n} = \partial_{i_1 \ldots i_n} \Gamma^i_\alpha .$$

Si possono poi trovare le equazioni di trasformazione delle componenti della derivata covariante basica $D_\alpha \Omega^{\mathcal{A}}$

$$D_\alpha \Omega^{\mathcal{A}}(p, \varkappa_2) = \tag{9.3}$$

$$= \mathcal{F}^{\mathcal{A}}_{\mathcal{B}}(\Omega^{\Gamma}(p, \varkappa_1), A^{\gamma}_{\gamma_1 \ldots \gamma_n}(p, \varkappa_1, \varkappa_2), A^{n+i}_{n+i_1, \ldots n+i_n}(p, \varkappa_1, \varkappa_2)) A^{\beta}_{\alpha}(p, \varkappa_2, \varkappa_1) \cdot$$

$$\cdot D_\beta \Omega^{\mathcal{B}}(p, \varkappa_1) +$$

$$+ \sum_{n=1}^{v} \mathcal{F}^{\mathcal{A}\gamma_1 \ldots \gamma_n}_{\gamma}\left(\Omega^{\mathcal{B}}(p, \varkappa_1), A^{\delta}_{\delta_1 \ldots \delta_n}(p, \varkappa_1, \varkappa_2), A^{n+i}_{n+i_1 \ldots n+i_n}(p, \varkappa_1, \varkappa_2)\right) A^{\gamma}_{\gamma_1 \ldots \gamma_n \beta}(p, \varkappa_1, \varkappa_2) A^{\beta}_{\alpha}(p, \varkappa_2, \varkappa_1)$$

donde possiamo vedere che la derivata covariante basica di un oggetto differenziale geometrico riducibile è un oggetto differenziale riducibile benchè in generale non geometrico.

Se un oggetto differenziale $\overset{\mathcal{A}_2}{\Omega_2}$ è un comitante locale di un oggetto differenziale riducibile $\overset{\mathcal{A}_1}{\Omega_1}$

$$(9.4)\qquad \overset{\mathcal{A}_2}{\Omega_2}(p,\varkappa) = f^{\mathcal{A}_2}\left(\xi_0^{\alpha}(p),\ \eta_0^{i}(p),\ \overset{\mathcal{A}_1}{\Omega_1}(p,\varkappa)\right)$$

allora secondo (7.13) avremo

$$(9.5)\qquad D_\alpha \overset{\mathcal{A}_2}{\Omega_2}(p,\varkappa) = f^{\mathcal{A}_2}_{A_1}\left(\xi_0^{\alpha}(p),\ \eta_0^{i}(p), \overset{\mathcal{B}_1}{\Omega_1}(p,\varkappa)\right) D_\alpha \overset{\mathcal{A}_1}{\Omega}(p,\varkappa) +$$

$$+ f^{\mathcal{A}_2}_{\beta}\left(\xi_0^{\alpha}(p),\ \eta_0^{i}(p),\ \overset{\mathcal{B}_1}{\Omega_1}(p,\varkappa)\right) D_\alpha \xi_0^{\beta}(p,\varkappa) + f^{\mathcal{A}_2}_{j}\left(\xi_0^{\alpha}(p),\ \eta_0^{i}(p), \overset{\mathcal{B}_1}{\Omega_1}(p,\varkappa)\right) D_\alpha \eta_0^{j}(p,\varkappa)$$

donde si deduce che se $D_\alpha \overset{\mathcal{A}_1}{\Omega_1}$ è un oggetto differenziale riducibile allora $D_\alpha \overset{\mathcal{A}_2}{\Omega_2}$ è ancora un oggetto differenziale riducibile.

Giacchè ogni oggetto differenziale riducibile può essere espresso come un comitante locale della parte riducibile dell'oggetto differenziale universale di ordine corrispondente e quest'ultimo è un oggetto differenziale geometrico riducibile, usando i risultati precedenti, otterremo che la derivata covariante basica di un oggetto differenziale riducibile qualunque è ancora un oggetto differenziale riducibile.

Usando la proprietà della derivata di Lie espressa da (7.14) si prova che le operazioni di derivazione covariante basica e di derivazione ordinaria rispetto a coordinate della fibra sono permutabili

$$(9.6)\qquad D_\alpha \partial_i \Omega^{\mathcal{A}} = \partial_i D_\alpha \Omega^{\mathcal{A}} \quad .$$

Considerando la differenza di derivate covarianti basiche del secondo ordine $D_\beta D_\alpha \Omega^A - D_\alpha D_\beta \Omega^A$ di un oggetto differenziale riducibile Ω^A, possiamo applicare la formula generale (7.15) per la derivata di Lie.

Fissando gl'indici α, β abbiamo che la differenza $D_\beta D_\alpha \Omega^A - D_\alpha D_\beta \Omega^A$ uguaglia la derivata di Lie rispetto al commutatore dei campi vettoriali $(\delta^\gamma_\beta, -\Gamma^i_\beta)$ e $(\delta^\gamma_\alpha, -\Gamma^i_\alpha)$. E' facile calcolare che questo commutatore è uguale a $(0, -R_{\beta\alpha}^{\cdot\cdot i})$. Indicando la derivata di Lie rispetto al campo vettoriale $(0, R_{\beta\alpha}^{\cdot\cdot i})$ col simbolo $\mathcal{R}_{\beta\alpha}$ otterremo la formula

$$2 D_{[\beta} D_{\alpha]} \Omega^A = -\mathcal{R}_{\beta\alpha} \Omega^A \tag{9.7}$$

la quale è una generalizzazione dell'identità di Ricci.

In relazione a ciò, chiameremo $\mathcal{R}_{\beta\alpha} \Omega^A$ la derivata di Ricci di un oggetto differenziale Ω^A. Se Ω^A è un oggetto differenziale geometrico otterremo l'espressione seguente della sua derivata di Ricci

$$\mathcal{R}_{\beta\alpha} \Omega^A = R_{\beta\alpha}^{\cdot\cdot i} \partial_i \Omega^A - \sum_{n=1}^{v} R_{\beta\alpha\cdot i_1 \ldots i_n}^{\cdot\cdot i} \overset{0}{\mathcal{F}}{}^{An+i_1 \ldots \ldots n+i_n}_{n+i}(\Omega^B) \tag{9.8}$$

ove

$$R_{\beta\alpha\cdot i_1 \ldots i_n}^{\cdot\cdot i} = \partial_{i_1 \ldots i_n} R_{\beta\alpha}^{\cdot\cdot i}$$

Applicando l'identità di Jacobi ai tre campi vettoriali $(\delta^\delta_\gamma, -\Gamma^i_\gamma)$ $(\delta^\delta_\beta, -\Gamma^i_\beta)$, $(\delta^\delta_\alpha, -\Gamma^i_\alpha)$, secondo (7.17), abbiamo

$$D_\gamma R_{\beta\alpha}^{\cdot\cdot i} + D_\beta R_{\gamma\alpha}^{\cdot\cdot i} + D_\alpha R_{\beta\gamma}^{\cdot\cdot i} = 0 \tag{9.9}$$

ossia

(9.10) $$D_{[\gamma} R^{\cdot\cdot i \cdot\cdot}_{\beta\alpha]} = 0 .$$

L'ultima formula rappresenta una generalizzazione dell'identità di Bianchi pel caso di una connessione parziale arbitraria.

Se Ω^{A} è un oggetto differenziale di una fibra possiamo costruire mediante la sua derivata covariante basica un oggetto differenziale della fibra

(9.11) $$D\, \Omega^{A} = D_{\alpha}\, \Omega^{A} d\, \xi^{\alpha}$$

i cui componenti sono forme di Pfaff rispetto ai differenziali $d\xi^{\alpha}$. Questo oggetto differenziale della fibra si dice il <u>differenziale assoluto basico</u> dello oggetto differenziale Ω^{A} della fibra.

Sia Ω^{A} un oggetto differenziale di una fibra i cui componenti sono forme differenziali emisimmetriche di ordine r rispetto ai differenziali $d\xi^{\alpha}$

(9.12) $$\Omega^{A} = \Omega^{A}_{\alpha_1 \ldots\ldots \alpha_r} \left[d\, \xi^{\alpha_1} \ldots\ldots d\, \xi^{\alpha_r} \right] .$$

Possiamo ottenere dalla derivata covariante basica di un oggetto differenziale riducibile $\Omega^{A}_{\alpha_1 \ldots \alpha_r}$ l'oggetto differenziale fibrale

(9.13) $$\left[D\, \Omega^{A} \right] = D_{[\alpha}\, \Omega^{A}_{\alpha_1 \ldots\ldots \alpha_r]} \left[d\, \xi^{\alpha}\, d\, \xi^{\alpha_1} \ldots\ldots d\, \xi^{\alpha_r} \right]$$

i cui componenti sono forme differenziali emisimmetriche d'ordine $r+1$.

Questo oggetto differenziale fibrale si dice il <u>differenziale assoluto basico esterno</u> di Ω^{A} .

Se Ω^{A} è un oggetto differenziale fibrale i cui componenti sono forme differenziali emisimmetriche di differenziali $d\xi^{\alpha}$ l'identità di Ricci per differenziazione assoluta basica esterna può essere rappresentata in modo seguente

(9.14) $$2\left[D^2\,\Omega^{\mathfrak{A}}\right] = -\left[\mathcal{R}\,\Omega^{\mathfrak{A}}\right]$$

dove a sinistra si trova il differenziale assoluto basico esterno del secondo ordine $\left[D^2\Omega^{\mathfrak{A}}\right]$ e a destra la derivata esterna di Ricci

(9.15) $$\left[\mathcal{R}\,\Omega^{\mathfrak{A}}\right] = \mathcal{R}_{[\beta\alpha}\,\Omega^{\mathfrak{A}}_{\alpha_1\ldots\alpha_r]}\left[d\xi^{\beta}d\xi^{\alpha}d\xi^{\alpha_1}\ldots d\xi^{\alpha_r}\right].$$

In particolare quanto $\Omega^{\mathfrak{A}}$ è un oggetto differenziale della fibra : ordinaria l'identità (9.15) può essere trascritta nella forma seguente

(9.16) $$2\left[D^2\Omega^{\mathfrak{A}}\right] = -\mathcal{R}\Omega^{\mathfrak{A}} .$$

III - PSEUDOMETRICHE E METRICHE VETTORIALI

§ 10. Superficie negli spazi lineari.

Come già detto, uno spazio lineare L_n è uno spazio di Klein semplice il cui gruppo coordinato è un sottogruppo del gruppo centro affine $\mathfrak{A}_n$. E' evidente che lo spazio lineare L_n si può considerare come uno spazio vettoriale. In questo spazio vettoriale ad ogni sistema coordinato dello spazio lineare corrisponde una base di vettori indipendenti che chiameremo una base speciale. Consideriamo ora lo spazio vettoriale coniugato. Come è noto ad ogni base di uno spazio vettoriale corrisponde una base nello spazio coniugato. Dunque ad ogni base speciale corrisponde una base nello spazio coniugato che chiameremo pure una base speciale. L'insieme di tutte le basi speciali dello spazio vettoriale coniugato vi definisce un atlante coordinato nello spazio. Dotando di coordinate lo spazio coniugato vettoriale con l'aiuto di questo atlante otterremo uno spazio dotato di coordinate lineare L_n^* che si dice lo spazio coniugato dello spazio lineare L_n.

Il punto di uno spazio lineare L_n avente le coordinate tutte nulle, il che è una proprietà invariante, indipendente dalla scelta del sistema coordinato, si dice il centro dello spazio lineare.

Punti dello spazio lineare differenti dal centro si dicono punti propri.

Analogamente i piani nello spazio lineare che non passano pel centro si dicono piani propri.

Si costituisce una corrispondenza biunivoca fra iperpiani propri dello spazio lineare L_n e punti propri dello spazio lineare coniugato L_n^* dove ad un iperpiano

$$y_\alpha x^\alpha = 1 \tag{10.1}$$

determinato dal covettore y_α corrisponde questo covettore considerato come un punto dello spazio coniugato L_n^*.

Questa corrispondenza si dice la corrispondenza <u>duale propria</u> dello spazio lineare. Abbiamo anche la cosidetta corrispondenza duale <u>centrale</u> dove ad un iperpiano centrale dello spazio lineare L_n corrisponde la retta centrale dello spazio coniugato L_n^* i cui punti propri corrispondono a iperpiani propri di L_n paralleli all'iperpiano centrale.

Una superficie in L_n si dice <u>centralmente semplice</u> se ogni semiretta centrale passa non più che per un suo punto.

Una superficie in L_n si dice <u>propria</u> se tutti i suoi piani tangenti sono propri. Ne segue che in particolare tutti i punti di una superficie propria sono punti propri cioè che la superficie non passa pel centro.

Se

$$X^\alpha = \ell^\alpha(\eta^i) \tag{10.2}$$

sono equazioni parametriche di una superficie propria ad s dimensioni allora gli $s+1$ vettori $\ell_i^\alpha, \ell^\alpha$ sono indipendenti.

E' facile vedere che ogni superficie propria è localmente centralmente semplice.

Ogni iperpiano che passa per un piano tangente della superficie si chiama un iperpiano tangente nel punto corrispondente della superficie.

L'insieme di tutti i punti nello spazio coniugato L_n^* che corrispondono a iperpiani tangenti propri di una superficie si dice l'immagine tangenziale propria di quella. Se abbiamo una superficie ad s dimensioni ove $s<n-1$ allora la sua immagine tangenziale propria è un'unione di piani a $(n-1-s)$-dimensioni che sono le immagini dei fasci di iperpiani tangenti alla superficie.

Consideriamo in uno spazio lineare L_n una superficie propria ad s

dimensioni S e una ipersuperficie propria I che abbiano un punto comune p_o . Sia p un punto sulla S vicino al punto p_o tale che la semiretta centrale 0p passi per un punto della I e sia q il punto comune alla I più vicino a p . Allora in un intorno corrispondente del punto p_o sulla S possiamo definire una funzione di punto $\delta(p)$ ponendo

$$\delta(p) = \frac{\overrightarrow{pq}}{\overrightarrow{0q}} \tag{10.3}$$

che chiameremo la <u>deviazione</u> della superficie S dall'ipersuperficie I nell'intorno del punto comune p_o .

Indicando con η^i coordinate curvilinee sulla S possiamo considerare la deviazione δ come una funzione delle η^i .

Se S e I hanno nel punto comune un contatto d'ordine ν tutte le derivate parziali della deviazione d'ordine fino a ν incluso sono uguali a zero nel punto p_o . Se ν è l'ordine di contatto preciso allora le derivate parziali d'ordine $\nu+1$ della deviazione δ non saranno tutte nulle nel punto p_o .

Se S e I hanno nel punto comune p_o un contatto d'ordine ν allora le derivate parziali d'ordine $\nu+1$ nel punto p_o sono componenti di un tensore covariante di valenza $\nu+1$, $D_{i_1 \ldots i_{\nu+1}}$ che si dice il <u>tensore di deviazione d'ordine</u> $\nu+1$ della superficie S dall'ipersuperficie I .

Se ν è l'ordine preciso del contatto allora il tensore di deviazione d'ordine $\nu+1$ non sarà nullo.

In questo caso la deviazione della S della I in un intorno infinitesimale del punto di contatto può essere espressa a meno di infinitesimi di ordine superiore come una forma differenziale d'ordine $\nu+1$:

(10.4) $$\delta \approx \frac{1}{(\nu+1)!} D_{i_1 \ldots\ldots i_{\nu+1}} d\eta^{i_1} \ldots\ldots d\eta^{i_{\nu+1}} \quad .$$

Un interesse speciale ha il tensore di deviazione del secondo ordine di una superficie da un suo iperpiano tangente, che chiameremo il <u>tensore tangenziale</u> di questo iperpiano tangente. Se un iperpiano tangente è determinato dal covettore $\overset{\circ}{y}_\alpha$ allora la deviazione della superficie dall'iperpiano è espressa dalla formula

(10.5) $$\delta = 1 - \overset{\circ}{y}_\alpha \ell^\alpha(\eta^i)$$

dalla quale si deduce l'espressione del tensore tangenziale

(10.6) $$g_{ij} = -\overset{\circ}{y}_\alpha \ell^\alpha_{ij}(\eta^k_\circ) \quad .$$

Un iperpiano proprio tangente ad una superficie si dice <u>regolare</u> se il suo tensore tangenziale è non degenere , cioè se il suo discriminante è differente da zero.

Una superficie si dice <u>regolare</u> se in ogni suo punto esiste un iperpiano tangente regolare.

Un punto dell'immagine tangenziale propria di una superficie si dice regolare se corrisponde ad un iperpiano tangente regolare.

Dimostreremo che nell'intormo di ogni punto regolare dell'immagine tangenziale propria di una superficie possiamo prendere una parte di questa che è una ipersuperficie.

Infatti l'immagine tangenziale propria di una superficie si può determinare dal seguente sistema di equazioni :

$$
\begin{cases}
\ell_i^{\alpha}(\eta^j)y_\alpha = 0 & (10.7)\\
\ell^{\alpha}(\eta^j)y_\alpha = 1 & (10.8)
\end{cases}
$$

Sia $\overset{o}{y}_\alpha$ un covettore che determina nel punto $p_o(\eta^j_o)$ di una superficie un iperpiano tangente regolare. I valori $\eta^j_o, \overset{o}{y}_\alpha$ soddisfano il sistema (10.7). Prendendoli come valori iniziali ed utilizzando la regolarità dell'iperpiano tangente $\overset{o}{y}_\alpha$ applichiamo al sistema (10.7) il teorema delle funzioni implicite. Secondo questo teorema possiamo risolvere il sistema (10.7) rispetto alle η^j in un intorno dei valori iniziali, esprimendo le η^j come funzioni delle y_α. E' evidente che queste funzioni $\eta^j(y_\alpha)$ saranno positivamente omogenee di grado 0. Con l'aiuto di queste funzioni formiamo la funzione positivamente omogenea di grado 1:

$$
H(y_\alpha) = \ell^\alpha(\eta^j(y_\alpha))y_\alpha . \tag{10.9}
$$

Indicando con H^α la derivata della H rispetto a y_α avremo

$$
H^\alpha = \ell^\alpha + \ell_i^\beta y_\beta \frac{\partial \eta^i}{\partial y_\alpha} ;
$$

quindi, utilizzando (10.7), otterremo

$$
H^\alpha = \ell^\alpha . \tag{10.10}
$$

Per la (10.8), una parte della immagine tangenziale propria della superficie nell'intorno del punto p_o può essere determinata dall'equazione

$$
H(y_\alpha) = 1 \tag{10.11}
$$

e perciò è una ipersuperficie.

Una superficie S si dice positivamente (negativamente) convessa rispetto a un iperpiano tangente proprio T, se S si trova dalla stessa parte (o da parte opposta) del centro dello spazio rispetto a T. Una superficie S ad s dimensioni si dice localmente positivamente (negativamente) convessa rispetto a un iperpiano tangente proprio T, se una sua parte, la quale sia ancora una superficie ad s dimensioni avente T come un iperpiano tangente, è positivamente (negativamente) convessa rispetto a T.

Considerando la deviazione della superficie S dall'iperpiano tangente T come una funzione definita su tutta la superficie abbiamo : Affinchè una superficie sia positivamente (negativamente) convessa rispetto a un iperpiano tangente proprio è necessario e sufficiente che la sua deviazione da questo iperpiano tangente non prenda valori negativi (positivi)

(10.12) $$\delta \geqslant 0 \qquad (\delta \leqslant 0) .$$

Poi con l'aiuto della formula (10.4) otteniamo: Affinchè una superficie sia localmente positivamente (negativamente) convessa rispetto ad un iperpiano tangente proprio regolare è necessario e sufficiente che la forma differenziale $g_{ij}\, d\eta^i\, d\eta^j$ sia positivamente (negativamente) definita

(10.13) $$g_{ij}\, d\eta^i\, d\eta^j \geqslant 0 \qquad (g_{ij}\, d\eta^i\, d\eta^j \leqslant 0) .$$

Si vede che rinunziando alla condizione di regolarità dell'iperpiano tangente la condizione (10.13) è necessaria ma non sufficiente per la convessità locale della superficie.

Passiamo ora dallo studio di suprficie proprie allo studio di superficie non proprie. Superficie completamente improprie sono i semiconi centrali.

Un insieme puntuale di uno spazio lineare L_n si dice centralmente semiconico se esso è l'unione di semirette centrali.[11)]

Una superficie ad ℓ dimensioni si dice un semicono centrale ad ℓ dimensioni se essa è un insieme centralmente semiconico.

Sarà poi comodo chiamare una regione aperta centralmente semiconica un semicono centrale ad n dimensioni.

Su ogni semicono centrale C ad ℓ dimensioni possiamo intrinsecamente definire un atlante coordinato ℓ-dimensionale K consistente di tutti i sistemi coordinati rispetto ai quali il semicono è determinato da un sistema di equazioni parametriche

$$X^\alpha = L^\alpha(x^\lambda) \tag{10.14}$$

ove L^α sono funzioni positivamente omogenee del grado 1. E' facile vedere che lo spazio dotato di coordinate (C, K) è uno spazio semiconico H_ℓ.

Se S è una superficie propria e centralmente semplice ad $\ell-1$ dimensioni l'insieme di tutte le semirette centrali intersecanti la superficie è un semicono C ad ℓ dimensioni. La superficie S si dice una direttrice di questo semicono.

Un semicono con una data direttrice si dice un semicono diretto da questa direttrice.

Se

$$X^\alpha = \ell^\alpha(\eta^i) \tag{10.15}$$

sono le equazioni della superficie direttrice allora il semicono da essa diret-

---

11) Notiamo che una semiretta centrale non contiene il centro stesso.

to può essere determinato dalle equazioni

$$X^{\alpha} = X \ell^{\alpha}(\eta^{i}) \qquad (X > 0) \tag{10.16}$$

Considerando una direttrice di un semicono ad ℓ dimensioni come un insieme puntuale di uno spazio $\mathfrak{X}_{\ell-1}$ possiamo considerare il semicono stesso come l'insieme puntuale dello spazio prodotto $\mathfrak{X}_{\ell-1} \boxtimes R_1^+$ ove R^+ è l'insieme di tutti i numeri reali positivi e R_1^+ è lo spazio dotato di coordinate $(R^+, \{\Delta_{R^+}\})$ ad una dimensione con un solo sistema coordinato Δ_{R^+}.

Come nel caso delle superficie proprie ogni iperpiano che passa per un piano tangente del semicono centrale è un iperpiano tangente. E' evidente che tutti gli iperpiani tangenti ad un semicono centrale sono iperpiani centrali. Ogni iperpiano proprio parallelo ad un iperpiano tangente del semicono centrale si dice un iperpiano <u>pseudotangente.</u>

L'insieme di tutti i punti dello spazio coniugato L_n^* che corrispondono ad iperpiani pseudotangenti di un semicono centrale si dice <u>l'immagine tangenziale</u> di quello. In altri termini l'immagine tangenziale di un semicono centrale è l'immagine dell'insieme di tutti gli iperpiani tangenti centrali del semicono nella corrispondenza duale centrale, ed è l'unione di punti propri di rette centrali.

Sia $\overset{0}{y}_{\alpha}$ un covettore che determina un iperpiano pseudotangente $\mathcal{S}$ del semicono centrale. La funzione di punto sul semicono definita dalla formula

$$\Pi = - \overset{0}{y}_{\alpha} L^{\alpha}(\mathfrak{X}^{\lambda}) \tag{10.17}$$

si dice la <u>pseudodeviazione</u> del semicono centrale dall'iperpiano pseudotangente $\mathcal{S}$.

E' evidente che la pseudodeviazione e le sue derivate del primo ordine sono nulle lungo la semiretta generatrice del semicono per cui passa l'iperpiano tangente centrale parallelo all'iperpiano pseudotangente $\mathcal{P}$. Le derivate del secondo ordine della pseudodeviazione lungo questa semiretta generatrice sono componenti di un tensore covariante simmetrico di valenza due :

$$G_{\lambda\mu} = - \overset{\circ}{y}_{\alpha} L^{\alpha}_{\lambda\mu} \tag{10.18}$$

il quale si dice il <u>tensore tangenziale</u> di questo iperpiano pseudotangente $\mathcal{P}$.

E' facile vedere che i componenti del tensore tangenziale in due punti della semiretta generatrice sono proporzionali.

Poichè pel tensore tangenziale vale sempre l'identità

$$G_{\lambda\mu} X^{\mu} = 0 \tag{10.19}$$

il suo discriminante è sempre nullo :

$$\operatorname{Det} \left| G_{\lambda\mu} \right| = 0 \quad . \tag{10.20}$$

Un iperpiano pseudotangente si dice regolare se il rango della matrice

$$\left\| G_{\lambda\mu} \right\| \tag{10.21}$$

è uguale ad ℓ-1 .

E' evidente che i tensori tangenziali degli iperpiani pseudotangenti paralleli sono proporzionali con un fattore di proporzionalità costante. Dunque tutti gli iperpiani pseudotangenti paralleli a un iperpiano pseudotangente regolare sono pure regolari. Un iperpiano tangente si dice <u>regolare</u> se sono regolari gl'iperpiani pseudotangenti paralleli.

Un semicono centrale si dice regolare se per ogni semiretta generatrice esiste un iperpiano tangente regolare.

Una retta generatrice dell'immagine tangenziale del semicono centrale si dice regolare se essa corrisponde ad un iperpiano tangente regolare del semicono.

Dimostreremo che in vicinanza di ogni retta generatrice regolare dell'immagine tangenziale del semicono centrale possiamo prendere una parte di questa che sia un ipercono.

Infatti l'immagine tangenziale del semicono centrale (10.14) si può determinare dal seguente sistema di equazioni

$$(10.22) \qquad L^{\alpha}_{\lambda}(x^{\mu})y_{\alpha} = 0$$

Sia $\overset{0}{y}_{\alpha}$ un covettore che determina un iperpiano pseudotangente regolare e siano x^{λ}_{0} le coordinate di un punto della semiretta generatrice del semicono. Badando al fatto che il rango della matrice (10.21) vale ℓ-1 per i valori iniziali x^{λ}_{0}, $\overset{0}{y}_{\alpha}$ possiamo eliminare le variabili x^{λ} dalle equazioni (10.22) e in vicinanza dei valori iniziali $\overset{0}{y}_{\alpha}$ sostituire al sistema (10.22) un equazione

$$(10.23) \qquad \varphi(y_{\alpha}) = 0$$

ove φ è una funzione omogenea di grado 0 .

Un semicono centrale C si dice <u>convesso</u> rispetto ad un iperpiano tangente T se C si trova tutto da una parte rispetto a T .

E' facile vedere che questo equivale a dire che la pseudodeviazione di C da ogni iperpiano pseudotangente parallelo a T è di segno costante e si può scegliere un iperpiano pseudotangente in tal modo che la pseudodeviazione sia non negativa

$$\Pi \geqslant 0 \quad . \tag{10.24}$$

In questo caso il semicono si troverà tutto da una parte rispetto all'iperpiano pseudotangente scelto.

Analogamente un semicono centrale ad ℓ dimensioni si dice <u>localmente convesso</u> rispetto ad un iperpiano tangente T se una sua parte, che sia pure un semicono centrale ad ℓ dimensioni avente T come un iperpiano tangente, è convessa rispetto a T .

Affinchè un semicono centrale sia localmente convesso rispetto ad un iperpiano tangente regolare è necessario e sufficiente che la forma differenziale determinata dal tensore tangenziale di un iperpiano pseudotangente parallelo sia definita.

Consideriamo ora un semicono centrale (10.16) diretto da una superficie ad ℓ-1 dimensioni (10.15) .

La pseudodeviazione del semicono centrale da un iperpiano pseudotangente può ora essere considerata come una funzione delle coordinate η^i, X . Fra le componenti del tensore tangenziale di un iperpiano pseudotangente possono essere differenti da zero solamente le componenti

$$G_{ij} = X\,\gamma_{ij} \quad , \quad \gamma_{ij} = -\overset{o}{y}_\alpha\,\ell_{ij} \quad . \tag{10.25}$$

ove γ_{ij} sono le componenti di un tensore sulla superficie direttrice. La regolarità dell'iperpiano pseudotangente è equivalente al fatto che il discriminante del tensore γ_{ij} sia differente da zero.

Considerando il sistema di equazioni dall'immagine tangenziale del semicono centrale diretto

(10.26) $$\ell_1^{\alpha}(\eta^j)\, y_\alpha = 0$$

(10.27) $$\ell^{\alpha}(\eta^j)\, y_\alpha = 0$$

in vicinanza di valori iniziali corrispondenti ad un iperpiano pseudotangente regolare possiamo definire la funzione $H(y_\alpha)$ nello stesso modo come abbiamo fatto in (10, 9) . Allora la parte corrispondente dell'immagine tangenziale del semicono diretto sarà determinata dall'equazione

(10.28) $$H(y_\alpha) = 0 \quad .$$

§ 11. Pseudometriche e metriche vettoriali in uno spazio lineare.

Come è noto da ogni spazio vettoriale W_n ad n dimensioni possiamo ottenere lo spazio centro-affine prendendo l'atlante che consiste di tutti i sistemi coordinati definiti dalle basi di W_n .

Un insieme di vettori di W_n si dice un fascio vettoriale semiconico ad ℓ dimensioni se esso è un semicono centrale ad ℓ dimensioni dello spazio centro-affine corrispondente.

Se questo semicono è determinato da equazioni parametriche

$$X^{\alpha} = L^{\alpha}(x^{\lambda}) \tag{11.1}$$

e un vettore $\bar{X}$ appartenente al fascio semiconico è un punto del semicono con coordinate curvilinee x^{λ} , allora le x^{λ} si dicono coordinate curvilinee del vettore $\bar{X}$.

Una pseudometrica ad ℓ dimensioni nello spazio vettoriale W_n è una funzione numerica μ di vettore, la quale considerata come una relazione binaria fra vettori di W_n e numeri reali è un fascio vettoriale semiconico ad ℓ dimensioni dello spazio vettoriale ad n+1 dimensioni $W_n \times R$.

Ne consegue che l'insieme $pr_1 \mu$ di vettori sul quale la funzione μ è definita sarà un fascio vettoriale semiconico ad ℓ dimensioni e che la funzione stessa sarà positivamente omogenea di grado +1 , cioè per ogni $c > 0$ e $X \in pr_1 \mu$ avremo $\mu(c\bar{X}) = c\mu(\bar{X})$.

Vettori appartenenti al fascio semiconico $pr_1 \mu$ si dicono vettori misurabili della pseudometrica μ e le direzioni di questi vettori si dicono le direzioni misurabili.

Il numero $\mu(\bar{X})$ ove $\bar{X} \in pr_1(\mu)$ si dice lo pseudomodulo del vettore misurabile $\bar{X}$.

Un vettore misurabile $\bar{X}$ si dice positivo, negativo, isotropo secondo quale delle tre condizioni seguenti $\mu(\bar{X}) > 0$, $\mu(\bar{X}) < 0$, $\mu(\bar{X}) = 0$ è soddisfatta.

La pseudometrica μ si dice semisimmetrica se dalla misurabilità del vettore $\bar{X}$ segue la misurabilità del vettore $-\bar{X}$. E' facile vedere che questo è equivalente al fatto che il semicono $pr_1\mu$ di vettori misurabili è un cono dello spazio centro-affine corrispondente.

Una pseudometrica semisimmetrica si dice una pseudometrica simmetrica se per tutti i vettori misurabili è : $\mu(-\bar{X}) = \mu(\bar{X})$.

Una pseudometrica si dice completa se ogni vettore non nullo è misurabile. E' evidente che la dimensione di una pseudometrica completa coincide con quella dello spazio vettoriale stesso.

Notiamo che secondo la nostra definizione il vettore nullo è sempre non misurabile. In certi casi e in particolare nel caso di pseudometrica completa, è comodo attribuire al vettore nullo pseudomodulo zero.

Una pseudometrica si dice una metrica se lo pseudomodulo è sempre positivo. In questo caso lo pseudomodulo si dice il modulo. Una pseudometrica si dice una metrica impropria se lo pseudomodulo è sempre non negativo ma prende il valore zero.

Una pseudometrica si dice regolare se μ è un semicono regolare.

Siano

$$X^\alpha = L^\alpha(\chi^\lambda) \quad , \quad X = L(\chi^\lambda) \tag{11.2}$$

le equazioni coordinate parametriche di una pseudometrica μ come di un fascio vettoriale semiconico ad ℓ dimensioni nello spazio vettoriale $W_n \times R$.

Le prime n di queste equazioni determinano il fascio semiconico ad ℓ dimensioni $pr_1\mu$ di vettori misurabili nello spazio W_n e l'ultima equazione definisce lo pseudomodulo di un vettore misurabile $\bar{X}$ come una funzione delle sue coordinate curvilinee x^λ .

Se μ è una pseudometrica ad n dimensioni allora il sistema di equazioni (11.2) può essere ridotto ad una sola equazione

$$X = L(X^\alpha) \tag{11.3}$$

se scegliamo le coordinate ordinarie X^α di un vettore misurabile come un caso particolare di coordinate curvilinee.

Un vettore misurabile si dice <u>stazionario</u> se il suo pseudomodulo è un valore stazionario della funzione μ , cioè $L_\lambda = 0$.

E' evidente che se $\overline{X}$ è un vettore stazionario allora tutti i vettori colla stessa direzione sono stazionari. Per conseguenza possiamo parlare di una direzione stazionaria. Utilizzando l'identità di Eulero per la funzione L cioè $L_\lambda \dot{x}^\lambda = L$ otterremo che ogni direzione stazionaria è isotropa. L'inverso non è vero in generale ma è vero nel caso di una metrica impropria perchè in questo caso ogni vettore isotropo corrisponde al valore minimale della funzione μ .

Diremo che in uno spazio lineare L_n è data una pseudometrica vettoriale se una pseudometrica è definita nello spazio vettoriale corrispondente. Dunque una pseudometrica vettoriale in L_n può essere considerata come un semicono centrale nello spazio $L_n \boxtimes R_1$, ove $R_1 = (R, \{\Delta_R\})$ è lo spazio aritmetico dotato di coordinate ad una dimensione.

Consideriamo le sezioni del semicono centrale μ nello spazio $L_n \boxtimes R_1$ con tre iperpiani $\pi_1(L_n) \times \{+1\}$, $\pi_1(L_n) \times \{-1\}$, $\pi_1(L_n) \times \{0\}$ e indichiamo le proiezioni di queste sezioni nello spazio L_n stesso rispettivamente con $\mathfrak{J}_+$, $\mathfrak{J}_-$, $\mathfrak{J}_o$.

$\mathfrak{J}_+$, $\mathfrak{J}_-$, $\mathfrak{J}_o$ si dicono rispettivamente l'indicatrice positiva, negativa e isotropa della pseudometrica μ.

Se il semicono μ è dato in $L_n \otimes R_1$ dalle equazioni (11.2) allora le indicatrici possono essere definite rispettivamente dalle equazioni

$$(11.4) \qquad L(\chi^\lambda) = 1 \quad , \qquad L(\chi^\lambda) = -1 \; , \qquad L(\chi^\lambda) = 0 \quad .$$

Benchè le indicatrici determinino completamente la pseudometrica μ in L_n, ciò nonostante è più comodo nel caso generale di studiare il semicono μ stesso. Le cose stanno diversamente nel caso di una metrica vettoriale. In questo caso $\mathfrak{J}_-$ e $\mathfrak{J}_o$ sono insiemi vuoti e l'indicatrice positiva da sola determina la metrica vettoriale. Per brevità chiameremo $\mathfrak{J}_+$ l'indicatrice della metrica e la indicheremo semplicemente con $\mathfrak{J}$.

E' facile vedere che l'indicatrice di una metrica vettoriale ad ℓ dimensioni è una superficie ad ℓ-1 dimensioni propria e centralmente semplice. Inversamente ogni superficie propria e centralmente semplice ad ℓ-1 dimensioni data in uno spazio lineare L_n vi definisce una metrica vettoriale ad ℓ dimensioni della quale essa è l'indicatrice.

Poi è facile verificare che la regolarità di una metrica vettoriale è equivalente alla regolarità dell'indicatrice.

Invece di determinare l'indicatrice di una metrica vettoriale in L_n mediante l'equazione implicita

$$(11.5) \qquad L(\chi^\lambda) = 1 \quad ,$$

considerandola come una direttrice del semicono centrale

$$(11.6) \qquad X^\alpha = L^\alpha(\chi^\lambda) \quad ,$$

possiamo determinarla immediatamente mediante un sistema di equazioni parametriche

(11.7) $$x^{\alpha} = \ell^{\alpha}(\eta^{i}) \quad .$$

Una pseudometrica μ nello spazio vettoriale W_n si dice lineare se il fascio vettoriale semiconico μ appartiene ad un sottospazio proprio dello spazio vettoriale $W_n \times R$ non contenente il sottospazio ad una dimensione $\{0\} \times R$.

La condizione di linearità della pseudometrica μ in termini coordinati è che esista un covettore costante C_{α} tale che

(11.8) $$L(x^{\lambda}) = C_{\alpha} L^{\alpha}(x^{\lambda}) \quad .$$

L'insieme M di tutte le pseudometriche di uno spazio lineare con vettori misurabili comuni è evidentemente uno spazio vettoriale, del quale è un sottospazio l'insieme di tutte pseudometriche lineari con gli stessi vettori misurabili.

Una trasformazione nello spazio M della forma

(11.9) $$\bar{\mu} = \varepsilon(\mu - \lambda)$$

ove λ è una pseudometrica lineare di M e $\varepsilon = \pm 1$, si dice una trasformazione di Caratheodory nello spazio M. Se $\varepsilon = +1$ la trasformazione di Caratheodory si dice propria e se $\varepsilon = -1$ impropria.

Per mezzo delle coordinate una trasformazione di Caratheodory può essere espressa dall'equazione

(11.10) $$\bar{L}(x^{\lambda}) = \varepsilon\,(L(x^{\lambda}) - C_{\alpha} L^{\alpha}(x^{\lambda})) \quad ,$$

ove le funzioni L^α determinano il semicono (11.6) di vettori misurabili comuni a tutte le pseudometriche di M, il covettore C_α determina la pseudometrica lineare λ e le funzioni L e $\overline{L}$ corrispondono rispettivamente alle pseudometriche μ e $\overline{\mu}$.

Ogni trasformazione di Caratheodory di una pseudometrica vettoriale μ in uno spazio lineare L_n può essere considerata come indotta da una trasformazione puntuale nello spazio $L_n \boxtimes R_1$

$$(11.11) \qquad \overline{X}^\alpha = X^\alpha \quad , \quad \overline{X} = \varepsilon (X - C_\alpha X^\alpha) \quad .$$

Per determinare questa trasformazione è sufficiente dare l'iperpiano centrale

$$(11.12) \qquad X - C_\alpha X^\alpha = 0$$

il quale sarà trasformato nell'iperpiano $\pi_1(L_n) \times \{0\}$, e fissare un valore di ε. Chiameremo l'iperpiano (11.12) l'<u>iperpiano determinante</u> della trasformazione (11.11).

Una pseudometrica vettoriale μ si dice riducibile ad una metrica se esiste una trasformazione di Caratheodory che trasforma questa pseudometrica in una metrica. E' facile vedere che per questo è necessario e sufficiente che nello spazio $L_n \boxtimes R_1$ esista un iperpiano centrale $\mathcal{P}$ non contenente la semiretta $\{0\} \times R^+$ tale che μ si trovi da una stessa parte di questo avendo in comune l'unico punto 0. Scegliendo $\mathcal{P}$ come l'iperpiano determinante della trasformazione (11.11) e prendendo $\varepsilon = +1$ se la semiretta $\{0\} \times R^+$ si trova dalla stessa parte di μ rispetto a $\mathcal{P}$ e $\varepsilon = -1$ in caso contrario, otterremo una trasformazione di Caratheodory che trasforma la pseudometrica μ in una metrica.

Sia T un iperpiano tangente rispetto al quale il semicono μ è convesso. Diremo che μ è <u>positivamente</u> convesso rispetto a T se μ e la semiretta $\{0\} \times R^+$ si trovano da una stessa parte di T e che μ è <u>negativamente</u> convesso rispetto a T in caso contrario.

Consideriamo una trsformazione di Caratheodory di una pseudometrica vettoriale μ, indotta da una trasformazione (11.11) tale che μ è positivamente (negativamente) convessa rispetto all'iperpiano determinante (11.12) e $\varepsilon = +1$ ($\varepsilon = -1$). In questo caso otterremo dalla pseudometrica vettoriale μ una metrica vettoriale impropria. Infatti la proiezione nello spazio L_n della semiretta generatrice del semicono μ, lungo la quale l'iperpiano determinante è tangente a μ, è una semiretta dello spazio L_n la quale è isotropa rispetto alla pseudometrica trasformata.

Chiameremo tali trasformazioni di Caratheodory di pseudometriche a metriche improprie trasformazioni di Caratheodory speciali.

Consideriamo ora una trasformazione di Cartheodory di una metrica vettoriale μ in una metrica vettoriale $\bar{\mu}$.

E' facile vedere che l'indicatrice $\bar{\mathfrak{J}}$ della metrica trasformata $\bar{\mu}$ è l'immagine dell'indicatrice $\mathfrak{J}$ della metrica data

$$X^\alpha = \ell^\alpha(\eta^i) \tag{11.13}$$

mediante una trasformazione puntuale dello spazio L_n

$$\bar{X}^\alpha = \frac{X^\alpha}{\varepsilon(1 - C_\beta X^\beta)} \tag{11.14}$$

ove è soddisfatta la condizione

$$\varepsilon(1 - C_\alpha \ell^\alpha(\eta^i)) > 0 \qquad , \tag{11.15}$$

che è necessaria e sufficiente affichè $\bar{\mu}$ sia anche una metrica vettoriale.

Le trasformazioni puntuali della forma (11. 14) dello spazio L_n le chiameremo <u>traslazioni duali</u> perchè le trasformazioni duali corrispondenti a (11. 14) nello spazio coniugato L_n^* sono

$$\bar{y}_\alpha = \varepsilon (y_\alpha - C_\alpha) \tag{11.16}$$

Se $\varepsilon = +1$ la traslazione duale (11. 14) si dice propria e se $\varepsilon = -1$ impropria.

Se il covettore C_α non è nullo la traslazione duale (11. 14) è una trasformazione parziale che è definita sul complemento dell'iperpiano

$$C_\alpha X^\alpha = 1 \tag{11.17}$$

nello spazio L_n. Chiameremo l'iperpiano (11. 17) l'iperpiano determinante della traslazione (11. 14).

La condizione (11. 15) esprime che l'indicatrice $\mathfrak{J}$ si trova da una stessa parte rispetto all'iperpiano determinante e non ha con esso punti comuni. Con questo se $\varepsilon = +1$ l'indicatrice si trova dalla stessa parte del centro dello L_n rispetto all'iperpiano determinante e se $\varepsilon = -1$ da parte opposta.

Considerando una trasformazione di Caratheodory speciale di una metrica vettoriale μ dobbiamo considerare una traslazione duale il cui iperpiano determinante è tangente all'indicatrice della metrica μ.

§ 12. Spazio m-vettoriale e cono di Grassmann.

Un caso particolare importante di spazio lineare è uno spazio m-vettoriale, che è comodo considerare come associato ad un dato spazio centro-affine A_n per il quale esso è lo spazio di tutti gli m-vettori.

Considereremo un m-vettore $X^{\alpha_1 \ldots \ldots \alpha_m}$ nello spazio centro-affine A_n come un oggetto coordinato con $\binom{n}{m}$ componenti prendendo come componenti essenziali solamente le componenti $X^{\{\alpha_1 \ldots \ldots \alpha_m\}}$, ove le parentesi graffe indicano che sono ammissibili solo i sistemi crescenti di indici $\alpha_1 < \ldots \ldots < \alpha_m$. Dunque lo spazio di tutti gli m-vettori sarà uno spazio ad $\binom{n}{m}$ dimensioni, che indicheremo con $M_{\binom{n}{m}}$.

Il gruppo coordinato dello spazio $M_{\binom{n}{m}}$ è un sottogruppo del gruppo $\mathfrak{A}_{\binom{n}{m}}$ consistente di trasformazioni della forma seguente

$$\overline{X}^{\{\alpha_1 \ldots \ldots \alpha_m\}} = A^{\{\alpha_1 \ldots \ldots \alpha_m\}}_{\{\beta_1 \ldots \beta_m\}} X^{\{\beta_1 \ldots \ldots \beta_m\}} \tag{12.1}$$

dove i coefficienti hanno la struttura speciale :

$$A^{\{\alpha_1 \ldots \alpha_m\}}_{\{\beta_1 \ldots \beta_m\}} = m! \, A^{\{[\alpha_1}_{\{[\beta_1} \ldots A^{\alpha_m]\}}_{\beta_m]\}} \tag{12.2}$$

E' facile vedere che lo spazio coniugato $\overset{*}{M}_{\binom{n}{m}}$ di $M_{\binom{n}{m}}$ può essere identificato con lo spazio di tutti gli m-covettori $y_{\alpha_1 \ldots \alpha_m}$ dello spazio A_n considerati come oggetti geometrici con $\binom{n}{m}$ componenti $y_{\{\alpha_1 \ldots \alpha_m\}}$.

Come è noto la condizione necessaria e sufficiente affinchè un m-vettore $X^{\alpha_1 \ldots \alpha_m}$ sia semplice, cioè sia il prodotto alternato di m vettori, può essere espressa dal sistema di equazioni

(12.3) $$x^{[\alpha_1 \ldots \alpha_m} x^{\beta_1] \ldots \beta_m} = 0 \quad .$$

Questo sistema di equazioni definisce nello spazio $M_{\binom{n}{m}}$ un cono ad $m(n-m)+1$ dimensioni si dice il <u>cono di Grassmann.</u>

Analogamente il sistema di equazioni

(12.4) $$y_{[\alpha_1 \ldots \alpha_m} y_{\beta_1] \ldots \beta_m} = 0 \quad ,$$

che esprime la condizione di semplicità di un m-covettore $y_{\alpha_1 \ldots \alpha_m}$ definisce nello spazio coniugato $\overset{*}{M}_{\binom{n}{m}}$ un cono ad $m(n-m)+1$ dimensioni che chiameremo pure cono di Grassmann.

I punti del cono di Grassmann si dicono punti semplici dello spazio $M_{\binom{n}{m}}$. Analogamente gli iperpiani che corrispondono ai punti semplici dello spazio coniugato $\overset{*}{M}_{\binom{n}{m}}$ si dicono iperpiani semplici.

Poi si dicono semplici le semirette consistenti di punti semplici e gli iperpiani centrali paralleli a iperpiani propri semplici.

Come è noto ogni piano centrale ad m-dimensioni in A_n può essere considerato come l'insieme puntuale di uno spazio A_m il cui atlante coordinato è definito in modo intrinseco. Tale spazio A_m si dice <u>intrinsecamente immerso</u> nello spazio A_n.

Ogni spazio lineare subordinato ad uno spazio A_m intrinsecamente immerso in A_n si dice semi-intrinsecamente immerso in A_n.

Considereremo poi, in particolare, spazi centro-affini orientati $\vec{A}_m$ e spazi centro-equiaffini orientati $\vec{U}_m$ semi-intrinsecamente immersi in A_n.

Sia A_m intrinsecamente immerso in A_n e siano B^{α}_{a} $(a = 1, \ldots m)$ il tensore di connessione unitario. Com'è noto ([20], [22]) le componenti

$B^{\alpha}_{a}(x, \chi)$ del tensore di connessione unitario eguagliano le componenti dei vettori basici che definiscono il sistema coordinato χ in A_m rispetto al sistema coordinato x in A_n.

Poniamo :

$$B^{\alpha_1 \ldots \alpha_m} = \mathcal{E}^{a_1 \ldots a_m} B^{\alpha_1}_{a_1} \ldots\ldots B^{\alpha_m}_{a_m} \tag{12.5}$$

ove $\mathcal{E}^{a_1 \ldots a_m}$ è la densità m-vettoriale unitaria in A_m definita dall'equazione

$$\mathcal{E}^{1 \ldots, m} = 1 \quad . \tag{12.6}$$

Poichè la densità m-vettoriale unitaria $\mathcal{E}^{a_1 \ldots a_m}$ è di peso +1 l'oggetto coordinato di connessione $B^{\alpha_1 \ldots \alpha_m}$ sarà un m-vettore semplice rispetto ad A_n e l'unione di densità scalare di peso +1 rispetto ad A_m mediante la corrispondenza duale centrale l'immagine di un piano centrale ad m dimensioni in A_n è un piano centrale ad n-m dimensioni nello spazio coniugato A^*_n. Sia $\hat{A}^*_{n-m}$ lo spazio centro-affine intrinsecamente immerso nell' A^*_n il quale corrisponde a questo piano ad n-1 dimensioni. Questo $\hat{A}^*_{n-m}$ si dice lo spazio <u>duale</u> dello spazio A^*_m corrispondente.

Indicheremo con C^p_{α} $(p = m+1 \ldots n)$ il tensore di connessione unitario dello spazio duale $\hat{A}^*_{n-m}$.

Abbiamo

$$C^p_{\alpha} B^{\alpha}_{a} = 0 \quad . \tag{12.7}$$

Come è noto un piano centrale ad m dimensioni nell' A_n si dice "<u>allestito</u>" se è dato un piano centrale ad n-m dimensioni avente con il precedente il centro come unico punto comune. Questo piano si dice il <u>piano alle-</u>

stente e lo spazio corrispondente $\hat{A}_{n-m}$ intrinsecamente immerso in A_n si dice lo spazio allestente.

Lo spazio duale $\hat{A}^*_{n-m}$ può essere identificato con lo spazio coniugato dello spazio allestente $\hat{A}_{n-m}$. In questo caso indicando con C^α_p il tensore di connessione unitario dello spazio allestente $\hat{A}_{n-m}$ avremo

$$C^q_\alpha C^\alpha_p = \delta^q_p \quad . \tag{12.8}$$

A sua volta lo spazio duale A^*_m dello spazio allestente $\hat{A}_{n-m}$ può essere identificato con lo spazio coniugato dello spazio A_m stesso.

In questo caso indicando con B^a_α il tensore di connessione unitario dello spazio A^*_m avremo

$$B^a_\alpha C^\alpha_p = 0 \quad , \quad B^b_\alpha B^\alpha_a = \delta^b_a \quad . \tag{12.9}$$

E' evidente che lo spazio A^*_m è uno spazio allestente per lo spazio duale $\hat{A}^*_{n-m}$. Dunque l'allestimento di uno spazio A_m intrinsecamente immerso in A_n è equivalente all'allestimento dello spazio duale $\hat{A}^*_{n-m}$ in A^*_n.

Poniamo

- 78 -

$$B_{\alpha_1 \dots \alpha_m} = \mathfrak{U}_{a_1 \dots a_m} B^{a_1}_{\alpha_1} \dots B^{a_m}_{\alpha_m} \tag{12.10}$$

dove $\mathfrak{U}_{a_1 \dots a_m}$ è una densità m-covettoriale unitaria nell'A_m definita dall'equazione

$$\mathfrak{U}_{1 \dots m} = 1 \quad . \tag{12.11}$$

(12.8) $C^q_\alpha C^\alpha_p = \delta^q_p$.

Poichè la densità m-covettoriale unitaria $\mathcal{U}_{a_1 \ldots a_m}$ è di peso -1 l'oggetto coordinato di connessione $B_{\alpha_1 \ldots \alpha_m}$ sarà un m-covettore semplice rispetto allo spazio A_n e l'unione di densità scalari di peso -1 rispetto all'A_m.

Da (12.9) consegue

$$(12.12) \qquad B_{\alpha_1 \ldots \alpha_m} B^{\alpha_1 \ldots \alpha_m} = m!$$

Ad ogni A_m intrinsecamente immerso in A_n corrisponde in modo biunivoco la retta centrale semplice nello spazio $M_{\binom{n}{m}}$ determinata dalle equazioni parametriche

$$(12.13) \qquad X^{\alpha_1 \ldots \alpha_m} = B^{\alpha_1 \ldots \alpha_m} \mathcal{T}$$

ove $\mathcal{T}$ è una densità scalare variabile di peso +1 nell' A_m.

Consideriamo ora uno spazio centro-affine orientato $\vec{A}_m$ semi-intrinseco immerso nello spazio A_n. In questo caso la disuguaglianza

$$(12.14) \qquad \mathcal{T} > 0$$

ove $\mathcal{T}$ è una densità scalare di peso +1, ha un senso invariante e perciò possiamo fare corrispondere allo spazio $\vec{A}_m$ la semiretta della retta (12.13) caratterizzata dai valori positivi del parametro.

Dunque otteniamo una corrispondenza biunivoca fra gli spazi $\vec{A}_m$ semi-intrinsecamente immersi in A_n e le semirette centrali semplici dello spazio $M_{\binom{n}{m}}$ associato.

Consideriamo infine uno spazio centro-equiaffine orientato $\vec{U}_m$ semi-intrinsecamente immerso in A_n. In questo caso $B^{\alpha_1 \dots \alpha_m}$ è un m-vettore semplice in A_n, il quale non dipende dal sistema coordinato nello spazio A_m e perciò $\vec{U}_m$ determina un punto semplice nello spazio $M_{\binom{n}{m}}$. Dunque otteniamo una corrispondenza biunivoca fra spazi $\vec{U}_m$ semi-intrinsecamente immersi in A_n e punti semplici propri dello spazio $M_{\binom{n}{m}}$ associato.

L'allestimento di uno spazio $\vec{U}_m$ semintrinsecamente immerso in A_n si riduce all'allestimento dello spazio duale con l'aiuto di uno spazio $\vec{U}^*_m$ semi-intrinsecamente immerso in A^*_n il quale è identificato con lo spazio coniugato di $\vec{U}_m$.

A questo spazio $\vec{U}^*_m$ semi-intrinsecamente immerso nell' A^*_n corrisponde biunivocamente l'iperpiano semplice nello spazio $M_{\binom{n}{m}}$.

Dalla relazione (12.12) si deduce che questo iperpiano passa per il punto che rappresenta in $M_{\binom{n}{m}}$ lo spazio U_m stesso.

Dunque l'allestimento di uno spazio $\vec{U}_m$ semi-intrinsecamente immerso nello spazio A_n si può determinare scegliendo un iperpiano semplice nello spazio $M_{\binom{n}{m}}$ passante per il punto corrispondente allo spazio $\vec{U}_m$.

Consideriamo il fascio Π_p di tutti gli iperpiani propri dello spazio $M_{\binom{n}{m}}$ che passano per un punto semplice p. Sia $\vec{U}_m$ lo spazio semi-intrinsecamente immerso nello spazio A_n, che corrisponde al punto p.

Se un m-covettore $y_{\alpha_1 \dots \alpha_m}$ determina un iperpiano $\mathcal{P}$ di Π_p allora è facile dimostrare che le quantità

$$B^a_\alpha = \frac{1}{(m-1)!} \, y_{\alpha \alpha_2 \dots \alpha_m} \, \mathcal{E}^{a a_2 \dots a_m} B^{\alpha_2}_{a_2} \dots B^{\alpha_m}_{a_m} \tag{12.15}$$

ove B^α_a sono componenti del tensore di connessione unitario dello spazio

$\vec{U}_m$ soddisfanno alle equazioni

(12.16) $$B^{b}_{\alpha} B^{\alpha}_{a} = \delta^{b}_{a}$$

e perciò sono componenti del tensore di connessione unitario di un $\vec{U}^*_m$ nello spazio A^*_n il quale determina un allestimento di $\vec{U}_m$.

A questo allestimento dell' $\vec{U}_m$ corrisponde un iperpiano semplice del fascio Π_p che indichiamo con $\alpha_p(\mathfrak{P})$.

In tal modo abbiamo ottenuto la trasformazione α_p del fascio Π_p , che ad ogni iperpiano $\mathfrak{P}$ fa corrispondere l'iperpiano semplice $\alpha_p(\mathfrak{P})$. Una coppia di iperpiani del fascio Π_p si dice α_p-equivalente se ad ambedue gli iperpiani corrisponde lo stesso iperpiano semplice. E' evidente che questa relazione di α_p-equivalenza è il nucleo della trasformazione α_p .

Segue che possiamo allestire uno spazio $\vec{U}_m$ semi-intrinsecamente immerso in A_n scegliendo nello spazio $M_{\binom{n}{m}}$ associato un iperpiano arbitrario che passa per il punto semplice corrispondente ad $\vec{U}_m$.

Con questo gli iperpiani α_p-equivalenti determineranno lo stesso allestimento.

Sia $\vec{U}_m$ uno spazio centro-equiaffine semi-intrinsecamente immerso nello spazio A_n e allestito da uno spazio $\hat{A}_{m-m}$.

Consideriamo i tensori di connessione seguenti, formati dai tensori di connessione unitari.

(12.17) $$B^{\alpha_1 \ldots \alpha_m a_1 \ldots a_r}_{p_1 \ldots p_r} = \frac{m!}{(m-r)!} \mathcal{E}^{a_1 \ldots a_r a_{r+1} \ldots a_m} C^{[\alpha_1}_{p_1} \ldots C^{\alpha_r}_{p_r} B^{\alpha_{r+1}}_{a_{r+1}} \ldots B^{\alpha_m]}_{a_m}$$

$$(12.18)\quad B^{p_1\dots p_r}_{\alpha_1\dots\alpha_m a_1\dots a_r} = \frac{m!}{(m-r)!} \mathcal{W}_{a_1\dots a_r a_{r+1}\dots a_m} C^{p_1}_{[\alpha_1}\dots C^{p_r}_{\alpha_r} B^{a_{r+1}}_{\alpha_{r+1}}\dots B^{a_m}_{\alpha_m]}$$

ove $r = 1,\dots,t$ e $t = \mathrm{Min}(m, n-m)$. Se $r = 0$ i tensori corrispondenti si riducono a $B^{\alpha_1\dots\alpha_m}$ e $B_{\alpha_1\dots\alpha_m}$.

Si può poi dimostrare che :

$$(12.19)\quad B^{p_1\dots p_r}_{\alpha_1\dots\alpha_m a_1\dots a_r} B^{\alpha_1\dots\alpha_m b_1\dots b_2}_{q_1\dots q_r} = m!\, \delta^{b_1\dots b_r}_{a_1\dots a_r} \delta^{p_1\dots p_r}_{q_1\dots q_r}$$

e se $r_1 \neq r_2$,

$$B^{p_1\dots p_{r_1}}_{\alpha_1\dots\alpha_m a_1\dots a_{r_1}} B^{\alpha_1\dots\alpha_m b_1\dots b_{r_2}}_{q_1\dots q_{r_2}} = 0$$

consegue quindi che questi tensori fanno corrispondere ad ogni coppia di si- sistemi coordinati negli spazi $\vec{U}_m$ e $\hat{A}_{n-m}$ una coppia di basi vettoriali coniugate negli spazi $M_{\binom{n}{m}}$ e $M^*_{\binom{n}{m}}$.

Ogni m-vettore $X^{\alpha_1\dots\alpha_m}$ ed ogni m-covettore $y_{\alpha_1\dots\alpha_m}$ possono essere rappresentati nel modo seguente

$$(12.20)\quad X^{\alpha_1\dots\alpha_m} = \sum_{r=0}^{t} B^{\alpha_1\dots\alpha_m \{a_1\dots a_r\}}_{\{p_1\dots p_r\}} X^{\{p_1\dots p_r\}}_{\{a_1\dots a_r\}}$$

$$(12.21) \qquad y_{\alpha_1 \cdots \alpha_m} = \sum_{r=0}^{t} B_{\alpha_1 \dots \alpha_m \{a_1 \dots a_r\}}^{\{p_1 \dots p_r\}} \, y_{p_1 \dots p_r}^{a_1 \dots a_r}$$

ove $X_{a_1 \dots a_r}^{p_1 \dots p_r}$ e $y_{p_1 \dots p_r}^{a_1 \dots a_r}$ sono tensori emisimmetrici di connessione negli spazi $\vec{U}_m$ e $\hat{A}_{n-m}$ che chiameremo il tensore coordinato d'ordine r dell' m-vettore $X^{\alpha_1 \dots \alpha_m}$ e il tensore coordinato d'ordine r dello m-covettore $y_{\alpha_1 \dots \alpha_m}$ rispetto allo spazio allestito $\vec{U}_m$.

I tensori coordinati d'ordine zero sono scalari X e y .

Notiamo che il prodotto scalare di un m-vettore $X^{\alpha_1 \dots \alpha_m}$ e un m-covettore $y_{\alpha_1 \dots \alpha_m}$ si può esprimere in termini dei loro tensori coordinati nel modo seguente

$$(12.22) \qquad y_{\{\alpha_1 \dots \alpha_m\}} X^{\{\alpha_1 \dots \alpha_m\}} = \sum_{r=0}^{t} y_{\{p_1 \dots p_r\}}^{\{a_1 \dots a_r\}} X_{\{a_1 \dots a_r\}}^{\{p_1 \dots p_r\}} .$$

Consideriamo il fascio di iperpiani propri Π_p col centro in un punto semplice p che corrisponde ad uno spazio $\vec{U}_m$ semi-intrinsecamente immerso in A_n . Supponiamo che $\vec{U}_m$ sia allestito con l'aiuto di un iperpiano semplice P del fascio Π_p .

Determineremo ogni iperpiano di Π_p con l'aiuto dei tensori coordinati di m-covettori corrispondenti rispetto allo spazio $\vec{U}_m$ allestito.

Notiamo che affinchè un iperpiano proprio appartenga a Π_p è necessario e sufficiente che sia

$$(12.23) \qquad y = 1 \quad .$$

Poi è facile dimostrare che coppie (P_1, P_2) α-equivalenti di iperpiani di Π_p sono caratterizzati dall'uguaglianza dei tensori coordinati del primo ordine

$$\overset{1}{y}{}^a_p = \overset{2}{y}{}^a_p \quad . \tag{12.24}$$

In particolare tutti gli iperpiani di Π_p che determinano l'allestimento dato dello spazio $\overrightarrow{U}_m$ sono determinati dall'equazione

$$y^a_p = 0 \quad . \tag{12.25}$$

§ 13. Superficie sul cono di Grassmann in uno spazio m-vettoriale.

Sia S una superficie propria ad s dimensioni sul cono di Grassmann nello spazio $M_{\binom{n}{m}}$ determinata dal sistema di equazioni parametriche

$$X^{\alpha_1 \dots \alpha_m} = \ell^{\alpha_1 \dots \alpha_m}(\eta^i) \;. \tag{13.1}$$

A questa superficie corrisponde un famiglia ad ℓ parametri di spazi $\vec{U}_m$ semi-intrinsecamente immersi nello spazio A_n pei quali si ha :

$$B^{\alpha_1 \dots \alpha_m} = \ell^{\alpha_1 \dots \alpha_m} \;. \tag{13.2}$$

Consideriamo uno di questi spazi $\vec{U}_m$ e supponiamo che esso sia allestito da uno spazio $\hat{A}_{n-m}$. Calcolando le derivate $\ell_i^{\alpha_1 \dots \alpha_m}$ nel punto corrispondente p della superficie S otteniamo

$$\ell_i^{\alpha_1 \dots \alpha_m} = m\,\mathscr{C}^{a_1 a_2 \dots a_m} B^{[\alpha_1}_{a_1 i} B^{\alpha_2}_{a_2} \dots B^{\alpha_m]}_{a_m} \tag{13.3}$$

donde vediamo che per gli m-vettori $\ell_i^{\alpha_1 \dots \alpha_m}$ i tensori coordinati di ordine superiore al primo sono nulli.

Indicando con ℓ_i e ℓ_{ia}^{p} rispettivamente i tensori coordinati d'ordine zero e d'ordine uno abbiamo

$$\ell_i = B^a_\alpha B^\alpha_{a\,i} \;, \qquad \ell_{ia}^{p} = C^p_\alpha B^\alpha_{ai} \;. \tag{13.4}$$

Dimostreremo che il rango della matrice

$$\left\| \ell_{ia}^{p} \right\| \tag{13.5}$$

ove i è l'indice delle righe e $\overset{p}{a}$ è l'indice delle colonne, è uguale ad s .

Per questo consideriamo dapprima la uguaglianza

$$\ell_i^{\alpha_1 \ldots \alpha_m} = B^{\alpha_1 \ldots \alpha_m} \ell_i + B^{\alpha_1 \ldots \alpha_m a}_{p} \ell^p_{ia} \tag{13.6}$$

Sia c^i una soluzione del sistema di equazioni lineari $c^i \ell^p_{ia} = 0$ allora otteniamo da (13.6)

$$c^i \ell_i^{\alpha_1 \ldots \alpha_m} = c^i \ell_i \ell^{\alpha_1 \ldots \alpha_m} \tag{13.7}$$

donde, in conseguenza dell'indipendenza lineare degli m-vettori $\ell_i^{\alpha_1 \ldots \alpha_m}$, $\ell^{\alpha_1 \ldots \alpha_m}$, si deduce che $c^i = 0$. Dunque il sistema di equazioni considerato ammette solamente la soluzione nulla e perciò il rango della matrice (13.5) è uguale ad s .

Gli iperpiani del fascio Π_p che sono tangenti alla superficie S nel punto p sono definiti da un m-covettore $Y_{\alpha_1 \ldots \alpha_m}$ che accanto alla condizione

$$\ell^{\alpha_1 \ldots \alpha_m} Y_{\alpha_1 \ldots \alpha_m} = m! \, \ell^{\{\alpha_1 \ldots \alpha_m\}} Y_{\{\alpha_1 \ldots \alpha_m\}} = m! \tag{13.8}$$

soddisfano alle equazioni

$$\ell_i^{\alpha_1 \ldots \alpha_m} Y_{\alpha_1 \ldots \alpha_m} = 0 \tag{13.9}$$

Secondo (12.22) in termini dei tensori coordinati di $Y_{\alpha_1 \ldots \alpha_m}$ rispetto allo spazio $\vec{U}_m$ allestito le equazioni (13.9) sono equivalenti alle equazioni

$$(13.10) \qquad \ell_i + \ell_{ia}^{p} Y_p^a = 0 \ .$$

Siccome il rango della matrice (13.5) è uguale ad s, il sistema (13.10) di equazioni lineari è compatibile. Sia Y_p^a una soluzione di questo sistema; allora il vettore semplice

$$(13.11) \qquad Y_{\alpha_1 \dots \alpha_m} = \mathcal{U}_{a_1 \dots a_m} (B_{\alpha_1}^{a_1} + C_{\alpha_1}^{p_1} Y_{p_1}^{a_1}) \dots (B_{\alpha_m}^{a_m} + C_{\alpha_m}^{p_m} Y_{p_m}^{a_m})$$

$$= B_{\alpha_1 \dots \alpha_m} + \sum_{r=1} B_{\alpha_1 \dots \alpha_m}{}^{\{p_1 \dots p_r\}}_{\{a_1 \dots a_r\}} Y_{\{[p_1}^{\{[a_1} \dots Y_{p_r]\}}^{a_r]\}}$$

ha per suoi tensori coordinati d' ordine zero e uno rispettivamente 1 e Y_p^a e perciò determina un iperpiano tangente semplice della superficie S nel punto p .

Supponiamo ora che lo spazio $\vec{U}_m$ sia allestito con l'aiuto di un iperpiano tangente semplice. Questo è equivalente al fatto che il sistema (13.10) ammetta la soluzione $Y_p^a = 0$ cioè al fatto che

$$(13.12) \qquad \ell_i = 0 \ .$$

Ora possiamo scrivere il sistema (13.10) nella forma :

$$(13.13) \qquad \ell_{ia}^{p} Y_p^a = 0 \ .$$

Ad ogni soluzione Y^a_p del sistema (13.13) corrisponde una classe di iperpiani tangenti che sono α-equivalenti fra loro.

Ogni classe contiene un unico iperpiano tangente semplice.

Siccome il sistema (13.13) ammette $m(n-m) - s$ soluzioni linearmente indipendenti in ogni punto della superficie abbiamo una famiglia di iperpiani tangenti semplici ad $m(n-m) - s$ parametri.

Notiamo che in ogni punto di una superficie ad $m(n-m)$ dimensioni abbiamo solamente un iperpiano tangente semplice.

L'insieme di tutti i punti dello spazio coniugato $\overset{*}{M}_{\binom{n}{m}}$, che corrispondono ad iperpiani tangenti propri semplici di una superficie S sul cono di Grassmann nello spazio $M_{\binom{n}{m}}$ si dice l'<u>immagine tangenziale grassmanniana propria</u> di questa superficie S.

Consideriamo ora gli iperpiani tangenti centrali in un punto della superficie S.

Le equazioni

$$\ell_i^{\alpha_1 \dots \alpha_m} Y_{\alpha_1 \dots \alpha_m} = 0\,, \quad \ell^{\alpha_1 \dots \alpha_m} Y_{\alpha_1 \dots \alpha_m} = 0 \tag{13.14}$$

esprimono che un m-covettore $Y_{\alpha_1 \dots \alpha_m}$ determina un iperpiano tangente centrale della superficie S. Secondo (12.22) le equazioni (3..14) sono equivalenti alle seguenti in termini di tensori coordinati di $Y_{\alpha_1 \dots \alpha_p}$

$$\ell^p_{ia} Y^a_p = 0 \quad , \quad Y = 0 \; . \tag{13.15}$$

Supponiamo che S sia una superficie ad $m(n-m)$ dimensioni. In questo caso tutti gli iperpiani tangenti centrali coincidono cogli iperpiani tangenti del cono di Grassmann. D'altra parte in questo caso il sistema $\ell^p_{ia} Y^a_p = 0$

ammette l'unica soluzione $Y^a_p = 0$ e perciò il sistema (13.15) è equivalente al sistema seguente

$$(13.16) \qquad Y^a_p = 0 \qquad , \qquad Y = 0 \quad .$$

Utilizzando il risultato ottenuto si può dimostare che tutti gli iperpiani pseudo-tangenti in un punto p del cono di Grassmann possono essere determinati dal sistema di equazioni

$$(13.17) \qquad \overset{}{\underset{\circ}{X}}{}^{\beta\alpha_2 \ldots \alpha_m} Y_{\alpha\alpha_2 \ldots \alpha_m} = 0$$

ove $\underset{\circ}{X}{}^{\beta\alpha_2 \ldots \alpha_m}$ sono coordinate del punto p .

Possiamo dare una interpretazione geometrica della relazione dell' α-equivalenza fra iperpiani di un fascio Π_p con centro in punto semplice p dello spazio $M_{\binom{n}{m}}$.

E' facile vedere che da (12.24) e (13.16) consegue che $\ell' \alpha$ -equivalenza di due iperpiani del fascio Π_p significa che la differenza $\overset{2}{Y}_{\alpha_1 \ldots \alpha_m} - \overset{1}{Y}_{\alpha_1 \ldots \alpha_m}$ degli m-covettori, i quali determinano questi iperpiani, determina un iperpiano pseudotangente del cono di Grassmann.

Una funzione numerica $f(\|X^\alpha_a\|)$ della matrice $\|X^\alpha_a\|$ ove $\alpha = 1, \ldots, n; a = 1, \ldots, m < n$ si dice positivamente omogenea di grado k se essa soddisfa alla condizione

$$(13.18) \qquad f(\|C^b_a\| \, \|X^\alpha_a\|) = (\mathrm{Det}\, \|C^b_a\|)^k \cdot f(\|X^\alpha_a\|) \quad ,$$

dove $\|C^b_a\|$ è una matrice arbitraria col determinante positivo

(13.19) $$\mathrm{Det} \left\| C_a^b \right\| > 0 \quad .$$

Considerando la funzione della matrice $\| X_a^\alpha \|$ come una funzione $f(X_a^\alpha)$ degli elementi della matrice cioè di mn variabili X_a^α, possiamo riscrivere la condizione (13.18) nella forma seguente :

(13.20) $$f(C_a^b X_b^\alpha) = (\mathrm{Det} \| C_a^b \|)^k f(X_b^\alpha) \quad .$$

Si può dimostrare che la condizione (13.20) equivale alle equazioni differenziali

(13.21) $$f_\alpha^b X_a^\alpha = k \delta_a^b f \quad .$$

E' facile vedere che una funzione numerica positivamente omogenea di grado k definita sul cono di Grassmann mediante le coordinate può essere espressa come una funzione positivamente omogenea di grado k della matrice $\| X_a^\alpha \|$ ove le X_a^α sono coordinate di m vettori in termini dei quali sono espresse le coordinate di punto sul cono di Grassmann e che sono componenti essenziali dell' m-vettore

(13.22) $$X^{\alpha_1 \ldots \alpha_m} = m! \, X_1^{[\alpha_1} \ldots X_m^{\alpha_m]} \quad .$$

Gli elementi della matrice $\| X_a^\alpha \|$ possono essere considerati come coordinate soprannumerarie di punto sul cono di Grassmann, le quali sono definite a meno di una trasformazione

(13.23) $$\overline{X}_a^\alpha = u_a^b X_b^\alpha \quad , \text{ ove } \quad \mathrm{Det} \| u_a^b \| = 1$$

Rappresentando un punto del cono di Grassmann come uno spazio $\vec{U}_m$ semi-intrinsecamente immerso nello spazio A_n possiamo interpretare le coordinate X_a^α come componenti del tensore di connessione unitario di questo $\vec{U}_m$.

Sia $L(X_a^\alpha)$ una funzione positivamente omogenea di grado $+1$ della matrice $\| X_a^\alpha \|$. Supponiamo anche che L prenda solo valori positivi : $L > 0$.

Considerando L come una funzione definita sul cono di Grassmann abbiamo che l'equazione

$$L(X_a^\alpha) = 1 \tag{13.24}$$

determina una superficie propria ad $m(n-m)$ dimensioni sul cono di Grassmann. Come abbiamo visto, in ogni punto di una superficie ad $m(n-m)$ dimensioni sul cono di Grassmann esiste un solo iperpiano tangente semplice. Per ottenere questo iperpiano tangente semplice consideriamo dapprima nello spazio corrispondente A_n la famiglia ad $m(n-m)$ parametri η^i di spazi $\vec{U}_m$ corrispondenti ai punti della superficie (13.24) . Le componenti B_a^α del tensore di connessione unitario di questi spazi $\vec{U}_m$ le considereremo come funzioni dei parametri η^i .

Sostituendo $B_a^\alpha(\eta^i)$ al posto di X_a^α nell'equazione (13.24) e derivando l'identità ottenuta avremo

$$L_\alpha^a \partial_i B_a^\alpha = 0 \tag{13.25}$$

D'altra parte, in virtù di (13.21) , abbiamo

$$L_\alpha^b B_a^\alpha = \delta_a^b \tag{13.26}$$

Usando (13.4), (13.12), (13.13) otterremo da (13.25), (13.26) che

$$B^a_\alpha = L^a_\alpha(B^\beta_b) \qquad (13.27)$$

sono componenti del tensore di connessione unitario di uno $\vec{U}^*_m$ nello spazio A^*_n, il quale determina un allestimento dell' $\vec{U}_m$ corrispondente all'iperpiano tangente semplice della superficie (13.24) .

E' facile vedere che un semicono centrale ad ℓ dimensioni sul cono di Grassmann nello spazio $M_{\binom{n}{m}}$ può essere determinato da un sistema di equazioni

$$\overset{z}{f}(X^\alpha_a) = 0 \qquad (z = \ell+1, \ldots, m(n-m)+1) \qquad (13.28)$$

ove $\overset{z}{f}$ sono funzioni positivamente omogenee della matrice $\| X^\alpha_a \|$.

Consideriamo ora una superficie ad ℓ-1 dimensioni S sul cono di Grassmann come l'intersezione di un semicono centrale ad ℓ dimensioni (13.28) con una superficie ad m(n-m) dimensioni (13.24).

Un iperpiano tangente semplice della superficie (13.24) già trovata evidentemente è tangente alla superficie S . Per trovare tutti gli iperpiani tangenti semplici della S procediamo come segue. Consideriamo la famiglia ad ℓ-1 parametri η^i degli spazi $\vec{U}_m$ corrispondenti ai punti della superficie S . Consideriamo le componenti B^α_a del tensore di connessione unitario di questi spazi come funzioni dei parametri η^i . Sostituendo $B^\alpha_a(\eta^i)$ al posto di X^α_a nelle equazioni (13.28) e derivando le identità ottenute avremo

$$\overset{z}{f}{}^a_\alpha \, \partial_i B^\alpha_a = 0 \qquad (13.29)$$

D'altra parte in virtù di (13.21) abbiamo

$$\text{(13.30)} \qquad \overset{z}{f}{}^{b}_{\alpha} \, B^{\alpha}_{a} = 0 \quad .$$

Ora è facile vedere che tutti gli iperpiani tangenti semplici della superficie S in un punto dato p , corrispondono agli allestimenti dello spazio $\vec{U}_m$ corrispondente al punto p mediante i tensori di connessione unitari

$$\text{(13.31)} \qquad B^{a}_{\alpha} = L^{a}_{\alpha}(B^{\beta}_{b}) + \lambda_{z} \overset{z}{f}{}^{a}_{\alpha} (B^{\beta}_{b})$$

ove λ_z sono parametri arbitrari.

Una superficie propria sul cono di Grassmann si dice <u>regolare</u> in senso stretto se ad ogni punto esiste un iperpiano tangente regolare semplice.

Se un m-covettore $\overset{\circ}{y}_{\alpha_1 \ldots \alpha_m}$ determina un iperpiano tangente regolare di una superficie S sul cono di Grassmann possiamo definire nello intorno dei valori $\overset{\circ}{y}_{\alpha_1 \ldots \alpha_m}$ la funzione $H(y_{\{\alpha_1 \ldots \alpha_m\}})$ secondo (10.9) . L'equazione

$$\text{(13.32)} \qquad H(y_{\{\alpha_1 \ldots \ldots \alpha_m\}}) = 1$$

determinerà una parte corrispondente dell'immagine tangenziale propria della superficie S .

Se l' m-covettore $\overset{\circ}{y}_{\alpha_1 \ldots \alpha_m}$ è semplice $\overset{\circ}{y}_{\alpha_1 \ldots \alpha_m} = m! \, \overset{1}{y}_{[\alpha_1} \cdots \overset{m}{y}_{\alpha_m]}$ allora possiamo definire nell'intorno dei valori iniziali $\overset{\circ}{y}{}^{a}_{\alpha}$ la funzione

$$(13.33) \qquad Q(y_\alpha^a) = H(m!\, y^1_{\{[\alpha_1} \cdots y^m_{\alpha_m]\}})$$

e la parte corrispondente dell'immagine tangenziale grassmanniana propria della superficie S può essere determinata dall'equazione

$$(13.34) \qquad Q(y_\alpha^a) = 1$$

ove y_α^a sono coordinate soprannumerarie di punto sul cono di Grassmann nello spazio coniugato $M^*_{\binom{n}{m}}$.

Derivando la funzione Q otteniamo

$$(13.35) \qquad Q_a^\alpha (B_\beta^b) = B_a^\alpha$$

ove B_a^α e B_α^a sono componenti dei tensori di connessione unitari di uno spazio $\overrightarrow{U}_m$ allestito con l'aiuto dell'iperpiano tangente semplice corrispondente.

§ 14. Pseudometriche e metriche m-vettoriali grassmaniane in $\mathfrak{X}_n$.

Come abbiamo detto, ogni superficie ad m dimensioni in uno spazio $\mathfrak{X}_n$ può essere considerata come l'insieme puntuale di uno spazio $\mathfrak{X}_m$ il cui atlante coordinato è definito in modo intrinseco e che si dice intrinsecamente immerso in $\mathfrak{X}_n$.

Una superficie ad m dimensioni in $\mathfrak{X}_n$ si dice orientata se lo spazio corrispondente $\mathfrak{X}_m$ è orientato e perciò ridotto a uno spazio $\vec{\mathfrak{X}}_m$, che chiameremo semi-intrinsecamente immerso nello spazio $\mathfrak{X}_n$.

E' evidente che gli spazi tangenti del primo ordine di uno spazio $\vec{\mathfrak{X}}_m$ sono spazi centro-affini orientati $\vec{A}_m$. Dunque gli spazi tangenti di uno spazio $\mathfrak{X}_m$ semi-intrinsecamente immerso in uno spazio $\mathfrak{X}_n$ sono spazi $\vec{A}_m$ semi-intrinsecamente immersi negli spazi tangenti corrispondenti dello spazio $\mathfrak{X}_n$.

Diremo che in uno spazio $\mathfrak{X}_n$ è dato un campo grassmanniano di ℓ-direzioni semiconiche m-vettoriali se in ogni spazio $M_{\binom{n}{m}}$ associato ad un punto di $\mathfrak{X}_n$ è dato un semicono centrale ad ℓ dimensioni sul cono di Grassmann.

Dal punto di vista della teoria di spazi fibrati un campo grassmanniano di ℓ-direzioni semiconiche m-vettoriali in uno spazio $\mathfrak{X}_n$ si può considerare come una superficie ℓ-secante S nello spazio fibrato $M_{n+(\binom{n}{m})}$ di tutti gli m-vettori dello spazio $\mathfrak{X}_n$, la quale taglia ogni fibra $M_{\binom{n}{m}}$ dello spazio fibrato $M_{n+(\binom{n}{m})}$ secondo un semicono centrale ad ℓ dimensioni sul cono di Grassmann.

Una superficie orientata ad m dimensioni si dice ammissibile rispetto al campo grassmanniano di ℓ-direzioni semiconiche m-vettoriali se in ogni suo punto il suo spazio tangente $\vec{A}_m$ corrisponde a una semiretta generatrice del semicono.

Un campo grassmanniano di ℓ-direzioni semiconiche m-vettoriali si dice olonomo se per ogni punto p dello spazio $\mathfrak{X}_n$ e semiretta generatrice g del semicono corrispondente a p , esiste una superficie ammissibile che passa per p e il cui spazio tangente $\vec{A}_m$ nel punto p corrisponde alla g .

Sia

$$X^{\alpha_1 \dots \alpha_m} = L^{\alpha_1 \dots \alpha_m}(\xi^\beta, \chi^\lambda) \tag{14.1}$$

un sistema di equazioni parametriche di un campo grassmaniano di ℓ-direzioni semiconiche m-vettoriali in $\mathfrak{X}_n$.

Se una superficie orientata ad m-dimensioni

$$\xi^\alpha = \xi^\alpha(t^a) \tag{14.2}$$

è ammissibile, il sistema di equazioni

$$\mathcal{E}^{a_1 \dots a_m} \xi^{\alpha_1}_{a_1} \dots \xi^{\alpha_m}_{a_m} = L^{\alpha_1 \dots \alpha_m}(\xi^\alpha, \chi^\lambda) \tag{14.3}$$

definisce χ^λ come funzioni di t^a

$$\chi^\lambda = \chi^\lambda(t^a) \quad . \tag{14.4}$$

Queste funzioni χ^λ sono componenti funzionali di ℓ densità scalari di peso 1 nello spazio $\vec{\mathfrak{X}}_m$ che corrisponde alla superficie orientata.

Se un campo grassmanniano di ℓ-direzioni semiconiche m-vettoriali è determinato dal sistema di equazioni implicite in funzione di coordinate soprannumerarie X^α_a

(14.5) $$\overset{z}{f}(\xi^{\alpha}, X_a^{\alpha}) = 0 \qquad z = \ell+1, \ldots, m(n-m)+1$$

allora le superficie ammissibili sono superficie integrali del sistema di equazioni differenziali

(14.6) $$\overset{z}{f}(\xi^{\alpha}, \xi_a^{\alpha}) = 0 \quad .$$

L'olonomia del campo grassmanniano di ℓ-direzioni semiconiche m-vettoriali è equivalente all'integrabilità completa di questo sistema di equazioni differenziali nel senso che ad ogni sistema di valori iniziali ξ^{α}, ξ_a^{α} soddisfacenti al sistema corrisponde una soluzione.

Un insieme γ di superficie connesse orientate, ad m dimensioni di uno spazio $\mathfrak{X}_n$ si dice completo se sono soddisfatte le due condizioni seguenti:

1) Se l'unione di due superficie di γ è ancora una superficie connessa orientata allora questa superficie appartiene a γ,

2) Se una superficie connessa orientata è una parte di una superficie di γ allora questa superficie appartiene a γ.

Una pseudometrica m-areale in uno spazio $\mathfrak{X}_n$ è una funzione $\mathfrak{a}$ definita su un insieme completo di superficie connesse orientate ad m dimensioni, il cui valore $\mathfrak{a}(S)$ è una densità scalare di peso 1 data nello spazio corrispondente $\vec{\mathfrak{X}}_n$ la quale soddisfa alla condizione seguente: se una superficie S_2 è una parte di una superficie S_1 di γ, allora il campo di densità scalare $\mathfrak{a}(S_2)$ è una restrizione del campo di densità scalare $\mathfrak{a}(S_1)$.

Superficie di $pr_1\mathfrak{a}$ si dicono superficie misurabili rispetto alla pseudometrica m-areale data.

Se $\mathcal{G}$ è una regione integrabile di una superficie misurabile S l'integrale della $\mathfrak{A}(S)$ su $\mathcal{G}$ si chiama la pseudoarea della $\mathcal{G}$. Una superficie misurabile S si dice isotropa se $\mathfrak{A}(S) = 0$ cioè la pseudoarea di ogni regione integrabile di questa superficie vale zero.

Una pseudometrica m-areale $\mathfrak{A}$ si dice una metrica m-areale se per ogni superficie misurabile S la densità scalare $\mathfrak{A}(S)$ è dappertutto positiva. In questo caso per ogni superficie misurabile S possiamo scegliere il sotto atlante coordinato K caratterizzato dalla condizione $\mathfrak{A}(S) = 1$ e allora (S, K) è uno spazio "equivoluminale" orientato $\vec{V}_m$ subordinato allo spazio corrispondente $\vec{\mathfrak{X}}_m$.

Sistemi coordinati di K si dicono sistemi coordinati "equiareali" della superficie misurabile S.

Una pseudometrica m-areale $\mathfrak{A}$ si dice una metrica m-areale impropria se per ogni superficie misurabile S la densità scalare $\mathfrak{A}(S)$ è dappertutto non negativa, ma può essere nulla.

Una pseudometrica m-areale $\mathfrak{A}$ si dice completa se $\mathfrak{A}$ è definita per tutte le superficie connesse orientate ad m dimensioni.

Una pseudometrica m-areale per m = 1 si dice una pseudometrica lineare e la pseudo area in questo caso si dice la pseudolunghezza dell'arco di curva misurabile corrispondente.

Una pseudometrica grassmanniana m-vettoriale ad ℓ dimensioni locali in uno spazio $\mathfrak{X}_n$ è una funzione numerica μ del punto dello spazio fibrato $M_{n+(\binom{n}{m})}$ di tutti gli m-vettori dello spazio $\mathfrak{X}_n$, la cui restrizione rispetto ad ogni fibra $M_{\binom{n}{m}}^{(p)}$ è una pseudometrica vettoriale μ_p nello spazio $M_{\binom{n}{m}}^{(p)}$, ove $pr_1\mu_p$ è un semicono centrale ad ℓ dimensioni sul cono di Grassmann.

Una pseudometrica grassmanniana m-vettoriale ad ℓ dimensioni locali μ si dice olonoma se $pr_1 \mu$ è un campo grassmanniano di ℓ-direzioni m-vettoriali olonomo.

Nel seguito ci limiteremo a considerare solo pseudometriche grassmaniane m-vettoriali olonome.

Se μ è una pseudometrica grassmaniana m-vettoriale ad ℓ dimensioni locali in uno spazio $\mathfrak{X}_n$, ogni superficie ad m dimensioni ammissibile rispetto al campo grassmanniano di ℓ-direzioni semiconiche m-vettoriali $pr_1\mu$ si dice una superficie misurabile della pseudometrica grassmanniana m-vettoriale μ.

E' facile vedere che l'insieme di tutte le superficie ammissibili rispetto ad un campo grassmanniano di ℓ-direzioni semiconiche m-vettoriali è sempre un insieme completo di superficie connesse orientate. Da questo segue che l'insieme di tutte le superficie misurabili è un insieme completo.
Secondo la definizione una pseudometrica grassmanniana m-vettoriale ad ℓ dimensioni locali μ è un sottoinsieme puntuale dello spazio $M_{n+(\binom{n}{m})} \boxtimes R_1$.
E' facile vedere che lo spazio $M_{n+(\binom{n}{m})} \boxtimes R_1$ può essere considerato come lo spazio fibrato di un prolungamento differenziale dello spazio $\mathfrak{X}_n$ le cui fibre sono spazi $M_{\binom{n}{m}} \boxtimes R_1$. Dunque μ può essere considerata come una superficie ℓ-secante dello spazio fibrato la quale seca ogni fibra $M_{\binom{n}{m}} \boxtimes R_1^{(p)}$ secondo un semicono centrale μ_p che è la pseudometrica vettoriale locale nello spazio $M_{\binom{n}{m}}^{(p)}$.

Sia

$$X^{\alpha_1 \dots \alpha_m} = L^{\alpha_1 \dots \alpha_m}(\xi^\alpha, \chi^\lambda) \quad , \qquad X = L(\xi^\alpha, \chi^\lambda) \tag{14.7}$$

un sistema di equazioni parametriche della μ nello spazio $M_{n+(\binom{n}{m})} \boxtimes R_1$.

Consideriamo una superficie misurabile S determinata dal sistema di equazioni

(14.8) $$\xi^{\alpha} = \xi^{\alpha}(t^a) \quad .$$

Come in (14.4) possiamo determinare χ^{λ} come funzioni delle t^a

(14.9) $$\chi^{\lambda} = \chi^{\lambda}(t^a) \quad .$$

Allora

(14.10) $$L(\xi^{\alpha}(t^a), \chi^{\lambda}(t^a))$$

sarà una densità scalare di peso +1 definita nello spazio $\vec{\mathfrak{X}}_m$ corrispondente alla superficie S .

Dunque vediamo che ogni pseudometrica grassmanniana m-vettoriale in uno spazio $\mathfrak{X}_n$ ne definisce una pseudometrica m-areale.

La pseudoarea σ di una regione φ di una superficie misurabile (14.8) può essere espressa come l'integrale m-uplo :

(14.11) $$\sigma = \int \cdots \int_{\varkappa(\varphi)} L(\xi^{\alpha}(t^a), \chi^{\lambda}(t^a))\, dt^1 \ldots \ldots dt^a$$

se φ è rappresentata nel sistema coordinato $\varkappa$, cioè $\varphi \subset pr_1 \varkappa$.

Una pseudometrica grassmanniana m-vettoriale in uno spazio $\mathfrak{X}_n$ per m=1 si dice una <u>pseudometrica vettoriale</u> in $\mathfrak{X}_n$. Essa definisce una pseudometrica lineare nello $\mathfrak{X}_n$.

Una pseudometrica grassmanniana m-vettoriale μ in uno $\mathfrak{X}_n$ si dice una metrica grassmanniana m-vettoriale se μ prende solo valori positivi. E' evidente che una metrica grassmanniana m-vettoriale definisce una metrica m-areale.

Analogamente definiamo la nozione di metrica grassmanniana m-vettoriale impropria la quale definisce una metrica m-areale impropria.

Una metrica grassmanniana m-vettoriale ad ℓ dimensioni locali può essere determinata dalla superficie ℓ-secante $\mathfrak{I}$ nello spazio fibrato $M_{n+\left(\binom{n}{m}\right)}$ la quale taglia ogni fibra $M_{\binom{n}{m}}^{(p)}$ secondo la superficie ad $\ell-1$ dimensioni sul cono di Grassmann che è l'indicatrice della metrica vettoriale fibrale μ_p .

Questa superficie $\mathfrak{I}$ si dice l'indicatrice globale della metrica grassmanniana m-vettoriale .

L'indicatrice globale $\mathfrak{I}$ può essere determinata o dall'equazione

$$L(\xi^\alpha, \chi^\lambda) = 1 \tag{14.12}$$

o dal sistema di equazioni parametriche

$$X^{\alpha_1 \dots \alpha_m} = f^{\alpha_1 \dots \alpha_m}(\xi^\alpha, \eta^i) \quad . \tag{14.13}$$

Una pseudometrica grassmanniana m-vettoriale μ in $\mathfrak{X}_n$ si dice lineare se tutte le pseudometriche fibrali μ_p sono lineari cioè se la funzione L è della forma seguente

$$L(\xi^\alpha, \chi^\lambda) = W_{\{\alpha_1 \dots \alpha_m\}}(\xi^\alpha) \, L^{\{\alpha_1 \dots \alpha_m\}}(\xi^\alpha, \chi^\lambda) \tag{14.14}$$

ove $W_{\alpha_1 \dots \alpha_m}$ è un m-covettore nello spazio $\mathfrak{X}_n$.

Una pseudometrica grassmanniana vettoriale lineare si dice un <u>gradiente</u> se l' m-covettore $W_{\alpha_1 \ldots \alpha_m}$ è una derivata alternata di un (m-1)-covettore

(14.15) $$W_{\alpha_1 \ldots \alpha_m} = \partial_{[\alpha_1} \varphi_{\alpha_2 \ldots \alpha_m]} \quad .$$

In virtù del teorema di Stokes la pseudoarea σ di una regione integrabile $\mathcal{G}$ di una superficie misurabile rispetto ad un gradiente è uguale all'integrale della forma differenziale esterna

(14.16) $$\varphi = \varphi_{\alpha_1 \ldots \alpha_m} \left[d\xi^{\alpha_1} \ldots d\xi^{\alpha_m} \right]$$

moltiplicata per $\frac{1}{m}$ esteso alla frontiera $\mathfrak{f}$ della regione $\mathcal{G}$

(14.17) $$\sigma = \frac{1}{m} \int_{\mathfrak{f}} \varphi \quad .$$

Consideriamo l'insieme M di tutte le pseudometriche grassmanniane m-vettoriali ad ℓ dimensioni locali definite sullo stesso campo di ℓ-direzioni semiconiche m-vettoriali in uno spazio $\mathcal{X}_n$

(14.18) $$X^{\alpha_1 \ldots \alpha_m} = L^{\alpha_1 \ldots \alpha_m}(\xi^\alpha, \chi^\lambda)$$

E' facile vedere che M può essere considerato come uno spazio vettoriale.

Una trasformazione nello spazio vettoriale M

(14.19) $$\bar{\mu} = \varepsilon (\mu - \lambda) \qquad (\varepsilon = + 1) \quad ,$$

ove λ è una pseudometrica grassmanniana m-vettoriale lineare, fissata nello spazio M, si dice una trasformazione di Caratheodory. Se $\varepsilon = +1$ la trasformazione di Caratheodory si dice propria e se $\varepsilon = -1$ impropria.

Notando che le funzioni L e $\bar{L}$ corrispondenti alle pseudometriche grassmanniane m-vettoriali μ e $\bar{\mu}$ stanno nella relazione

$$(14.20) \qquad \bar{L} = \varepsilon\,(L - W_{\{\alpha_1 \dots \alpha_m\}} L^{\{\alpha_1 \dots \alpha_m\}}) \quad ,$$

per due superficie misurabili colla frontiera comune S_1, S_2 si ha :

$$(14.21) \qquad \bar{\sigma}_2 - \bar{\sigma}_1 = \varepsilon\,(\sigma_2 - \sigma_1) \quad ,$$

ove σ_1, σ_2 sono le pseudoaree di S_1 e S_2 rispetto alla μ e $\bar{\sigma}_1$, $\bar{\sigma}_2$ sono le pseudoaree di S_1 e S_2 rispetto alla $\bar{\mu}$.

Dalla relazione (14.21) si deduce che superficie misurabili di pseudoarea estrema conservano questa proprietà se assoggettate ad una trasformazione di Caratheodory arbitraria di pseudometrica grassmanniana m-vettoriale. Se la trasformazione di Caratheodory è propria si conserva anche la specie dell'estremo (minimo o massimo). Se la trasformazione di Caratheodory è impropria la specie dell'estremo cambia.

E' evidente che una superficie isotropa rispetto ad una metrica grassmanniana m-vettoriale impropria è sempre una superficie di area minima. Dunque per trovare superficie di pseudoarea minima (massima) rispetto ad una pseudometrica grassmanniana m-vettoriale data μ possiamo cercare le trasformazioni di Caratheodory proprie (improprie) le quali trasformano μ in una metrica grassmanniana m-vettoriale impropria e poi trovare le superficie isotrope rispetto a quest'ultima.

Una trasformazione di Caratheodory di una pseudometrica grassmanniana m-vettoriale in uno spazio $\mathfrak{X}_n$ determina una trasformazione di Caratheodory (nel senso del § 11) della pseudometrica vettoriale fibrale μ_p in ogni spazio $M_{\binom{n}{m}}$. Se tutte queste trasformazioni di Caratheodory delle pseudometriche fibrali sono speciali la trasformazione di Caratheodory della pseudometrica grassmanniana m-vettoriale data si dice <u>speciale</u>. Ogni trasformazione di Caratheodory speciale di una pseudometrica grassmanniana m-vettoriale la trasforma in una metrica grassmanniana m-vettoriale impropria.

Più tardi considereremo la teoria geometrica delle trasformazioni di Caratheodory speciali di metriche grassmanniane m-vettoriali.

IV - LA CONDIZIONE DI EULERO PER LE CURVE DI PSEUDOLUNGHEZZA ESTREMALE IN UNO SPAZIO $\mathfrak{X}_n$ DOTATO DI UNA PSEUDOMETRICA VETTORIALE.

§ 15. Campo di ℓ-direzioni semiconiche.

Diremo che in uno spazio $\mathfrak{X}_n$ è dato un campo di ℓ-direzioni semiconiche se in ogni spazio tangente associato ad un punto dello spazio $\mathfrak{X}_n$ è dato un semicono centrale ad ℓ dimensioni.[12)]

Una curva orientata nello spazio $\mathfrak{X}_n$ si dice ammissibile rispetto al campo di ℓ-direzioni semiconiche se in ogni suo punto la sua semitangente è una generatrice del semicono corrispondente.

Dal punto di vista della teoria degli spazi fibrati un campo di ℓ-direzioni semiconiche in uno spazio $\mathfrak{X}_n$ si può considerare come una superficie ℓ-secante S nello spazio fibrato $T_{n+(n)}$ del prolungamento tangente del primo ordine dello spazio $\mathfrak{X}_n$, la quale seca ogni fibra di $T_{n+(n)}$ secondo un semicono centrale ad ℓ dimensioni.

Siano

$$X^\alpha = L^\alpha(\xi^\beta, \mathfrak{X}^\lambda) \qquad \begin{matrix}(\alpha, \beta = 1, \dots n)\\ (\lambda = 1, \dots, \ell)\end{matrix} \tag{15.1}$$

le equazioni si S.

Consideriamo una curva ammissibile C

---

12) Secondo il paragrafo precedente un campo di ℓ-direzioni semiconiche può essere considerato come una caso particolare di un campo grassmanniano di ℓ-direzioni semiconiche m-vettoriali per m=1.

$$\xi^{\alpha} = \xi^{\alpha}(t) \tag{15.2}$$

e supponiamo che essa possa essere immersa in una famiglia ad un parametro a di curve ammissibili

$$\xi^{\alpha} = \xi^{\alpha}(t, a) \tag{15.3}$$

ove essa corrisponde al valore $a = 0$ del parametro.

Allora le equazioni

$$X^{\alpha} = \xi^{\alpha}_{a}(t, 0) \tag{15.4}$$

insieme con le equazioni (15, 2) determineranno una curva nello spazio fibrato $T_{n+(n)}$ la quale ricopre la curva C . Questa curva ricoprente si dice la curva variazionale della curva C corrispondente alla famiglia (15. 3) di curve ammissibili.

Poichè la famiglia (15. 3) sicompone di curve ammissibili le equazioni

$$\xi^{\alpha}_{t}(t, a) = L^{\alpha}(\xi^{\beta}(t, a), \chi^{\lambda}) \tag{15.5}$$

danno χ^{λ} come funzioni continue e differenziabili a tratti dei parametri t e a [13)]

$$\chi^{\lambda} = \chi^{\lambda}(t, a) \quad . \tag{15.6}$$

---

13) Consideriamo curve orientate con un numero finito di punti angolosi.

Posto :

$$Z^{\lambda} = \chi^{\lambda}_{a}(t, 0) \tag{15.7}$$

e derivando (15.4) rispetto al parametro a , e ponendo dopo la derivazione $a = 0$ si ottiene il sistema di equazioni differenziali

$$\dot{X}^{\alpha} = \partial_{\beta} L^{\alpha} X^{\beta} + L^{\alpha}_{\lambda} Z^{\lambda} \tag{15.8}$$

il quale è soddisfatto dalla curva variazionale.

Si può dimostrare ([3]) che il sistema di equazioni differenziali (15.8), ove Z^{λ} sono funzioni continue a tratti qualunque, esprime la condizione necessaria e sufficiente affinchè una curva nello spazio fibrato $T_{n+(n)}$, che ricopra una curva ammissibile C data nello spazio $\mathfrak{X}_n$, sia una curva variazionale per questa curva C ;cioè affinchè essa possa essere immersa in una famiglia di curve ammissibili (15.3) in modo tale che la curva ricoprente sia determinata dal sistema di equazioni (15.2), (15.4).

Una generalizzazione importante di questa proposizione è la seguente ([3]); Se sono date r curve integrali del sistema di equazioni differenziali (15.7)

$$X^{\alpha} = X^{\alpha}_{u}(t) \qquad (u = 1, \ldots, r) \tag{15.8}$$

si può trovare una famiglia ad r parametri a^{u} di curve ammissibile la quale contiene la curva data C per $a^{u} = 0$, tale che sia

$$X^{\alpha}_{u}(t) = \xi^{\alpha}_{a^{u}}(t, 0) \quad . \tag{15.10}$$

Possiamo dare al sistema di equazioni differenziali (15.8) un'altra forma eliminando le funzioni arbitrarie Z^{λ}. Per questo considereremo il tensore di connessione unitario C^{p}_{α} dell'immagine duale del piano tangente ℓ-dimensionale al semicono fibrale.

Notando che

(15.11) $$C^{p}_{\alpha} L^{\alpha}_{\lambda} = 0$$

e ponendo

(15.12) $$\Lambda^{p}_{\alpha} = -C^{p}_{\beta} \partial_{\alpha} L^{\beta}$$

si ha che il sistema (15.8) equivale al sistema

(15.13) $$C^{p}_{\alpha} \dot{X}^{\alpha} + \Lambda^{p}_{\alpha} X^{\alpha} = 0$$

il quale può essere rappresentato come il sistema di Pfaff

(15.14) $$C^{p}_{\alpha} dX^{\alpha} + \Lambda^{p}_{\alpha} X^{\alpha} dt = 0$$

Dunque ad ogni curva nello spazio $\mathfrak{X}_n$ ammissibile rispetto al campo di ℓ-direzioni semiconiche corrisponde il sistema di equazioni differenziali (15.13) oppure il sistema di Pfaff equivalente (15.14), il quale determina le curve variazionali per la curva C.

Indicheremo con ρ_C la relazione binaria fra punti delle fibre $A_n(p_1)$ e $A_n(p_2)$ associate ai punti terminali (p_1, p_2) di una curva ammissibile C la quale è l'insieme di tutte le coppie ordinate di punti terminali delle curve variazionali per la curva C.

Questa relazione binaria ρ_c si dice la relazione variazionale per la curva C.

Una curva ammissibile C nello spazio $\mathfrak{X}_n$ si dice normale se la sua relazione variazionale ρ_c coincide con la relazione binaria universale cioè se :

$$\rho_c = \theta^{-1} \langle p_1 \rangle \times \theta^{-1} \langle p_2 \rangle \qquad (15.15)$$

Ciò significa che due punti qualunque degli spazi $A_n(p_1)$ e $A_n(p_2)$ possono essere congiunti con una curva variazionale. In caso contrario la curva si dice anormale.

Per lo studio delle curve anormali, le quali sono interessanti dal punto di vista delle loro applicazioni al calcolo delle variazioni, dobbiamo cominciare con lo studio delle relazioni variazionali.

Dapprima dimostreremo che la relaziona variazionale ρ_c per ogni curva anormale è una relazione binaria lineare, cioè è un piano centrale dello spazio-prodotto $A_n(p_1) \times A_n(p_2)$.

Notando che le equazioni (15.14) sono lineari rispetto a X^α, dX^α abbiamo che se le equazioni $X^\alpha = X_1^\alpha(t)$ e $X^\alpha = X_2^\alpha(t)$ determinano due curve variazionali per una curva ammissibile data dalle equazioni $\xi^\alpha = \xi^\alpha(t)$ allora le equazioni $X^\alpha = C_1 X_1^\alpha(t) + C_2 X_2^\alpha(t)$, ove C_1, C_2 sono costanti qualunque, determinano pure una curva variazionale per la curva C. Ne consegue immediatamente che se due coppie di punti $(\overline{p}_{11}, \overline{p}_{12})$ e $(\overline{p}_{21}, \overline{p}_{22})$ con coordinate $X_{11}^\alpha, X_{12}^\alpha$ e $X_{21}^\alpha, X_{22}^\alpha$ appartengono a ρ_c allora la coppia di punti $(\overline{p}_1, \overline{p}_2)$ di coordinate $C_1X_{11}^\alpha + C_2X_{21}^\alpha$ e $C_1X_{12}^\alpha + C_2X_{22}^\alpha$ appartengono pure a ρ_c.

Si vede poi che la relazione variazionale ρ_c è completa cioè

(15.16) $$pr_1 \rho_c = \overset{-1}{\theta} \langle p_1 \rangle \quad , \quad pr_2 \rho_c = \overset{-1}{\theta} \langle p_2 \rangle \quad ,$$

perchè per ogni punto dello spazio $A_n \langle p_1 \rangle$ passa una curva variazionale e per ogni punto dello spazio $A_n \langle p_2 \rangle$ passa una curva variazionale.

Per dimostrare che ogni relaziona binaria lineare completa fra punti di due spazi centro-affini è una quasi-rappresentazione la cui rappresentazione fattoriale è un isomorfismo degli spazi fattoriali corrispondenti dobbiamo ricorrere alla teoria generale delle relazioni binarie fra elementi di due gruppi.

Dimostriamo che ogni relazione binaria $\rho \subset G \times H$ fra elementi di due gruppi G e H, che sia un sottogruppo del prodotto diretto $G \times H$ di questi gruppi, è una quasi-rappresentazione parziale.

Per elementi g, h qualunque si ha :

$$(g, h) \in \rho \circ \overset{-1}{\rho} \circ \rho \longleftrightarrow \bigvee_{(g_1, h_1)} (g, h_1), (g_1, h_1), (g_1, h), (g_1, h) \in \rho \; .$$

Dal fatto che ρ è un sottogruppo del prodotto diretto $G \times H$ consegue

$$(g, h_1), (g_1, h_1), (g_1, h) \in \rho \rightarrow (g\, g_1^{-1} g_1, \; h_1 h_1^{-1} h) \in \rho$$

oppure

$$(g, h_1), (g_1, h_1), (g_1, h) \in \rho \longrightarrow (g, h) \in \rho$$

donde

$$(g, h) \in \rho \circ \overset{-1}{\rho} \circ \rho \longrightarrow (g, h) \in \rho$$

ossia

$$\rho \circ \overset{-1}{\rho} \circ \rho \subset \rho \quad ,$$

ciò significa (cfr. § 3) che ρ è una quasirappresentazione parziale.

Si può poi dimostrare che le proiezioni $pr_1\rho$ e $pr_2\rho$ sono sottogruppi dei gruppi G e H rispettivamente e che le sezioni $\overset{-1}{\rho}\langle e_2\rangle$ e $\rho\langle e_1\rangle$ delle relazioni binarie $\overset{-1}{\rho}$ e ρ secondo gli elementi unitari e_1 e e_2 sono sottogruppi invarianti dei gruppi $pr_1\rho$ e $pr_2\rho$ rispettivamente. Con questo $\overset{-1}{\rho}\circ\rho$ e $\rho\circ\overset{-1}{\rho}$ sono relazioni di equivalenza fra elementi del gruppo $pr_1\rho$ e del gruppo $pr_2\rho$ determinati dai sottogruppi invarianti $\overset{-1}{\rho}\langle e_2\rangle, \rho\langle e_1\rangle$ e la rappresentazione fattoriale corrispondente alla quasirappresentazione ρ è un isomorfismo del gruppo fattoriale $\dfrac{pr_1\rho}{\overset{-1}{\rho}\langle e_2\rangle}$ sul gruppo fattoriale $\dfrac{pr_2\rho}{\rho\langle e_1\rangle}$.

Sia ρ una relazione binaria lineare fra gli elementi di due spazi vettoriali $\mathcal{X}$ e $\mathcal{Y}$, cioè un sottospazio dello spazio prodotto vettoriale $\mathcal{X}\times\mathcal{Y}$. Secondo la proposizione precedente abbiamo che ρ è una quasi-rappresentazione parziale. Si dimostra poi facilmente che $pr_1\rho$ e $pr_2\rho$ sono sottospazi degli spazi vettoriali $\mathcal{X}$ e $\mathcal{Y}$, così come le sezioni $\overset{-1}{\rho}\langle 0_2\rangle$ e $\rho\langle 0_1\rangle$ ove 0_1, 0_2 sono i vettori nulli. La rappresentazione fattoriale della quasi-rappresentazione ρ sarà un isomorfismo degli spazi vettoriali fattoriali $\dfrac{pr_1\rho}{\overset{-1}{\rho}\langle 0_2\rangle}$ e $\dfrac{pr_2\rho}{\rho\langle 0_1\rangle}$.

Si può anche dimostrare che la dimensione degli spazi vettoriali fattoriali $\dfrac{pr_1\rho}{\overset{-1}{\rho}\langle 0_2\rangle}$ e $\dfrac{pr_2\rho}{\rho\langle 0_1\rangle}$ è uguale alla codimensione del sottospazio ρ dello spazio vettoriale $pr_1\rho\times pr_2\rho$, la quale si dice la codimensione intrinseca di ρ e sarà indicata con $\operatorname{codim}_i\rho$

(15.17) $$\operatorname{codim}_i \rho = \dim \mathrm{pr}_1 \rho + \dim \mathrm{pr}_2 \rho - \dim \rho .$$

E' evidente che la relazione binaria inversa ad una relazione binaria lineare anche è lineare.

Si può poi dimostrare che il prodotto $\sigma \circ \rho$ di due relazioni binarie lineari $\rho \subset \mathcal{X} \times \mathcal{Y}$ e $\sigma \subset \mathcal{Y} \times \mathcal{Z}$ ove $\mathcal{X}$, $\mathcal{Y}$, $\mathcal{Z}$ sono tre spazi vettoriale, è pure una relazione binaria lineare.

Applicando tutti questi risultati ad una relazione variazionale ρ_c otteniamo che essa è una quasi rappresentazione dello spazio $A_n(p_1)$ sullo spazio $A_n(p_2)$ e che la rappresentazione fattoriale corrispondente è un isomorfismo dello spazio fattoriale $\dfrac{A_n(p_1)}{\overset{-1}{\rho}_c \langle 0_2 \rangle}$ sullo spazio fattoriale $\dfrac{A_n(p_2)}{\rho_c \langle 0_1 \rangle}$. Notiamo che, in particolare, ne consegue che

$$\dim \overset{-1}{\rho}_c \langle 0_2 \rangle = \dim \rho_c \langle 0_1 \rangle .$$

Per la relazione variazionale ρ_c come per una relazione binaria lineare completa la codimensione intrinseca coincide con la codimensione ordinaria $\operatorname{codim} \rho_c$ la quale si dice il <u>grado di anormalità</u> della curva ammissibile C. Dunque il grado di anormalità di una curva normale è uguale a zero.

Sia p un punto interno di una curva ammissibile C con i punti terminali (p_1, p_2). Indichiamo con C_1 e C_2 i suoi archi $\overset{\frown}{p_1 p}$ e $\overset{\frown}{p p_2}$. Si può dimostrare che il prodotto $\rho_{c_2} \circ \rho_{c_1}$ coincide con ρ_c

(15.18) $$\rho_{c_2} \circ \rho_{c_1} = \rho_c .$$

E' comodo definire la relazione variazionale anche per una curva degenerata in un punto p_o . Definiamo la relazione variazionale in questo caso come uguale alla relazione binaria identica $\Delta_{\pi_1(A_n(p_o))}$. Allora la formula (15.18) vale anche nel caso in cui il punto p coincide con uno dei punti terminali.

Consideriamo la relazione binaria

$$\nu_{(c,p)} = \overset{-1}{\rho}_{c_2} \circ \rho_c \tag{15.19}$$

ove p è un punto arbitrario della curva C . $\nu_{(c,p)}$ è una relazione binaria lineare come prodotto delle relazioni binarie lineari ρ_c e $\overset{-1}{\rho}_{c_2}$. E' facile dimostrare che $\overset{-1}{\nu}_{(c,p)}\langle 0\rangle = \overset{-1}{\rho}_c\langle 0_2\rangle$ e perciò $\nu_{(c,p)}$ determina un isomorfismo dello spazio fattoriale $\dfrac{A_n(p_1)}{\overset{-1}{\rho}_c\langle 0_2\rangle}$ sullo spazio fattoriale $\dfrac{A_n(p)}{\nu_{(c,p)}\langle 0_1\rangle}$. Poi si può dimostrare che l'insieme di tutte le curve variazionali per la curva C uscenti dai punti di un piano $\mathcal{S}$ nello spazio $A_n(p_1)$, il quale è un punto dello spazio fattoriale $\dfrac{A_n(p_1)}{\overset{-1}{\rho}_c\langle 0_2\rangle}$, taglia lo spazio $A_n(p)$ secondo il piano $\nu_{(c,p)}(\mathcal{S})$.

Scegliamo in tutti gli spazi fattoriali $\dfrac{A_n(p)}{\nu_{(c,p)}\langle 0_1\rangle}$ ove p è un punto arbitrario della curva C , i sistemi coordinati in tal modo che essi siano corrispondenti ad un sistema coordinato fisso nello spazio fattoriale $\dfrac{A_n(p_1)}{\overset{-1}{\rho}_c\langle 0_2\rangle}$ rispetto agli isomorfismi determinati dalle quasirappresentazioni $\nu_{(c,p)}$.

Supponiamo che le equazioni

$$X^p = Y^p_\alpha X^\alpha \qquad (p = 1, \ldots\ldots, s = \text{codim } \rho_c) \tag{15.20}$$

determinino la rappresentazione canonica dello spazio $A_n(p)$ sullo spazio fattoriale $\frac{A_n(p)}{\nu_{(c,p)}{<}0_1{>}}$ ove i coefficienti Y^p_α sono considerati come funzioni del parametro t il quale determina la posizione del punto p sulla curva C.

Allora le equazioni

$$Y^p_\alpha(t)\, X^\alpha = C^p \tag{15.21}$$

determinano una famiglia di insiemi puntuali che sono i luoghi di tutte le curve variazionali che congiungono punti di due piani negli spazi $A_n(p_1)$ e $A_n(p_2)$ i quali sono punti degli spazi fattoriali $\frac{A_n(p_1)}{\overset{-1}{\rho}_c{<}0_2{>}}$ e $\frac{A_n(p_2)}{\rho_c{<}0_1{>}}$ corrispondenti sotto l'isomorfismo determinato dalla relazione variazionale ρ_c. Ogni tale insieme puntuale taglia lo spazio $A_n(p)$ ove p è un punto arbitrario della curva C, secondo un piano ad $n-s$ dimensioni.

Per dimostrare che questi insiemi puntuali sono superficie ad $n-s+1$ dimensioni con spigoli, dobbiamo dimostrare che le funzioni Y^p_α sono differenziabili a tratti.

Consideriamo il sistema di equazioni differenziali

$$\dot{\lambda}_\alpha = -\partial_\alpha L^\beta \lambda_\beta \tag{15.22}$$

rispetto a funzioni λ_α del parametro t lungo la curva C .

Sia λ_α una soluzione delle equazioni (15.22) .

Derivando la somma $\lambda_\alpha X^\alpha$ ove $X^\alpha = X^\alpha(t)$ sono equazioni d'una curva variazionale ricoprente la curva C e usando le equazioni (15.7) e (15.22) otterremo

$$\frac{d}{dt}(\lambda_\alpha X^\alpha) = \lambda_\alpha L^\alpha_\lambda Z^\lambda \tag{15.23}$$

donde si trae integrando

$$\lambda_\alpha X^\alpha \Big|_{t_1}^{t_2} = \int_{t_1}^{t_2} \lambda_\alpha L^\alpha_\lambda Z^\lambda \, dt \quad . \tag{15.24}$$

Indichiamo con λ^p_α (p = 1, ... , s) un sistema di s soluzioni delle equazioni differenziali (15.22) soddisfacenti alle condizioni iniziali

$$\lambda^p_\alpha(t_1) = Y^p_\alpha(t_1) \quad . \tag{15.25}$$

Siano $X^\alpha = X^\alpha(t)$ le equazioni di una curva variazionale che passa per il centro dello spazio $A_n(p_2)$, allora il punto iniziale di questa curva si trova nel piano centrale $\overset{-1}{\rho}_c \langle 0_2 \rangle$ e per conseguenza

$$\lambda^p_\alpha(t_1) X^\alpha(t_1) = 0 \quad , \quad X^\alpha_2(t_2) = 0 \tag{15.26}$$

donde, secondo (15.24) , abbiamo per le funzioni arbitrarie Z^λ

$$\int_{t_1}^{t_2} \lambda_\alpha L^\alpha_\lambda Z^\lambda \, dt = 0 \quad . \tag{15.27}$$

In virtù dell'arbitrarietà delle funzioni Z^λ ne consegue

(15.28) $$\lambda_\alpha L^\alpha_\lambda = 0 \quad .$$

Secondo le condizioni iniziali (15.25) il rango della matrice $\| \lambda^p_\alpha \|$ è uguale a s . Allora dalle equazioni (15.28) risulta che il grado d'anormalità non può essere più grande $n-\ell$:

(15.29) $$\operatorname{codim} \rho_c < n-\ell \quad .$$

Ritornando alle equazioni (15.23) e utilizzando (15.28) otterremo che lungo una curva variazionale qualunque è

(15.30) $$\frac{d}{dt} (\lambda^p_\alpha X^\alpha) = 0$$

donde

(15.31) $$\lambda^p_\alpha X^\alpha = C^p \quad .$$

Vediamo dalle equazioni (15.21) e (15.31) che valgono per tutte le curve variazionali e delle condizioni iniziali (15.25) , che i due sistemi di funzioni Y^p_α e λ^p_α coincidono

(15.32) $$Y^p_\alpha = \lambda^p_\alpha \quad ;$$

dunque le funzioni Y^p_α sono continue lungo la curva e differenziabili dappertutto salvo nei punti angolosi della curva C .

Ne segue che le curve anormali sono caratterizzate dalla condizione che lungo ogni tale curva C esiste un campo di covettori Y_α tale che le sue componenti sono soluzioni del sistema di equazioni differenziali misto

$$\dot{Y}_\alpha = -\partial_\alpha L^\beta Y_\beta \quad , \quad L^\alpha_\lambda Y_\alpha = 0 \quad , \tag{15.33}$$

con questo il numero di tali covettori indipendenti è uguale al grado d'anormalità della curva .

Si deduce dalle equazioni in termini finiti che il covettore Y_α determina un iperpiano pseudotangente del semicono corrispondente il quale è parallelo all'iperpiano tangente lungo la semiretta generatrice del semicono che coincide con la semitangente alla curva C .

In un punto angoloso della curva C vi sono due semitangenti e per ciò l'iperpiano pseudotangente Y_α è parallelo ad un iperpiano tangente al semicono lungo due semirette generatrici.

Il problema di trovare tutte le curve anormali si riduce all'integrazione del sistema di equazioni differenziali misto

$$\dot{\xi}^\alpha = L^\alpha \quad , \quad \dot{Y}_\alpha = -\partial_\alpha L^\beta Y_\beta \quad , \quad L^\beta_\lambda Y_\beta = 0 \quad . \tag{15.34}$$

Supponiamo ora che il campo di ℓ-direzioni semiconiche sia determinato dal sistema di equazioni implicite

$$\overset{z}{f}(\xi^\alpha, X^\alpha) = 0 \qquad (z = \ell+1, \ldots, n) \tag{15.35}$$

dove $\overset{z}{f}$ sono funzioni positivamente omogenee delle variabili X^α .

Sostituendo in queste equazioni ad X^α , le sue espressioni dalle equazioni (15.1) e derivando le identità ottenute avremo

(15.36) $$\partial_\alpha \overset{z}{f} + \overset{z}{f}_\beta \partial_\alpha L^\beta = 0$$

E' evidente che ogni covettore che definisce un iperpiano pseudotangente del semicono (15.35) può essere rappresentato nella forma seguente

(15.37) $$Y_\alpha = \lambda_z \overset{z}{f}_\alpha \qquad ,$$

ove λ_z sono certi coefficienti.

Da (15.36) e (15.37) si deduce

(15.38) $$\partial_\alpha L^\beta \, Y_\beta = - \lambda_z \partial_\alpha \overset{z}{f} \qquad .$$

Secondo (15.37) e (15.38) otterremo dal sistema (15.34) il nuovo sistema di equazioni differenziali per curve anormali

(15.39) $$\overset{z}{f}(\xi^\beta, \dot{\xi}^\beta) = 0, \quad \frac{d}{dt}(\lambda_z \overset{z}{f}_\alpha(\xi^\beta, \dot{\xi}^\beta)) - \lambda_z \partial_\alpha \overset{z}{f}(\xi^\beta, \dot{\xi}^\beta) = 0.$$

Una curva ammissibile C del campo di ℓ-direzioni semiconiche si dice libera se esistono intorni $\sigma(p_1)$ e $\sigma(p_2)$ dei suoi punti terminali p_1, p_2 tali che ogni punto di $\sigma(p_1)$ può essere congiunto con ogni punto di $\sigma(p_2)$ per mezzo di una curva ammissibile. Nel caso contrario la curva C si dice vincolata .

Teorema. Ogni curva ammissibile normale di un campo di ℓ-direzioni semiconiche in uno spazio $\mathfrak{X}_n$ è libera.

Dimostrazione.

Sia C una curva ammissibile normale. Poichè la relazione variazionale di una curva normale è la relazione binaria universale, esistono 2n curve variazionali C_u $(u = 1, \ldots\ldots, 2n)$ tali che le coppie dei loro punti terminali formano un sistema di 2n punti linearmente indipendenti nello spazio $A_n(p_1) \boxtimes A_n(p_2)$.

Siano

$$\xi^\alpha = \xi^\alpha(t, a^u) \qquad (t_1 \leqslant t \leqslant t_2) \tag{15.40}$$

le equazioni di una famiglia a 2n parametri di curve ammissibili alle quali corrispondono queste curve variazionali. In virtù della nostra ipotesi che le coppie di punti terminali delle curve variazionali scelte formino un sistema di punti linearmente indipendenti nello spazio $A_n(p_1) \boxtimes A_n(p_2)$ abbiamo

$$\mathrm{Det} \left| \xi^\alpha_u(t_1, 0) \quad , \quad \xi^\alpha_u(t_2, 0) \right| \neq 0 \, . \tag{15.41}$$

Ne consegue che il sistema di equazioni

$$\xi^\alpha(t_1, a^u) = \xi^\alpha_1 \quad , \; \xi^\alpha(t_2, a^u) = \xi^\alpha_2 \tag{15.42}$$

può essere risoluto rispetto alle variabili a^u in un intorno dei valori iniziali. Questo significa che possiamo scegliere intorni $\mathcal{O}(p_1)$ e $\mathcal{O}(p_2)$ tali che le equazioni (15.42) possono essere risolute nell'intorno $\mathcal{O}(p_1) \times \mathcal{O}(p_2)$ del punto (p_1, p_2) dello spazio $A_n(p_1) \boxtimes A_n(p_2)$. Dunque ogni coppia di punti $(\bar{p}_1, \bar{p}_2)$ ove $\bar{p}_1 \in \mathcal{O}(p_1), \bar{p}_2 \in \mathcal{O}(p_2)$ può essere congiunta da una curva ammissibile della famiglia (15.40) .

Come una conseguenza importantissima del teorema dimostrato abbiamo: Affinchè una curva ammissibile di un campo di ℓ-direzioni semiconiche in $\mathfrak{X}_n$ sia vincolata è necessario che essa sia anormale.

§ 16. Estremali in uno spazio $\mathfrak{X}_n$ dotato di una pseudometrica vettoriale.

Supponiamo che in uno spazio $\mathfrak{X}_n$ sia data una pseudometrica vettoriale μ ad ℓ dimensioni locali. Considerando lo spazio $T_{n+(n)} \boxtimes R_1$ come lo spazio fibrato del prolungamento differenziale del primo ordine dello spazio $\mathfrak{X}_n$, otterremo che μ sarà una superficie ℓ-secante nello spazio fibrato $T_{n+(n)} \boxtimes R_1$ la quale taglia ogni fibra $A_n \boxtimes R_1$ di $T_{n+(n)} \boxtimes R_1$ secondo un semicono centrale ad ℓ dimensioni, che rappresenta la metrica μ_p fibrale nello spazio tangente associato al punto p.

Sia

$$X^\alpha = L^\alpha(\xi^\beta, \chi^\lambda) \quad , \quad X = L(\xi^\beta, \chi^\lambda) \tag{16.1}$$

un sistema di equazioni parametriche di μ nello spazio $T_{n+(n)} \boxtimes R_1$.

Le prime n equazioni di questo sistema sono le equazioni della superficie ℓ-secante in $T_{n+(n)}$ che rappresenta il campo $pr_1\mu$ di ℓ-direzioni semiconiche misurabili.

Una curva misurabile è una curva ammissibile rispetto al campo $pr_1\mu$ di ℓ-direzioni semiconiche.

Se

$$\xi^\alpha = \xi^\alpha(t) \qquad (t_1 \leq t \leq t_2) \tag{16.2}$$

sono le equazioni di una curva misurabile C allora possiamo esprimere le coordinate curvilinee χ^λ del vettore tangente come funzioni del parametro t

$$\chi^\lambda = \chi^\lambda(t) \qquad (t_1 \leq t \leq t_2) \tag{16.3}$$

e determinare la pseudolunghezza $\mathcal{S}_c$ della curva C

$$(16.4) \qquad \mathcal{S}_c = \int_{t_1}^{t_2} L(\xi^\alpha(t), \mathcal{X}^\lambda(t))\, dt \quad .$$

Consideriamo lo spazio $\mathfrak{X}_n \boxtimes R_1$ come lo spazio fibrato del prolungamento lineare dello spazio $\mathfrak{X}_n$ le cui fibre sono i prodotti $\{p\} \times R$, ove p è un punto dello spazio base $\mathfrak{X}_n$.

Gli spazi tangenti dello spazio $\mathfrak{X}_n \boxtimes R_1$ possono essere considerati come i prodotti $A_n \boxtimes R_1$ di spazi tangenti degli spazi $\mathfrak{X}_n$ e R_1 .

Lo spazio fibrato $T_{n+(n)} \times R_1$ si può considerare come immerso nello spazio fibrato del prolungamento tangenziale di primo ordine dello spazio $\mathfrak{X}_n \boxtimes R_1$ e possiamo identificare le fibre $A_n \boxtimes R_1$ dello spazio $T_{n+(n)} \times R_1$ con gli spazi tangenti dello spazio $\mathfrak{X}_n \boxtimes R_1$ associati con i punti del tipo (p, 0) .

Dunque la pseudometrica μ definisce un campo di ℓ-direzioni semiconiche nello spazio $\mathfrak{X}_n \boxtimes R_1$ il quale era definito solamente sulla superficie secante $\Pi_1(\mathfrak{X}_n) \times \{0\}$. E' facile vedere che abbiamo un isomorfismo naturale fra gli spazi tangenti dello spazio $\mathfrak{X}_n \boxtimes R_1$ associati coi punti (p, 0) e (p, r) ove r è un numero reale arbitrario. Con l'aiuto di questi isomorfismi possiamo estendere su tutto lo spazio $\mathfrak{X}_n \boxtimes R_1$ il campo di ℓ-direzioni semiconiche inizialmente definite solamente sulla superficie $\Pi_1(\mathfrak{X}_n) \times \{0\}$.

Questo campo di ℓ-direzioni semiconiche nello spazio $\mathfrak{X}_n \boxtimes R_1$ sarà determinato dallo stesso sistema di equazioni (16. 1).

Indicando con ξ^α , ξ le coordinate di punto nello spazio $\mathfrak{X}_n \boxtimes R_1$ otteniamo il sistema di equazioni differenziali di curve ammissibili rispetto al campo di ξ-direzioni semiconiche nella forma seguente

$$\dot{\xi}^{\alpha} = L^{\alpha}(\xi^{\beta}, \mathfrak{X}^{\lambda}) \quad . \tag{16.5}$$

Vediamo che ogni curva misurabile (16.2) nello spazio $\mathfrak{X}_n$ è ricoperta da una famiglia ad un parametro di curve ammissibili in $\mathfrak{X}_n \boxtimes R_1$ la quale è definita dal sistema (16.2) e dall'equazione

$$\xi = \int_{t_1}^{t} L(\xi^{\alpha}(t), \mathfrak{X}^{\lambda}(t))dt + \xi_o \quad . \tag{16.6}$$

La curva di questa famiglia che passa per il punto $(p_1, 0)$ ove p_1 è il punto iniziale della curva C definita da (16.2) si dice <u>la curva ricoprente principale</u> della curva C . Dunque la curva ricoprente principale della curva C è definita dal sistema (16.2) cui va aggiunta l'equazione

$$\xi = \int_{t_1}^{t} L(\xi^{\alpha}(t), \mathfrak{X}^{\lambda}(t))\, dt \quad . \tag{16.7}$$

E' facile vedere che affinchè una curva misurabile nello spazio $\mathfrak{X}_n$ sia una curva di pseudolunghezza estremale è necessario che la sua curva ricoprente principale nello spazio $\mathfrak{X}_n \boxtimes R_1$ sia vincolata.

Infatti se $\widetilde{C}$ è la curva ricoprente principale di una curva misurabile C di pseudolunghezza estremale allora il punto iniziale $(p_1, 0)$ della curva $\widetilde{C}$ non può essere congiunto, da due curve ammissibili, con un punto $(p_2, s_c - \varepsilon)$ e con un punto $(p_2, s_c + \varepsilon)$ rispettivamente, ove $\varepsilon > 0$.

Per il risultato del § 15 si ha : Affinchè una curva misurabile nello spazio $\mathfrak{X}_n$ dotato di una pseudometrica vettoriale sia una curva di pseudolunghezza estremale è necessario che la sua curva ricoprente principale nello spazio $\mathfrak{X}_n \boxtimes R_1$ sia anormale.

Ogni curva misurabile nello spazio $\mathfrak{X}_n$ la cui curva ricoprente principale nello spazio $\mathfrak{X}_n \boxtimes R_1$ è anormale sarà chiamata una <u>estremale.</u>

In virtù di (15.34) otterremo il sistema di equazioni differenziali misto per estremali

$$\dot{\xi}^\alpha = L^\alpha \quad , \quad \dot{\xi} = L \quad , \quad \dot{y}_\alpha = -\partial_\alpha L^\beta y_\beta - \partial_\alpha L\, y \quad , \quad \dot{y} = 0 \; , \tag{16.8}$$

ove y_α , y sono componenti di un covettore nello spazio $\mathfrak{X}_n \boxtimes R_1$.

Da (16.8) segue che y = cost. Se $y \neq 0$ senza perdere la generalità possiamo porre $y = -1$ e trascrivere il sistema (16.8) nel modo seguente

$$\dot{\xi}^\alpha = L^\alpha \quad , \quad \dot{y}_\alpha = \lambda_0 \partial_\alpha L - \partial_\alpha L^\beta y_\beta \quad , \quad L^\alpha_\lambda y_\alpha = \lambda_0 L_\lambda \; , \tag{16.9}$$

ove $\lambda_0 = 1, 0$ e y_α è un covettore nello spazio $\mathfrak{X}_n$.

Notiamo dapprima che ogni curva ammissibile anormale del campo di ℓ-direzioni misurabili $pr_1\mu$ nello spazio $\mathfrak{X}_n$ è una estremale poichè essa soddisfa il sistema (16.9) per $\lambda_0 = 0$.

Se il grado di anormalità di un estremale anormale C è uguale a α , il grado di anormalità della sua ricoprente principale può essere eguale ad α o ad $\alpha + 1$. In questo ultimo caso l'estremale anormale C soddisfa al sistema (16.8) anche per $\lambda_0 = 1$.

Tali estremali anormali si dicono debolmente anormali.

Per estremali normali e debolmente anormali il sistema (16.9) può essere scritto nel modo seguente

$$\dot{\xi}^\alpha = L^\alpha \quad , \quad \dot{y}_\alpha = \partial_\alpha L - \partial_\alpha L^\beta y_\beta \quad , \quad L^\alpha_\lambda y_\alpha = L_\lambda \; . \tag{16.10}$$

Se una pseudometrica vettoriale in $\mathcal{X}_n$ è definita dal sistema di equazioni implicite

(16.11) $$\overset{z}{f}(\xi^\alpha, X^\alpha) = 0 \ , \quad L(\xi^\alpha, X^\alpha) - X = 0 \qquad (z = \ell+1, \ldots . n)$$

che corrisponde alla definizione ordinaria del problema di Lagrange nel calcolo delle variazioni, secondo (15.39) otterremo il sistema di equazioni differenziali di estremali nella forma ben conosciuta

(16.12) $$\overset{z}{f}(\xi^\beta, \dot{\xi}^\beta) = 0$$

$$\frac{d}{dt}\left(\lambda_0 L_\alpha(\xi^\beta, \dot{\xi}^\beta) + \lambda_z \partial_\alpha \overset{z}{f}(\xi^\beta, \dot{\xi}^\beta)\right) - \left(\lambda_0 \partial_\alpha L(\xi^\beta, \dot{\xi}^\beta) + \lambda_z \partial_\alpha \overset{z}{f}(\xi^\beta, \dot{\xi}^\beta)\right) = 0$$

ove $\lambda_0 = 1, 0$.

Supponiamo ora che la pseudometrica vettoriale μ nello spazio $\mathcal{X}_n$ sia una metrica definita dall'indicatrice globale

(16.13) $$X^\alpha = \ell^\alpha(\xi^\beta, \eta^i) \qquad .$$

Dimostreremo che scegliendo la lunghezza d'arco s come un parametro lungo gli estremali possiamo trasformare il sistema (16.9) nella forma seguente

(16.14) $$\frac{d\xi^\alpha}{ds} = \ell^\alpha \ , \quad \frac{dy_\alpha}{ds} = -\partial_\alpha \ell^\beta y_\beta \quad , \quad \ell_i^\alpha y_\alpha = 0 \ , \quad \ell^\alpha y_\alpha = \lambda_0 \quad ,$$

ove $\lambda_0 = 1, 0$.

Infatti, notiamo che se la metrica μ è data dal sistema (16.1) l'indacatrice può essere determinata dal sistema di equazioni

(16.15) $$\chi^\lambda = \chi^\lambda(\xi^\alpha, \eta^i) \quad ,$$

e valgono le identità

(16.16) $$L^\alpha\left(\xi^\beta, \chi^\lambda(\xi^\beta, \eta^i)\right) = \ell^\alpha(\xi^\beta, \eta^i) \quad , \quad L\left(\xi^\beta, \chi^\lambda(\xi^\beta, \eta^i)\right) = 1$$

Derivando queste identità otteniamo

(16.17) $$\partial_\alpha L^\beta + L^\alpha_\lambda \partial_\alpha \chi^\lambda = \partial_\alpha \ell^\beta \quad , \quad \partial_\alpha L + L_\lambda \partial_\alpha \chi^\lambda = 0 .$$

Ricavando $\partial_\alpha L^\beta$ e $\partial_\alpha L$ dalle espressioni ottenute e sostituendole nelle equazioni (16.9) , si ottiene il sistema (16.14).

Per estremali normali e debolmente anormali il sistema (16.14) può essere scritto nel modo seguente

(16.18) $$\frac{d\xi^\alpha}{ds} = \ell^\alpha \quad , \quad \frac{dy_\alpha}{s} = -\partial_\alpha \ell^\beta y_\beta \quad , \quad \ell^\alpha_i y_\alpha = 0 \quad , \quad \ell^\alpha y_\alpha = 1 \quad .$$

Dalle equazioni in termini finiti di questo sistema segue che il covettore y_α in ogni punto di una estremale definisce l'iperpiano tangente dell'indicatrice locale nel punto corrispondente alla direzione tangente della estremale. Se l'estremale è normale il covettore y_α è definito in modo unico.

Per estremali anormali o in altri termini per curve anormali rispetto al campo $pr_1\mu$ di ℓ-direzioni semiconiche misurabili il sistema (16.14) può essere riscritto nel modo seguente

(16.19) $$\frac{d\xi^\alpha}{ds} = \ell^\alpha \quad , \quad \frac{dy_\alpha}{ds} = -\partial_\alpha \ell^\beta y_\beta \quad , \quad \ell^\alpha_i y_\alpha = 0 \quad , \quad \ell^\alpha y_\alpha = 0.$$

Supponiamo ora che la metrica μ sia regolare e che l'equazione

$$H(\xi^\alpha, y_\alpha) = 1 \tag{16.20}$$

determini le immagini tangenziali proprie delle indicatrici locali, ove, secondo (10.9) la funzione H è definita dalla formula

$$H(\xi^\alpha, y_\alpha) = \ell^\beta(\xi^\beta, \eta^j(\xi^\beta, y_\beta))\, y_\beta \tag{16.21}$$

e le funzioni η^i sono definite implicitamente dalle equazioni

$$\ell^\alpha_i(\xi^\beta, \eta^j)\, y_\alpha = 0 \quad . \tag{16.22}$$

Derivando (16.22) rispetto a ξ^α otteniamo in virtù di (16.22)

$$\partial_\alpha H = \partial_\alpha \ell^\beta y_\beta \quad . \tag{16.23}$$

Oltre a questo abbiamo da (10.10)

$$H^\alpha = \ell^\alpha \tag{16.24}$$

Sostituendo le espressioni ottenute (16.23), (16.24) nel sistema (16.18) otteniamo il sistema differenziale di estremali normali e debolmente anormali nella forma nuova

$$\frac{d\xi^\alpha}{ds} = H^\alpha \quad , \qquad \frac{dy_\alpha}{ds} = -\partial_\alpha H \quad . \tag{16.25}$$

E' facile vedere che questo sistema coincide col sistema di equazioni differenziali delle curve caratteristiche dell'equazione differenziale

$$H(\xi^\alpha, \partial_\alpha \varphi) = 1 \tag{16.26}$$

nella funzione incognita $\varphi(\xi^\alpha)$. Il significato geometrico di una soluzione φ di questa equazione è che il covettore gradiente $\partial_\alpha \varphi$ definisce in ogni punto dello spazio $\mathcal{X}_n$ un iperpiano tangente proprio della corrispondente indicatrice locale.

Dal § 10 segue che se i semiconi delle direzioni misurabili della metrica μ sono regolari possiamo determinare le loro immagini tangenziali mediante l'equazione

$$H(\xi^\alpha, \ {}_\alpha) = 0 \quad , \tag{16.27}$$

ove la funzione H è definita dalla stessa formula (16.21) .

Allora il sistema (16.19) di equazioni differenziali delle curve misurabili anormali può essere trasformato in modo da assumere la stessa forma (16.25) , ma in questo caso il sistema deve essere considerato come il sistema di equazioni differenziali delle curve caratteristiche dell'equazione differenziale

$$H(\xi^\alpha, \partial_\alpha \varphi) = 0 \quad . \tag{16.28}$$

Una soluzione φ di questa equazione differenziale è una funzione le cui ipersuperficie di livello inviluppano il campo di ℓ-direzioni semiconiche misurabili cioè in ogni punto di quella ipersuperficie il suo iperpiano tangente è tangente al semicono corrispondente.

V - COVETTORE DI CURVATURA EULERIANA

§ 17. Geometria di un campo allestito di m-direzioni in $\mathfrak{X}_n$.

Dal punto di vista della teoria degli spazi fibrati un campo di m-direzioni in uno spazio $\mathfrak{X}_n$ si può considerare come una superficie m-secante dello spazio fibrato $T_{n+(n)}$ del prolungamento tangenziale del primo ordine dello spazio $\mathfrak{X}_n$, la quale taglia ogni fibra di $T_{n+(n)}$ secondo un piano centrale ad m dimensioni. Questa superficie m-secante può essere considerata come l'insieme puntuale dello spazio fibrato $A_{n+(m)}$ di un prolungamento centro-affine dello spazio $\mathfrak{X}_n$. Lo spazio $A_{n+(m)}$ è immerso intrinsecamente nello spazio $T_{n+(n)}$.

Come è noto un campo di m-direzioni di $\mathfrak{X}_n$ si dice <u>allestito</u> se è dato nello spazio $\mathfrak{X}_n$ un campo di (n-m)-direzioni che sono complementari alle m-direzioni del campo iniziale. Questo campo di (n-m)-direzioni allestenti può essere anche considerato come una superficie (n-m)-secante $\hat{S}$ nello spazio fibrato $T_{n+(n)}$, la quale seca ogni fibra di $T_{n+(n)}$ secondo un piano centrale ad (n-m)-dimensioni. $\hat{S}$ sarà poi considerato come l'insieme puntuale dello spazio fibrato $\hat{A}_{n+(n-m)}$, il quale è immerso intrinsecamente nello spazio $T_{n+(n)}$.

Siano B^{α}_{a} , B^{a}_{α} , C^{α}_{p} , C^{p}_{α} $(a = 1, \dots, m \ , \ p = m+1, \dots, n)$ gli unici tensori di connessione delle fibre centro-affini degli spazi fibrati $A_{n+(m)}$ e $\hat{A}_{n+(n-m)}$.

Consideriamo un campo di m-direzioni in $\mathfrak{X}_n$ come un caso particolare di un campo di ℓ-direzioni semiconiche studiato nel § 15 .

La superficie m-secante S nello spazio $T_{n+(n)}$ può essere determinata dal sistema di equazioni

(17.1) $$X^{\alpha} = B^{\alpha}_{a}(\xi^{\beta})\, X^{a} \quad .$$

Se

(17.2) $$\xi^{\alpha} = \xi^{\alpha}(t)$$

sono le equazioni di una curva ammissibile del campo di m-direzioni, dalle equazioni

(17.3) $$\dot{\xi}^{\alpha} = B^{\alpha}_{a}\, X^{a}$$

possiamo esprimere le componenti X^{a} del vettore tangenziale come funzioni del parametro t

(17.4) $$X^{a} = X^{a}(t) \quad .$$

Il sistema di Pfaff (15.14) per curve variazionali puo essere ora riscritto nella forma seguente

(17.5) $$C^{p}_{\alpha}\, d\, X^{\alpha} + \Lambda_{\alpha}{}^{p}_{a}\, X^{a} X^{\alpha} dt = 0$$

ove

(17.6) $$\Lambda_{\alpha}{}^{p}_{a} = - C^{p}_{\beta}\, \partial_{\alpha}\, B^{\beta}_{a} \quad .$$

La superficie (n-m)-secante $\hat{S}$ può essere determinata dal sistema di equazioni

$$X^{\alpha} = C^{\alpha}_{p} (\xi^{\beta}) X^{p} \quad . \tag{17.7}$$

Consideriamo ora le curve variazionali per curve ammissibili di S le quali appartengano alla superficie allestente $\hat{S}$.

Sostituendo l'espressione (17.7) di X^{α} nel sistema di Pfaff (17.5) otteniamo un sistema di Pfaff

$$d X^{p} + \Gamma^{p}_{aq} \; X^{a} X^{q} dt = 0 \quad , \tag{17.8}$$

ove

$$\Gamma^{p}_{aq} = -B^{\alpha}_{a} C^{\beta}_{q} \partial_{[\alpha} C^{p}_{\beta]} \quad . \tag{17.9}$$

Dunque l'insieme di tutte le curve variazionali per curve ammissibili rispetto ad $A_{n+(m)}$, le quali appartengano allo spazio fibrato allestente $\hat{A}_{n+(n-m)}$, è una connessione isomorfa in quest'ultimo spazio le cui curve ammissibili della base sono le curve ammissibili rispetto a $A_{n+(m)}$. Questa connessione si dice <u>connessione variazionale</u>.

Scambiando $A_{n+(m)}$ e $\hat{A}_{n+(n-m)}$ otterremo nello spazio fibrato $A_{n+(m)}$ la connessione variazionale determinata dal sistema di Pfaff

$$dx^{a} + \Gamma^{a}_{pb} \, x^{p} x^{a} \, dt = 0 \quad , \tag{17.10}$$

ove

$$\Gamma^{a}_{pb} = - C^{\alpha}_{p} B^{\beta}_{b} \partial_{[\alpha} B^{a}_{\beta]} \quad . \tag{17.11}$$

Mediante le connessioni variazionali negli spazi fibrati $A_{n+(m)}$ e $\hat{A}_{n+(n-m)}$ possiamo definire le derivazioni covarianti di tensori e di densità tensoriali fibrali

$$\mathfrak{D}_p \mathfrak{T}_{a_1 \ldots}^{b_1 \ldots} = \partial_p \mathfrak{T}_{a_1 \ldots}^{b_1 \ldots} - \kappa \Gamma_{pc}^{c} \mathfrak{T}_{a_1 \ldots}^{b_1 \ldots} + \Gamma_{pc}^{b_1} \mathfrak{T}_{a_1 \ldots}^{c \ldots} + \tag{17.12}$$
$$+ \ldots - \Gamma_{p a_1}^{c} \mathfrak{T}_{c \ldots}^{b_1 \ldots} - \ldots$$

$$\mathfrak{D}_a \mathfrak{T}_{p_1 \ldots}^{q_1 \ldots} = \partial_a \mathfrak{T}_{p_1 \ldots}^{q_1 \ldots} - \kappa \Gamma_{ar}^{r} \mathfrak{T}_{p_1 \ldots}^{q_1 \ldots} + \Gamma_{ar}^{q_1} \mathfrak{T}_{p_1 \ldots}^{r \ldots} + \tag{17.13}$$
$$+ \ldots - \Gamma_{a p_1}^{r} \mathfrak{T}_{r \ldots}^{q_1 \ldots} ,$$

ove ∂_a e ∂_p sono gli operatori differenziali

$$\partial_a = B_a^{\alpha} \partial_{\alpha} , \qquad \partial_p = C_p^{\alpha} \partial_{\alpha} . \tag{17.14}$$

Chiameremo $\mathfrak{D}_p \mathfrak{T}_{a_1 \ldots}^{b_1 \ldots}$ e $\mathfrak{D}_a \mathfrak{T}_{p_1 \ldots}^{q_1 \ldots}$ le <u>derivate covarianti variazionali</u> delle densità tensoriali $\mathfrak{T}_{a_1 \ldots}^{b_1 \ldots}$ e $\mathfrak{T}_{p_1 \ldots}^{q_1 \ldots}$:

Dal punto di vista puramente formale le connessioni variazionali e le operazioni di derivazioni corrispondenti sono state introdotte da Schouten e van Kampen ([19]).

Supponiamo che nello spazio fibrato $A_{n+(m)}$ sia data una densità scalare fibrale $\breve{\mathfrak{U}}$ di peso $+1$ ovunque differente da zero. La condizione

$$\breve{\mathfrak{U}} = 1 \tag{17.15}$$

definisce un sottoatlante K dell'atlante coordinato dello spazio $A_{n+(m)}$ e lo spazio coordinizzato $\vec{U}_{n+(m)} = (S, K)$ sarà uno spazio fibrato subordinato allo spazio $A_{n+(m)}$ le cui fibre sono spazi centro-equiaffini orientati $\vec{U}_m$.

Questo spazio fibrato $\vec{U}_{n+(m)}$ si dice semi-intrinsecamente immerso nello spazio fibrato $T_{n+(n)}$.

Se $\vec{U}_{n+(m)}$ è allestito mediante lo spazio fibrato $\hat{A}_{n+(n-m)}$ il covettore fibrale nello spazio $\hat{A}_{n+(n-m)}$

$$h_p = - \frac{\mathfrak{D}\breve{\mathfrak{U}}}{\breve{\mathfrak{U}}} = \Gamma^c_{pc} \tag{17.16}$$

si dice il covettore di curvatura euleriana dello spazio $\vec{U}_{n+(m)}$.

Il covettore fibrale h_p dello spazio fibrato $\hat{A}_{n+(n-m)}$ può essere ridotto al covettore

$$h_\alpha = C^p_\alpha h_p \tag{17.17}$$

nello spazio base $\mathfrak{X}_n$; esso è pure detto il covettore di curvatura euleriana dello spazio $\vec{U}_{n+(m)}$.

Dopo alcuni passaggi si ottiene:

(17.18) $$h_\alpha = \partial_a B^a_\alpha + B^b_\beta \partial_\alpha B^\beta_b + 2B^a_\alpha \Omega_{ab}^{\ \ b}$$

o

(17.19) $$h_\alpha = \partial_a B^a_\alpha - B^\beta_b \partial_\alpha B^b_\beta + 2B^a_\alpha \Omega_{ab}^{\ \ b}$$

ove

(17.20) $$\Omega_{ab}^{\ \ c} = -\partial_{[a} B^\gamma_{b]} B^c_\gamma \quad .$$

Oltre a questo possiamo esprimere h_α in termini dell'm-vettore $B^{\alpha_1 \dots \alpha_m}$ e dell'm-covettore $B_{\alpha_1 \dots \alpha_m}$:

(17.21) $$h_\alpha = -\frac{m+1}{m!} \partial_{[\alpha} B_{\alpha_1 \dots \alpha_m]} B^{\alpha_1 \dots \alpha_m} .$$

L'allestimento dello spazio $\vec{U}_{n+(m)}$ si dice un <u>allestimento gradiente</u> se l'm-covettore $B_{\alpha_1 \dots \alpha_m}$ è un gradiente.

Si può dimostrare che

(17.22) $$\partial_{[\alpha} B_{\alpha_1 \dots \alpha_m]} = -h_{[\alpha} B_{\alpha_1 \dots \alpha_m]} - M_{pq}^{\cdot\cdot\, a} C^p_{[\alpha} B_{\alpha_1 \dots \alpha_m]a}^{\ \ q}$$

ove

(17.23) $$M_{pq}^{\cdot\cdot a} = - C_p^{\alpha} C_q^{\beta} \partial_{[\alpha} B_{\beta]}^a$$

è il tensore di anolonomia di Schouten del campo ad (n-m)-dimensioni allestente.

Se $B_{\alpha_1 \ldots \alpha_m}$ è un m-covettore gradiente, necessariamente risulta

(17.24) $$\partial_{[\alpha} B_{\alpha_1 \ldots \alpha_m]} = 0$$

che, secondo (17.22), equivale

(17.25) $$M_{pq}^{\cdot\cdot a} = 0 \quad , \quad h_{\alpha} = 0 \quad .$$

Dunque un allestimento gradiente di $\vec{U}_{n+(m)}$ è sempre olonomo e $\vec{U}_{n+(m)}$ allestito mediante gradiente è di curvatura euleriana nulla.

Come è noto dal punto di vista locale la condizione (17.24) è anche sufficiente affinchè l'm-covettore $B_{\alpha_1 \ldots \alpha_m}$ sia un gradiente.

Utilizzando i risultati del §12 possiamo determinare un allestimento dello spazio fibrato $\vec{U}_{n+(m)}$ mediante un m-covettore arbitrario $W_{\alpha_1 \ldots \alpha_m}$ dello spazio $\mathfrak{X}_n$ che soddisfa solamente alla condizione

(17.26) $$W_{\{\alpha_1 \ldots \alpha_m\}} B^{\{\alpha_1 \ldots \alpha_m\}} = 1$$

ma che non è semplice.

Un allestimento dello spazio fibrato $\vec{U}_{n+(m)}$ si dice un allestimento pseudogradiente se esso può essere determinato mediante un m-covettore gradiente $W_{\alpha_1 \ldots \alpha_m}$ in generale non semplice.

Una proprietà importante degli spazi fibrati $\vec{U}_{n+(m)}$ con un allestimento pseudogradiente è espressa dal teorema ([37]).

In uno spazio fibrato $\vec{U}_{n+(m)}$ con un allestimento pseudogradiente il covettore di curvatura euleriana è nullo su ogni superficie ad n dimensioni ammissibile.

§ 18. Superficie estremali in uno spazio $\mathfrak{X}_n$ dotato di una metrica grassmanniana m-vettoriale.

Sia data in uno spazio $\mathfrak{X}_n$ una metrica grassmanniana m-vettoriale μ ad ℓ dimensioni locali.

L'indicatrice globale $\mathfrak{J}$ della μ sarà considerata come lo spazio fibrato $\mathfrak{X}_{n+(\ell-1)}$ e sarà determinata dal sistema di equazioni

$$x^{\alpha_1 \dots \alpha_m} = \ell^{\alpha_1 \dots \alpha_m}(\xi^\alpha, \eta^i) \quad . \tag{18.1}$$

Una m-direzione orientata in un punto p di uno spazio $\mathfrak{X}_n$ è un piano centrale orientato ad m dimensioni nello spazio tangente A_n associato a p, cioè un piano centrale ad m dimensioni che è l'insieme puntuale di uno spazio $\vec{A}_m$ semintrisecamente immerso in A_n.

Una m-direzione orientata in un punto di uno spazio $\mathfrak{X}_n$ dotato di una metrica grassmanniana m-vettoriale si dice <u>misurabile</u> se questo $\vec{A}_m$ corrisponde ad una semiretta semplice misurabile rispetto alla metrica vettoriale corrispondente nello spazio $M_{\binom{n}{m}}$.

E' evidente che ogni m-direzione misurabile può essere considerata come un piano centrale ad m dimensioni che è l'insieme puntuale dello spazio $\vec{U}_m$ corrispondente ad un punto dell'indicatrice locale $\mathfrak{J}_p$. Ogni tale $\vec{U}_m$ si dice anche misurabile.

E' comodo considerare gli spazi $\vec{U}_m$ misurabili come le fibre dello spazio fibrato ad $n+\ell-1+m$ dimensioni $\vec{U}_{n+(\ell-1)+((m))}$ il cui spazio base è lo spazio fibrato $\mathfrak{X}_{n+(\ell-1)}$ corrispondente all'indicatrice globale. In questo caso i componenti dell'unico tensore di connessione B^α_a sono considerati come funzioni delle coordinate ξ^α, η^i di punto dello spazio $\mathfrak{X}_{n+(\ell-1)}$:

(18.2) $$B_a^\alpha = B_a^\alpha(\xi^\beta, \eta^i) \quad .$$

Una (n-m)-direzione si dice <u>trasversale</u> rispetto ad una m-direzione misurabile nello stesso punto p dello spazio $\mathcal{X}_n$ se questa (n-m)-direzione corrisponde ad un iperpiano proprio semplice nello spazio $M_{\binom{n}{m}}(p)$ il quale è tangente all'indicatrice nel punto corrispondente allo spazio $\vec{U}_m$.

Un campo di m-direzioni misurabili nello spazio $\mathcal{X}_n$ o uno spazio fibrato $\vec{U}_{n+(m)}$ semintrisecamente immerso nello $T_{n+(n)}$ le cui fibre sono $\vec{U}_m$ misurabili, può essere determinato mediante una superficie secante dello spazio fibrato $\mathcal{X}_{n+(\ell-1)}$ data dal sistema di equazioni

(18.3) $$\eta^i = \eta^i(\xi^\alpha) \quad .$$

Supponiamo ora che questo campo di m-direzioni misurabili sia allestito da un campo di (n-m)-direzioni trasversali.

I componenti del tensore di connessione B_α^a che determina questo allestimento trasversale sono dati come funzioni delle coordinate ξ^α di punto nello spazio $\mathcal{X}_n$

(18.4) $$B_\alpha^a = B_\alpha^a(\xi^\beta) \quad .$$

Consideriamo il covettore di curvatura euleriana dello spazio fibrato $\vec{U}_{n+(m)}$ allestito trasversalmente dallo spazio fibrato $\hat{A}_{n+(n-m)}$.

Da (13.4) e (13.12) otteniamo la condizione di trasversalità dell'allestimento nella forma seguente

(18.5) $$B_\beta^b(\xi^\alpha) \cdot B_{bi}^\beta(\xi^\alpha, \eta^j(\xi^\beta)) = 0 \quad .$$

Indicando con $B^{\beta}_{b\alpha}$ le derivate parziali delle funzioni (18.2) rispetto a ξ^{α} otteniamo per (18.5)

$$(18.6) \qquad B^{b}_{\beta} \partial_{\alpha} B^{\beta}_{b} = B^{b}_{\beta} B^{\beta}_{b\alpha} + B^{b}_{\beta} B^{\beta}_{bi} \partial_{\alpha} \eta^{i} = B^{b}_{\beta} B^{\beta}_{b\alpha} ,$$

ove $B^{\beta}_{b}(\xi^{\beta}, \eta^{j})$ sono considerate funzioni composte di ξ^{β} .

Sostituendo le espressioni (18.6) nella formula (17.18) otteniamo

$$(18.7) \qquad h_{\alpha} = \partial_{a} B^{a}_{\alpha} + B^{b}_{\beta} B^{\beta}_{b\alpha} + 2B^{a}_{\alpha} \Omega^{b}_{ab} .$$

Consideriamo ora una superficie ad m dimensioni ammissibile rispetto a un dato campo $\vec{U}_{n+(m)}$ di m-direzioni misurabili trasversalmente allestito, la quale è determinata dal sistema di equazioni

$$(18.8) \qquad \xi^{\alpha} = \xi^{\alpha}(s^{a}) ,$$

ove s^{a} sono coordinate "uniareali" su questa superficie misurabile. Possiamo scegliere un sistema coordinato nello spazio $\vec{U}_{n+(m)}$ in tal modo che sulla superficie sia

$$(18.9) \qquad B^{\alpha}_{a} = \frac{\partial \xi^{\alpha}}{\partial s^{a}}$$

e quindi

$$(18.10) \qquad \partial_{a} = B^{\alpha}_{a} \partial_{\alpha} = \frac{\partial}{\partial s^{a}} .$$

Notando che secondo (17.20), (18.9), (18.10) sulla superficie è identicamente

(18.11) $$\Omega_{ab}^{\ c} = 0 \quad ,$$

dalla formula (18.7) si ottiene la seguente espressione del covettore di curvatura euleriana sulla superficie

(18.12) $$h_\alpha = \frac{\partial B_\alpha^a}{\partial s^a} + B_\beta^b \, B_{b\alpha}^\beta \quad .$$

Da questa espressione si trae come conseguenza importantissima che il covettore di curvatura euleriana di uno spazio $\vec{U}_{n+(m)}$ misurabile, allestito trasversalmente su una superficie ad m dimensioni ammissibile dipende solamente dalla superficie stessa e dal suo allestimento trasversale e da niente altro.

Dunque possiamo introdurre la nozione di covettore di curvatura euleriana di una superficie misurabile arbitraria trasversalmente allestita definendolo mediante la formula (18.12).

Un campo di m-direzioni misurabili trasversalmente allestito o, in altri termini, uno spazio fibrato $\vec{U}_{n+(m)}$ misurabile trasversalmente allestito nello spazio $T_{n+(n)}$ si dice <u>geodetico</u> se il suo allestimento è uno pseudo-gradiente.

Consideriamo uno spazio $\vec{U}_{n+(m)}$ geodetico. Sia $W_{\alpha_1 \ldots \alpha_m}$ l'm-covettore gradiente che determina l'allestimento di $\vec{U}_{n+(m)}$.

E' facile vedere che in ogni punto dello spazio $\mathfrak{X}_n$ l'm-covettore $W_{\alpha_1 \ldots \alpha_m}$ determina un iperpiano tangente dell'indicatrice locale.

Questo segue immediatamente dal fatto che l'iperpiano tangente semplice che determina l'allestimento trasversale di una fibra $\vec{U}_m$ di $\vec{U}_{n+(m)}$

e l'iperpiano determinato dall'm-covettore $W_{\alpha_1 \dots \alpha_m}$ sono iperpiani α-equivalenti del fascio di iperpiani con centro nel punto semplice corrispondente allo spazio $\vec{U}_m$.

Uno spazio geodetico $\vec{U}_{n+(m)}$ si dice che soddisfa alla condizione minima (massima) di Weierstrass se in ogni punto dello spazio $\mathfrak{X}_n$ l'indicatrice locale è positivamente (negativamente) convessa rispetto all'iperpiano tangente determinato dall'm-covettore $W_{\alpha_1 \dots \alpha_m}$. In questo caso le traslazioni proprie $\mathcal{E} = +1$ (improprie $\mathcal{E} = -1$) negli spazi

(18.13) $$\bar{X}^{\{\alpha_1 \dots \alpha_m\}} = \frac{X^{\{\alpha_1 \dots \alpha_m\}}}{\mathcal{E}\left(1 - W_{\{\alpha_1 \dots \alpha_m\}}(\xi^{\alpha}) X^{\{\alpha_1 \dots \alpha_m\}}\right)}$$

determiniamo la trasformazione di Caratheodory speciale propria (impropria) della metrica grassmaniana m-vettoriale data.

Ogni superficie ad m dimensioni ammissibile rispetto ad uno spazio $\vec{U}_{n+(m)}$ geodetico soddisfacente alla condizione minima (massima) di Weierstrass rispetto alla metrica impropria trasformata sarà una superficie isotropa e perciò essa è una superficie di area minima (massima). Chiameremo tali superficie d'area estrema <u>virtualmente isotrope</u> . Secondo il teorema della fine del § 17 ogni superficie virtualmente isotropa può essere allestita trasversalmente in modo tale che essa diventi una superficie trasversalmente allestita di curvatura euleriana nulla.

Supponiamo ora che la metrica grassmanniana m-vettoriale considerata sia regolare cioè che tutte le indicatrici locali siano regolari.

Allora per i risultati del § 10 possiamo determinare le immagini tangenziali proprie delle indicatrici locali mediante l'equazione

(18.14) $$H(\xi^\alpha, Y_{\{\alpha_1 \dots \alpha_m\}}) = 1 \quad ,$$

ove

(18.15) $$H^{\{\alpha_1 \dots \alpha_m\}} = \ell^{\{\alpha_1 \dots \alpha_m\}} \quad .$$

Dalla (18.14) segue che per trovare gli spazi $\overrightarrow{U}_{n+(m)}$ geodetici dobbiamo integrare l'equazione differenziale

(18.16) $$H(\xi^\alpha, \partial_{\{[\alpha_1} \varphi_{\alpha_2 \dots \alpha_m]\}}) = 1 \quad .$$

Infatti se $\varphi_{\alpha_2 \dots \alpha_m}$ è una soluzione dell'equazione differenziale (18.16) allora l'm-vettore semplice

(18.17) $$H^{\alpha_1 \dots \alpha_m}(\xi^\alpha, \partial_{\{[\alpha_1} \varphi_{\alpha_2 \dots \alpha_m]\}})$$

determina lo spazio $\overrightarrow{U}_{n+(m)}$ geodetico il cui allestimento è determinato dall'm-covettore $\partial_{[\alpha_1} \varphi_{\alpha_2 \dots \alpha_m]}$.

Ritornando al caso della metrica grassmanniana m-vettoriale generale supponiamo che l'indicatrice globale sia data dal sistema di equazioni

(18.18) $$L(\xi^\alpha, X_a^\alpha) = 1$$

(18.19) $$\overset{z}{f}(\xi^\alpha, X_a^\alpha) = 0 \quad .$$

Secondo la (13, 31) un allestimento trasversale di una superficie misurabile (18, 8) si può determinare scegliendo funzioni

$$\lambda_z = \lambda_z(s) \tag{18.20}$$

che sono contenute nelle espressioni delle componenti dell'unico tensore di connessione

$$B^a_\alpha = L^a_\alpha\left(\xi^\beta, \frac{\partial \xi^\beta}{\partial s^b}\right) + \lambda_z \overset{z}{f}{}^a_\alpha\left(\xi^\beta, \frac{\partial \xi^\beta}{\partial s^b}\right). \tag{18.21}$$

Sostituendo le espressioni (18.2) di B^α_a nelle equazioni (18.18) e (18, 19) e derivando le identità ottenute rispetto a ξ^α avremo

$$\partial_\alpha L + L^b_\beta B^\beta_{b\alpha} = 0 \quad , \quad \partial_\alpha \overset{z}{f} + \overset{z}{f}{}^b_\beta B^\beta_{b\alpha} = 0 \tag{18.22}$$

donde in virtù di (18, 21), otteniamo

$$B^b_\beta B^\beta_{b\alpha} = -(\partial_\alpha L + \lambda_z \partial_\alpha \overset{z}{f}) \tag{18.23}$$

Sostituendo le espressioni (18.21) e (18.23) in (18.12) avremo

$$h_\alpha = \tag{18.24}$$

$$= \frac{\partial}{\partial s^a}\left(L^a_\alpha(\xi^\beta, \frac{\partial \xi^\beta}{\partial s^b}) + \lambda_z \overset{z}{f}{}^a_\alpha(\xi^\beta, \frac{\partial \xi^\beta}{\partial s^b})\right) - \left(\partial_\alpha L(\xi^\beta, \frac{\partial \xi^\beta}{\partial s^b}) + \lambda_z \partial_\alpha \overset{z}{f}(\xi^\beta, \frac{\partial \xi^\beta}{\partial s^b})\right).$$

Considerando il problema di trovare le superficie misurabili di area estrema rispetto a una metrica grassmanniana m-vettoriale data dal sistema di equazioni dell'indicatrice globale (18.18), (18.19) veniamo al problema di Lagrange per l'integrale multiplo nel calcolo delle variazioni in forma ordinaria.

Analogamente al caso del problema di Lagrange per l'integrale semplice è naturale considerare come condizione di Eulero per il problema di Lagrange per l'integrale multiplo la condizione che esistano funzioni λ_z tali che la superficie misurabile data soddisfi al sistema di equazioni differenziali

$$\frac{\partial}{\partial s^a}(\lambda_o L^a_\alpha + \lambda_z \overset{za}{f_\alpha}) - (\partial_\alpha L + \lambda_z \partial_\alpha \overset{z}{f}) = 0 \quad , \tag{18.25}$$

ove $\lambda_o = 1$ o $\lambda_o = 0$, chiamate equazioni di Eulero.

Chiameremo tali superficie misurabili le superficie estremali. Una superficie estremale si dice normale se essa soddisfa al sistema (18.25) solamente quando $\lambda_o = 1$. Nel caso contrario la superficie estremale si dice anormale. Se una superficie estremale anormale soddisfa il sistema (18.25) anche per $\lambda_o = 1$ essa si dice debolmente anormale. E' facile vedere che per una superficie estremale normale le funzioni λ_z sono univocamente determinate.

Finora la necessità della condizione di Eulero per l'estremo nel caso generale del problema di Lagrange per l'integrale multiplo non è dimostrata benchè essa sembri verosimile.

Confrontando il sistema di equazioni differenziali di Eulero (18.25) con l'espressione (18.24) del covettore di curvatura euleriana otteniamo l'interpretazione geometrica delle superficie estremali normali e debolmente anormali come superficie misurabili che ammettono un allestimento trasversale tale che esse diventano le superficie trasversalmente allestite di curvatura euleriana nulla. Con questo se una superficie estremale è normale questo allestimento trasversale è unico e se essa è debolmente anormale questo allestimento si trova con una certa arbitrarietà.

Come caso particolare, per $m = 1$ otteniamo l'interpretazione geometrica di estremali normali e debolmente anormali in uno spazio $\mathcal{X}_n$ dotato di una metrica vettoriale come le curve misurabili che ammettono un allestimento trasversale tale che esse diventino le curve trasversalmente allestite di curvatura euleriana nulla.

Per $\ell = m(n-m)$ in ogni punto dell'indicatrice locale esiste un unico iperpiano tangente semplice e perciò per ogni m-direzione misurabile esiste una unica (n-m)-direzione trasversale. Dunque per ogni superficie misurabile esiste un unico allestimento trasversale. In questo caso possiamo parlare del covettore di curvatura euleriana di una superficie misurabile supponendola allestita trasversalmente (nell'unico modo possibile).

Come è noto la necessità della condizione di Eulero in questo caso si deduce dalla condizione che la variazione dell'area sia nulla.

Ritornando di nuovo al caso generale di una metrica grassmanniana m-vettoriale in uno spazio $\mathcal{X}_n$ supponiamo che essa sia regolare in senso stretto cioè che tutte le indicatrici locali sono regolari in senso stretto.

Determiniamo le immagini tangenziali grassmanniane proprie delle indicatrici locali dall'equazione

$$Q(\xi^\alpha, Y^a_\alpha) = 1 \tag{18.26}$$

Consideriamo ora uno spazio fibrato $\overrightarrow{U}_{n+(m)}$ misurabile, trasversalmente allestito. Sostituendo nell'equazione (18.26) i componenti B^a_α dello unico tensore di connessione di $\overrightarrow{U}_{n+(m)}$ e derivando l'identità ottenuta avremo

(18.27) $$\partial_\alpha Q + Q_b^\beta \partial_\alpha B_\beta^b = 0 \quad .$$

D'altra parte da (13.35) abbiamo

(18.28) $$Q_b^\beta = B_b^\beta$$

e perciò da (18.27) possiamo dedurre,

(18.29) $$B_b^\beta \partial_\alpha B_\beta^b = - \partial_\alpha Q \quad .$$

Sostituendo le espressioni di $B_b^\beta \partial_\alpha B_\beta^b$ ricavate da (18.29) nella formula (17.18) otteniamo l'espressione seguente per il covettore di curvatura euleriana di $\vec{U}_{n+(m)}$

(18.30) $$h_\alpha = \partial_a B_\alpha^a + \partial_\alpha H + 2B_\alpha^a \Omega_{ab}^{\;\;b} \quad .$$

Per una superficie ammissibile di $\vec{U}_{n+(m)}$ da (18.30) si ottiene

(18.31) $$h_\alpha = \frac{\partial B_\alpha^a}{\partial s^a} + \partial_\alpha H \quad .$$

Poi è facile vedere da (18.28) e (28.31) che le superficie misurabili di curvatura euleriana nulla possono essere determinate dal seguente sistema di equazioni differenziali

(18.32) $$\frac{\partial \xi^\alpha}{\partial s^a} = H_a^\alpha \qquad \frac{\partial Y_\alpha^a}{\partial s^a} = - \partial_\alpha H \quad .$$

V. V. Wagner

VI - GEOMETRIA FINSLERIANA GENERALIZZATA

§19. La teoria delle ipersuperficie regolari nello spazio centro-affine .

Uno spazio finsleriano generalizzato è uno spazio $\mathfrak{X}_n$ in cui è data una metrica vettoriale regolare ad n dimensioni locali.

Da questa definizione segue che uno spazio finsleriano generalizzato è uno spazio finsleriano se la metrica vettoriale data è completa, simmetrica e convessa. Vedremo che il contenuto ordinario della geometria finsleriana sarà conservato da questa generalizzazione della nozione di spazio finsleriano.

Considerando la geometria finsleriana generalizzata come la teoria degli invarianti differenziali d'una metrica vettoriale regolare ad n dimensioni locali data nello spazio $\mathfrak{X}_n$ abbiamo che essa equivale alla teoria delle superficie $(n-1)$-secanti nello spazio fibrato $T_{n+(n)}$ le quali siano le indicatrici globali della metrica vettoriale nello spazio $\mathfrak{X}_n$ oppure in altre parole alla teoria di un campo di ipersuperficie negli spazi tangenti dello spazio $\mathfrak{X}_n$, le quali sono le indicatrici locali.

Naturalmente per lo studio della indicatrice globale dobbiamo cominciare dallo studio delle indicatrici locali, cioè dallo studio delle ipersuperficie regolari nello spazio centro-affine.

Sia S una ipersuperficie propria nello spazio centro-affine ad n-dimensioni A_n definita dalle equazioni parametriche

$$x^{\alpha} = \ell^{\alpha}(\eta^{i}) \qquad (19.1)$$

Ad ogni punto della ipersuperficie S abbiamo l'iperpiano tangente determinato dal covettore ℓ_{α}, che si trova mediante le equazioni

(19.2) $$\ell_i^\alpha \ell_\alpha = 0 \qquad \ell^\alpha \ell_\alpha = 1 .$$

Il tensore tangenziale

(19.3) $$g_{ij} = -\ell_\alpha \ell_{ij}^\alpha = -\ell_{\alpha\, ij} \ell^\alpha = \ell_{\alpha\, i} \ell_j$$

di questo iperpiano tangente si dice il primo tensore fondamentale della ipersuperficie S .

Rappresentando i vettori ℓ_{ji}^α come combinazioni lineari dei vettori indipendenti ℓ_i^α, ℓ^α abbiamo

(19.4) $$\ell_{ji}^\alpha = G_{ji}^k \ell_k^\alpha - g_{ji} \ell^\alpha \quad ,$$

ove G_{ji}^k sono componenti d'un oggetto della connessione affine senza torsione .

Scrivendo le equazioni (19.4) nella forma

(19.5) $$\nabla_j \ell_i^\alpha = - g_{ji} \ell^\alpha$$

otteniamo le loro condizioni d'integrabilità

(19.6) $$R_{kji}^{\cdot\cdot\cdot h} = 2 \delta_{[k}^h g_{j]i} \quad , \qquad \nabla_{[k} g_{j]i} = 0$$

le quali esprimono che la connessione definita da G_{ji}^k è equiaffine e proiettivamente euclidea.

La necessità di questa ultima condizione segue immediatamente dal fatto che le geodetiche rispetto alla connessione G_{ji}^k saranno le sezioni della ipersuperficie con gli iperpiani centrali donde avremo che la proiezione centrale darà una rappresentazione della parte corrispondente dell'ipersuperficie sull'iperpiano che preserva le geodetiche.

Dalla prima delle equazioni (19.6) si può ottenere che il tensore fondamentale differisca dal tensore di Ricci solamente per il fattore $\frac{1}{n-2}$ cioè

(19.7) $$g_{ji} = \frac{1}{n-2} R_{ji} \quad .$$

Dunque un'ipersuperficie propria data nello spazio centroaffine A_n a meno di un automorfismo arbitrario può essere considerata come uno spazio $\mathfrak{X}_{n-1}$ dotato di una connessione equiaffine e proiettivamente euclidea, senza torsione.

Se S è un'ipersuperficie regolare allora la sua immagine tangenziale S^* sarà un'ipersuperficie nello spazio coniugato A^*_n data dalle equazioni

(19.8) $$Y_\alpha = \ell_\alpha(\eta^i)$$

Rappresentano i vettori $\ell_{\alpha\, ji}$ come conbinazioni lineari dei vettori indipendenti $\ell_{\alpha\, i}$, ℓ_α abbiamo

(19.9) $$\ell_{\alpha\, ji} = {}^*G^k_{ji}\, \ell_{\alpha\, k} - g_{ji}\, \ell_\alpha \quad ,$$

ove ${}^*G^k_{ji}$ sono i componenti d'un oggetto della connessione affine senza torsione la quale è anche equiaffine e proiettivamente euclidea.

Questa connessione affine si dice <u>coniugata</u> rispetto alla connessione affine determinata da G^k_{ji} .

Calcolando le derivate covarianti del tensore fondamentale g_{ji} rispetto a G^k_{ji} e a $^*G^k_{ji}$ otterremo che essi differiscano tra loro solamente per il segno.

Porremo

$$\nabla_k g_{ji} = -{}^*\nabla_k g_{ji} = 2A_{kji} \tag{19.10}$$

ove A_{kji} è un tensore simmetrico, il quale si dice il secondo tensore fondamentale dell'ipersuperficie S .

Dalla (19.10) segue che la derivata covariante del primo tensore fondamentale g_{ji} rispetto all'oggetto della connessione affine

$$\overset{\circ}{G}{}^k_{ji} = \frac{1}{2}\left(G^k_{ji} + {}^*G^k_{ji}\right) \tag{19.11}$$

è uguale a zero, il che significa che la connessione affine determinata da $\overset{\circ}{G}{}^k_{ji}$ è una connessione riemanniana corrispondente al tensore g_{ji} .

Questa connessione riemanniana si dice la connessione affine media della ipersuperficie S .

Mediante l'oggetto della connessione affine media possiamo modificare le equazioni (19.4) e (19.9) e scriverle nella forma seguente

$$\overset{\circ}{\nabla}_j \ell^\alpha_i = -A_{ji}^{\cdot\cdot k}\,\ell^\alpha_k - g_{ji}\,\ell^\alpha \;, \quad \overset{\circ}{\nabla}_j \ell_{\alpha i} = A_{ji}^{\cdot\cdot k}\,\ell_{\alpha k} - g_{ji}\ell_\alpha \tag{19.12}$$

ove il tensore $A_{ji}^{\cdot\cdot k}$ è ottenuto dal tensore A_{jik} con l'aiuto della operazione di innalzamento degli indici definita dal tensore g^{kh} .

Le condizioni di integrabilità per ambedue i sistemi di equazioni differenziali (19.12) coincidono e possono essere ridotte alla forma seguente

(19.13) $\overset{\circ}{R}_{kjih} = 2g_{h[k}g_{j]i} - 2A_{\ell h[k}A_{j]i}^{\cdot\cdot\cdot\ell}, \quad \overset{\circ}{\nabla}_{[k}A_{j]ih} = 0$.

Per trovare un significato geometrico del secondo tensore fondamentale dell'ipersuperficie consideriamo l'iperquadrica centrale osculatrice.

L'iperquadrica centrale osculatrice in un dato punto p dell'ipersuperficie S è una iperquadrica centrale il cui centro coincide col centro dello spazio A_n e la quale nel punto p è osculatrice di secondo ordine all'ipersuperficie.

Se

(19.14) $$a_{\beta\alpha} \; x^{\beta} \; x^{\alpha} = 1$$

è l'equazione dell'iperquadrica centrale osculatrice possiamo ottenere le espressioni seguenti per le componenti del tensore $a_{\beta\alpha}$

(19.15) $$a_{\beta\alpha} = \ell^{j}_{\beta}\, \ell^{i}_{\alpha}\, g_{ji} + \ell_{\beta}\, \ell_{\alpha} \quad ,$$

donde abbiamo

(19.16) $$\text{Det} \left| a_{\beta\alpha} \right| = \text{Det} \left| g_{ji} \right| \cdot (\text{Det} \left| \ell^{i} \; , \ell_{\alpha} \right|)^{2}$$

ciò significa che la regolarità dell'ipersuperficie equivale alla non degenerazione delle iperquadriche centrali osculatrici.

Si può poi dimostrare che il secondo tensore fondamentale A_{kji} della ipersuperficie è uguale al tensore della deviazione centrale dell'ipersuperficie dalla sua iperquadrica centrale osculatrice moltiplicato per $\frac{1}{2}$.

Derivando (19.15) si può ottenere l'equazione

(19.17) $$\partial_i a_{\beta\alpha} = A_{ijk} \ell^j_\beta \ell^k_\alpha$$

dalla quale abbiamo che la condizione necessaria e sufficiente affinchè una ipersuperficie sia un'iperquadrica centrale, il cui centro coincida col centro dello spazio centro-affine, è che il secondo tensore fondamentale sia nullo :

(19.18) $$A_{ijk} = 0 \quad .$$

§ 20. La connessione lineare parziale intrinseca nello spazio finsleriano generalizzato.

Consideriamo l'indicatrice globale della metrica vettoriale nello spazio $\mathfrak{X}_n$, che definisce la geometria finsleriana come l'insieme puntuale di uno spazio fibrato generale $\mathfrak{X}_{n+(n-1)}$.

Il problema fondamentale della geometria finsleriana generalizzata è di trovare una connessione lineare parziale intriseca in questo spazio fibrato $\mathfrak{X}_{n+(n-1)}$.

Sia

$$x^{\alpha} = f^{\alpha}(\xi^{\beta}, \eta^{i}) \tag{20.1}$$

il sistema di equazioni dell'indicatrice globale.

Con l'aiuto dei covettori introdotti nella teoria delle ipersuperficie nello spazio centro-affine, determiniamo n forme di Pfaff indipendenti

$$\ell_i = \ell_{\alpha i} d\xi^{\alpha} \quad , \quad \ell = \ell_{\alpha} d\xi^{\alpha} \quad , \tag{20.2}$$

le prime n-1 delle quali sono componenti di un covettore fibrale nello spazio $\mathfrak{X}_{n+(n-1)}$ e l'ultima è una forma di Pfaff scalare.

Se nello spazio fibrato $\mathfrak{X}_{n+(n-1)}$ è data una connessione lineare parziale determinata da un sistema di equazioni di Pfaff

$$\delta\eta^i = d\eta^i + \Gamma^i = 0 \tag{20.3}$$

considerando la derivata covariante basica $\mathcal{D}_{\alpha}\,\Omega^{A}$ di un oggetto differenziale fibrale Ω^{A} costruiamo due nuovi oggetti differenziali fibrali $\mathcal{D}_i\Omega^{A}$ e $\mathcal{D}_0\Omega^{A}$

(20.4) $$\mathcal{D}_i \Omega^{\mathfrak{A}} = \ell_i^{\alpha} \mathcal{D}_{\alpha} \Omega^{\mathfrak{A}} \quad , \quad \mathcal{D}_o \Omega^{\mathfrak{A}} = \ell^{\alpha} \mathcal{D}_{\alpha} \Omega^{\mathfrak{A}}$$

i quali si dicono rispettivamente la derivata assoluta basica trasversale e la derivata assoluta basica radiale di questo oggetto differenziale fibrale $\Omega^{\mathfrak{A}}$.

Il differenziale assoluto basico di $\Omega^{\mathfrak{A}}$ s'esprime nel modo seguente

(20.5) $$\mathcal{D} \Omega^{\mathfrak{A}} = \ell_i \mathcal{D}^i \Omega^{\mathfrak{A}} + \ell \mathcal{D}_o \Omega^{\mathfrak{A}} \quad ,$$

ove

(20.6) $$\mathcal{D}^i \Omega^{\mathfrak{A}} = g^{ij} \mathcal{D}_j \Omega^{\mathfrak{A}} \quad .$$

La possibilità definire una connessione lineare parziale nello spazio $\mathfrak{X}_{n+(n-1)}$ in modo intrinseco segue dal teorema seguente :

Teorema. Nello spazio fibrato $\mathfrak{X}_{n+(n-1)}$ il cui insieme puntuale è una superficie (n-1)-secante nello spazio fibrato $T_{n+(n)}$ la quale taglia ogni fibra A_n secondo un'ipersuperficie propria regolare, esiste una e solamente una connessione lineare parziale che soddisfa alle condizioni seguenti :

(20.7) $$[\mathcal{D} \ell] = 0$$

(20.8) $$\mathcal{D}_o g_{ji} = 0$$

Dimostrazione.

Notiamo, dapprima, che la condizione (20.8) è equivalente alla condizione

(20.9) $$\mathcal{D}_o g^{ji} = 0 \quad .$$

Ponendo

$$\Gamma^i = \gamma^{ji}\ell_j + \gamma^i\ell \tag{20.10}$$

$$[\partial \ell] = \lambda^{ji}[\ell_j\ell_i] + 2\lambda^j[\ell_j\ell] \tag{20.11}$$

dopo alcuni passaggi si ottiene :

$$[\mathcal{D} \ell] = (\lambda^{ji} - \gamma^{ji})[\ell_j\ell_i] + (2\lambda^j + \gamma^j)[\ell_j\ell] \tag{20.12}$$

$$\mathcal{D}_o g^{ji} = \partial_o g^{ji} - \gamma^k \partial_k g^{ji} + 2\partial_k \gamma^{(j}g^{i)k} - 2\gamma^{(ji)} \ , \tag{20.13}$$

ove è posto per brevità $\partial_o = \ell^\alpha \partial_\alpha$.

Per le (20.12) e (20.13) possiamo modificare le equazioni (20.7) e (20.9) e ottenere un sistema di equazioni per i coefficienti γ^{ji} , γ^i .

Risolvendo questo ultimo sistema di equazioni e sostituendo le espressione ottenute per γ^{ji} , γ^i nella formula (20.10) abbiamo in definitiva l'espressione

$$\Gamma^i = \left(\lambda^{ji} + \frac{1}{2}\partial_o g^{ji} + \lambda^k \partial_k g^{ji} - 2\partial_k \lambda^{(j}g^{i)k}\right)\ell_j - 2\lambda^i\ell \tag{20.14}$$

per i componenti dell'oggetto di connessione lineare parziale nello spazio fibrato $\mathfrak{X}_{n+(n-1)}$ il quale è definito in modo intrinseco mediante le condizioni (20.7), (20.8).

Consideriamo ora una identità importante per la connessione parziale ottenuta.

Notiamo dapprima che in virtù della commutatività delle operazioni della differenziazione assoluta basica e della derivazione parziale fibrale nello spazio fibrato $\mathfrak{X}_{n+(n-1)}$ si ha :

(20. 15) $$\partial_i [\mathfrak{D}\, \ell] = [\mathfrak{D}\, \ell_i]$$

donde

(20. 16) $$[\mathfrak{D}\, \ell_i] = 0 \quad .$$

Dalla seconda equazione di (19. 12) abbiamo

(20. 17) $$\overset{o}{\nabla}_j \ell_i = A_{ji}^{\cdot\cdot k} \ell_k - g_{ji} \ell \quad .$$

Prendendo il differenziale assoluto basico esterno di ambedue i membri della equazione (20. 17) e utilizzando l'identità

(20. 18) $$[\mathfrak{D}\, \overset{o}{\nabla}_j \ell_i] = \overset{o}{\nabla}_j [\mathfrak{D}\, \ell_i] - [\mathfrak{D}\, \overset{o}{G}_{ji}^{k} \ell_k]$$

otteniamo

(20. 19) $$[(\mathfrak{D}\, \overset{o}{G}_{ji}^{k} + \mathfrak{D}\, A_{ji}^{\cdot\cdot k}) \ell_k] - [\mathfrak{D}\, g_{ji}\, \ell] = 0$$

donde segue

(20. 20) $$\mathfrak{D}_o \overset{o}{G}_{ji,k} + \mathfrak{D}_o A_{jik} + \mathfrak{D}_k g_{ji} = 0 \quad ,$$

ove $\overset{o}{G}_{ji,k}$ sono i simboli di Christoffel di prima specie rispetto al tensore g_{ji} .

Derivando l'equazione (20.8) rispetto a η^k otteniamo

$$\mathfrak{D}_o \partial_k g_{ji} = -\mathfrak{D}_k g_{ji} \tag{20.21}$$

da cui

$$\mathfrak{D}_o \overset{o}{G}_{ji,k} = \frac{1}{2}\left(\mathfrak{D}_o \partial_j g_{ik} + \mathfrak{D}_o \partial_i g_{jk} - \mathfrak{D}_o \partial_k g_{ji}\right) = \tag{20.22}$$

$$= \frac{1}{2}\left(\mathfrak{D}_k g_{ji} - \mathfrak{D}_j g_{ik} - \mathfrak{D}_i g_{jk}\right)$$

Sostituendo queste espressioni nelle equazioni (20.20) e alternando rispetto alla coppia di indici j, k ci si convince che il tensore $\mathfrak{D}_k g_{ji}$ è simmetrico e che si ha :

$$\mathfrak{D}_k g_{ji} = -2\,\mathfrak{D}_o A_{kji} \quad . \tag{20.23}$$

Consideriamo ora nello spazio fibrato $T_{n+(n)}$ del prolungamento tangenziale del primo ordine dello spazio dato $\mathfrak{X}_n$ una connessione lineare parziale arbitraria definita dalle forme di Pfaff

$$\delta x^\alpha = d x^\alpha + \Gamma^\alpha \qquad \text{ove} \quad \Gamma^\alpha = \Gamma^\alpha_\beta(\xi^\gamma, x^\gamma) d\xi^\beta \quad . \tag{20.24}$$

Associamo a questa connessione parziale una nuova connessione lineare parziale definita dalle forme di Pfaff

(20.25) $$\tilde{\delta} x^\alpha = \delta x^\alpha + d\xi^\alpha$$

che chiameremo la connessione parziale <u>eccentrica</u> per la connessione parziale (20.24) .

Il vettore S^α , che è la semidifferenza dei vettori di curvatura della connessione parziale eccentrica e della connessione parziale data, si dice il <u>vettore di torsione</u> della connessione parziale data

(20.26) $$S^\alpha = \frac{1}{2} (\tilde{R}^\alpha - R^\alpha) \quad .$$

Analogamente ai componenti del vettore di curvatura i componenti del vettore di torsione sono forme differenziali bilineari emisimmetriche

(20.27) $$S^\alpha = S_{\gamma\beta}^{\cdot\cdot\alpha} [d\xi^\gamma d\xi^\beta]$$

i cui coefficienti

(20.28) $$S_{\gamma\beta}^{\cdot\cdot\alpha} = \Gamma_{[\gamma\beta]}^{\alpha} \quad , \text{ove} \quad \Gamma_{\gamma\beta}^{\alpha} = \frac{\partial \Gamma_\gamma^\alpha}{\partial x^\beta}$$

sono componenti di un tensore del terzo ordine il quale si dice il <u>tensore di torsione.</u>

Supponendo che nello spazio $\mathfrak{X}_n$ sia data una metrica vettoriale regolare ad n dimensioni locali possiamo introdurre sistemi coordinati polari negli spzi centro-affine tangenti ponendo

(20.29) $$x^\alpha = x\ell^\alpha(\xi^\beta, \eta^i) \quad .$$

Sia V^α un campo vettoriale in uno spazio tangente, come di solito

V^i e V^o definiti dalle formule

(20.30) $$V^i = \frac{1}{x} l^i_\alpha V^\alpha \quad , \quad V^o = l_\alpha V^\alpha$$

si dicono le <u>componenti polari</u> del campo vettoriale V^α. Con questo, V^i si dicono componenti trasversali e V^o si dice la componente radiale.

Indicando con $\delta\eta^i$ le componenti trasversali e con δx la componente radiale del vettore δx^α si ha :

(20.31) $$\delta\eta^i = d\eta^i + \Gamma^i \quad , \quad \delta x = dx + \Gamma^o \quad ,$$

ove

(20.32) $$\begin{aligned} \Gamma^i &= l^i_\alpha \partial l^\alpha + \frac{1}{x} l^i_\alpha \Gamma^\alpha(\xi^\beta, x l^\beta) \\ \Gamma^o &= l_\alpha \partial l^\alpha + l_\alpha \Gamma^\alpha(\xi^\beta, x l^\beta) . \end{aligned}$$

Una connessione lineare parziale nello spazio fibrato $T_{n+(n)}$ si dice <u>vettoriale</u> se le rappresentazioni parziali delle fibre da essa definite conservano le semirette centrali e determinano sue rappresentazioni isomorfiche.

E' facile dimostrare che la condizione necessaria e sufficiente affinchè una connessione parziale data in $T_{n+(n)}$ sia vettoriale è che Γ^α_β siano funzioni positivamente omogenee di grado 1 rispetto alla variabile x^γ.

Data nello spazio $\mathfrak{X}_n$ una metrica finsleriana generalizzata possiamo determinare la connessione parziale vettoriale usando le forme di Pfaff corrispondenti riferite a un sistema di coordinate polari η^i, x .

In questo caso le forme di Pfaff $\delta\eta^i$ non dipendono da x e definiscono una connessione lineare parziale nello spazio fibrato $\mathfrak{X}_{n+(n-1)}$ e la forma di Pfaff δx si esprime così :

$$\delta x = dx + x\Gamma \qquad \text{, ove} \qquad \Gamma = \Gamma_\alpha(\xi^\gamma, \eta^i)d\xi^\alpha \tag{20.33}$$

Una connessione parziale vettoriale nello spazio finsleriano generalizzato si dice <u>isometrica</u> se le rappresentazioni isomorfe delle semirette centrali da essa definite sono isometriche, il che equivale a uguagliare a zero la forma di Pfaff Γ .

Dunque dare una connessione parziale vettoriale isometrica nello spazio finsleriano ad n dimensioni equivale a dare una connessione parziale nello spazio fibrato $\mathcal{X}_{n+(n-1)}$ il cui insieme puntuale è l'indicatrice globale.

Calcoliamo il vettore di torsione di una connessione parziale vettoriale isometrica nello spazio finsleriano la quale si definisce mediante forme di Pfaff $\delta\eta^i$.

Notiamo dapprima che la corrispondente connessione parziale eccentrica sarà determinata dalle forme di Pfaff

$$\tilde{\delta}\eta^i = \delta\eta^i + \frac{1}{x}\ell^i \quad , \quad \tilde{\delta}x = dx + \ell \quad . \tag{20.34}$$

Dopo alcuni calcoli si ottengono le seguenti espressioni per le componenti trasversali e radiale del vettore di torsione :

$$s^i = \frac{1}{x}[\mathcal{D}\ell^i] \qquad\qquad s^o = [\mathcal{D}\ell] \quad , \tag{20.35}$$

ove i differenziali assoluti basici esterni sono calcolati rispetto alla connessione parziale data nello spazio fibrato $\mathcal{X}_{n+(n-1)}$.

Dunque una connessione parziale nello spazio fibrato $\mathcal{X}_{n+(n-1)}$ la quale definisce una connessione parziale vettoriale isometrica senza torsione nello spazio finsleriano, deve soddisfare alle condizioni

(20.36) $$\left[\mathfrak{D}\,\ell^i\right] = 0 \qquad \left[\mathfrak{D}\,\ell\right] = 0 \quad .$$

Dimostriamo che le condizioni (20.36) sono equivalenti alle (20.7), (20.8). Derivando l'equazione $\ell^i = g^{ji}\ell_i$ si ottiene

(20.37) $$\left[\mathfrak{D}\ell^i\right] = \mathfrak{D}^{[k}g^{j]\,i}\left[\ell_k\ell_j\right] + \mathfrak{D}_o g^{ji}\left[\ell\ell_j\right] + g^{ji}\left[\mathfrak{D}\ell\right] \quad .$$

Dunque le condizioni (20.36) sono equivalenti alle condizioni

(20.38) $$\left[\mathfrak{D}\,\ell\right] = 0 \quad , \quad \mathfrak{D}_o g^{ji} = 0 \quad , \quad \mathfrak{D}^{[k}g^{j]\,i} \quad ,$$

le quali a loro volta sono equivalenti alle condizioni seguenti

(20.39) $$\left[\mathfrak{D}\,\ell\right] = 0 \quad , \quad \mathfrak{D}_o g_{ji} = 0 \quad , \quad \mathfrak{D}_{[k}g_{j]\,i} = 0 \quad .$$

Notiamo che in virtù della formula (20.23) l'ultima di queste condizioni è una conseguenza delle precedenti. Dunque abbiamo dimostrato che la connessione parziale nello spazio fibrato $\mathfrak{X}_{n+(n-1)}$ definita dalle condizioni (20.7), (20.8) coincide con quella che corrisponde alla connessione parziale vettoriale isometrica senza torsione nello spazio finsleriano.

Come è noto la connessione parziale vettoriale isometrica senza torsione nello spazio finsleriano può essere definita immediatamente dalle condizioni

(20.40) $$\mathfrak{D}\,L = 0 \qquad S_{\gamma\beta}^{\cdot\cdot\alpha} = 0 \quad ,$$

ove L è la funzione metrica.

Ponendo $\mathcal{F} = \frac{1}{2} L^2$ otteniamo che i componenti del tensore $a_{\beta\alpha}$, che definisce le iperquadriche centrali osculatrici delle indicatrici fibrali, coincidono con le derivate seconde della funzione $\mathcal{F}$ rispetto alle coordinate fibrali x^α

$$(20.41) \qquad a_{\beta\alpha} = \mathcal{F}_{\beta\alpha} \quad .$$

Indicando con $a^{\beta\alpha}$ i componenti del tensore reciproco rispetto al tensore $a_{\beta\alpha}$ possiamo scrivere le espressioni ben conosciute delle componenti dello oggetto della connessione parziale in termini della funzione metrica

$$(20.42) \qquad \Gamma^\alpha_\beta = \frac{\partial G^\alpha}{\partial x^\beta} \; , \quad G^\alpha = \frac{1}{2} a^{\alpha\gamma} (x^\beta \partial_\beta \mathcal{F}_\gamma - \partial_\gamma \mathcal{F}) \; .$$

Un caso interessante speciale è quando la connessione vettoriale (20.42) in uno spazio finsleriano è una connessione affine.

Questi spazi finsleriani furono studiati per la prima volta da Berwald, perciò possiamo chiamarli gli spazi di Berwald.

Definiamo <u>spazi di Berwald generalizzati</u> gli spazi finsleriani generalizzati per i quali la connessione parziale vettoriale (20.41) è una parte di una connessione affine. E' facile vedere che in questo caso la connessione affine è univocamente definita.

Si può dimostrare che affinchè uno spazio finsleriano generalizzato sia uno spazio di Berwald generalizzato è necessario e sufficiente che le derivate covarianti basiche dei tensori fibrali fondamentali g_{ji} e A_{kji} nello spazio fibrato corrispondente all'indicatrice globale, siano nulle

$$(20.43) \qquad \mathcal{D}_\alpha \, g_{ji} = 0$$

(20.44) $$\mathfrak{D}_\alpha A_{kji} = 0 \quad .$$

Dalle (20.8) e (20.23) otteniamo che la condizione (20.43) è una conseguenza della condizione (20.44) .

Uno spazio finsleriano generalizzato si dice <u>piano</u> se esso è uno spazio di Berwald generalizzato con la connessione affine di curvatura nulla.

E' evidente che la curvatura della connessione parziale vettoriale (20.42) sia nulla quando e solo quando è nulla la curvatura della connessione parziale (20.14) nello spazio fibrato $\mathfrak{X}_{n+(n-1)}$.

Dunque affinchè uno spazio finsleriano generalizzato sia piano è necessario e sufficiente che

(20.45) $$\mathfrak{D}_\alpha A_{kji} = 0 \quad , \quad R_{\beta\alpha}^{\cdot\cdot i} = 0 \quad .$$

Consideriamo ora il covettore di curvatura euleriana di una curva misurabile in uno spazio finsleriano generalizzato.

Sia

(20.46) $$\xi^\alpha = \xi^\alpha (s)$$

il sistema di equazioni di una curva misurabile, ove il parametro s è la lunghezza d'arco.

Applicando la formula generale (18.12) dobbiamo sostituirvi ℓ^α con B_a^α e ℓ_α con B_α^a . Abbiamo

(20.47) $$h_\alpha = \frac{d\ell_\alpha}{ds} + \ell_\beta \partial_\alpha \ell^\beta$$

o

(20.48) $$h_\alpha = \ell_{\alpha i} \frac{d\eta^i}{ds} + 2\ell^\beta \partial_{[\beta} \ell_{\alpha]}$$

Dalle formule (20.11) e (20.14) si ottiene

(20.49) $$2\ell^\beta \partial_{[\beta} \ell_{\alpha]} = \ell_{\alpha i} \Gamma^i_\beta \ell^\beta$$

e possiamo trascrivere (20.48) nel seguente modo

(20.50) $$h_\alpha = \ell_{\alpha i} \frac{\delta \eta^i}{ds}$$

ove

(20.51) $$\frac{\delta \eta^i}{ds} = \frac{d\eta^i}{ds} + \Gamma^i_\beta \ell^\beta$$

è il vettore di deviazione assoluta della curva

(20.52) $$\xi^\alpha = \xi^\alpha(s) \quad , \qquad \eta^i = \eta^i(s)$$

nello spazio fibrato $\mathfrak{X}_{n+(n-1)}$, la quale si dice la curva ricoprente tangenziale della curva misurabile (20.46).

In particolare le estremali sono le curve misurabili le cui ricoprenti tangenziali sono le curve di deviazione assoluta nulla.

BIBLIOGRAFIA

1) Barker, Ch. The lagrange multiplier rule for two dependent and two independent variables. Amer. J. Math. 67 (1945) 256-276 .

2) Berwald, L. Uber zweidimensionale allgemeine metrische Raume. J. reine angewandte Math. 156 (1927) 191-222 .

3) Bliss, G. A. Lectures on the calculus of variations. Chicago 1946.

4) Caratheodory, K. Uber die Variationsrechnung bei mehrfachen Integralen. Acta Szeged. 4 (1929) 193-216.

5) " " Variationsrechnung. Leipzig und Berlin, Teubner. 1935.

6) Cartan, E. Les espaces metriques fondés sua la notion d'aire. Actualités scientifiques et industrielles 72 Paris. Hermann. 1933.

7) " " Les espaces de Finsler. Actualités scientifiques et industrielles 79 Paris. Hermann. 1934.

8) Debever, R. Les champs de Mayer dans le calcul des variations des integrales multiples. Bull. de l'Academie royale de Belgique. 23 (1937) 809-815.

9) Ehresmann, Ch. Les connexions infinitésimales dans una espace fibré différentiable.
Colloque de topologie (Espaces fibrés). Tenu à Bruxelles du 5 au 8 juin 1950.

10) Kawaguchi, A. Eim metrischer Raum, der eine Verallgemeinung des Finslerischen Raumes ist.
Monatsh. Math. Phys. 43 (1936) 289-297.

11) " " Die Differentialgeometrie hoher Ordnung II Uber die n-dimensionalem metrischen Raume mit vom m-dimensionalen Flachenelement abhangigen Zusammehang.
J. Faculty of Science. Hokkaido Imper. Univ. 9 (1940) 153-188.

12) Kawaguchi, A. und Hokari, S. Die Grundlegung der Geometrie der n-dimensionalen metrischen Raume auf des Begriffs des k-dimensionalen Flacheninhalts.
Proceed. Imper. Acad. Japan. 26 (1940) 320-325.

13) Landers, A. W. Invariant multiple integrals in the calculus of variations.
Contributions to the calculus of variations 1938-1941. Chicago University Press. 179-206.

14) Lepage, Th. Sur les champs géodésiques du calcul des variations.
Bull. de l'Acad. roy. de Belgique 22 (1936) 716-729, 1036-1040.

15) Nyenhuis, A. Theory of the geometric object.
Thesis Univ. Amsterdam 1952.

16) Radon, J. Zum Problem von Lagrange.
Namburger Mathem. Einzelschriften 6 .
Leipzig Teubner 1928.

17) Riguet, J. Relations binaires, fermetures, correspondances de Galois.
Bull. Soc. Math. France 76 (1948) 114-155.

18) Rund, H. The differential geometry of Finsler space.
Berlin Springer 1959.

20) Schouten, J. A. und
van Kampen, E. R. Zur Einbettungs-und Krummungsthorie nichtholonomer Gebilde.
Math. Ann. 103 (1930) 752-783.

20) Schouten, J. A. und
Struik, D. J. Einfuhrung in die neueren Methoden der Differentialgeometrie I, II.
Groningen Noordhoff 1935, 1938.

21) Schouten, J. A. und
Haantjes, J. On the theory of the geometric object.
Proc. London Math. Soc. 42 (1937) 356-376.

22) Schouten, J. A. Ricci-Calculus
Berlin Springer 1954.

23) Veblen, O. and
Whitehead, J. H. C.-The foundations of differential geometry.
Cambr. Tracts. No. 29 1932.

24) Vessiot, E. Sur la théorie des multiplicités et le calcul des variations.
Bull. Soc. Math. France 40 (1912).

25) Wagner, V. V. Uber Berwaldsche Raume.
Recueil Math. 3(45) (1938) 655-662.

26) " " Les espaces de Finsler à deux dimensions à groupes d'holonomie finis et continus.
C. R. Acad. Sci. URSS 39 (1943) 210-212.

27) " " The absolute derivative of field of local geometric object in a compound manifold.
C. R. Acad. Sci. URSS 40 (1943) 94-97.

28) Wagner, V. V. Homological transformation of Finslerian metric
C. R. Acad. Sci. URSS 46 (1945) 263-265.

29) " " The generalization of Ricci's and Bianchi's identities for a connexion in the compound manifold.
C. R. Acad. Sci. URSS 46 (1945) 303-305.

30) " " Geometry of field of local curves in X_3 and the simplest case of Lagrange's problem in the calculus of variations.
C. R. Acad. URSS 48 (1945) 229-232.

31) Wagner, V. V. Geometry of field of local central plane curves in X_3. C. R. Acad. Sci. URSS 48 (1945) 382-384.

32) " " On a sufficient condition in the problem of Lagrange foe multiple integrals. C. R. Acad. Sci. URSS 54 (1946) 479-482.

33) " " On a geometric interpretation of Lagrange for multiple integrals. C. R. Acad. Sci. URSS 55(1946) 87-90.

34) " " Theory of field of local n-2-dimensional surfaces in X_n and its application to the problem of Lagrange in the calculus of variations Ann. of Math. 49 (1948) 141-188.

35) Winternitz, A. Uber die affine Grundlage der Metrik eines Variationsproblems. Sitzungsberichte Acad. Berlin (1930) 457-469.

36) Yano, K. The theory of Lie derivatives and its applications North-Holland Publishing Co.

37) Wagner, V. V. (Вагнер В. В.) La geometria di uno spazio dotato di metrica areale e le sue applicazioni al calcolo delle variazioni. (in russo) Mat. Sb. 19(61), (1946), 341-404 .

38) " " La geometria di uno spazio n-dimensionale con metrica riemanniana m-dimensionale e le sue applicazioni al calcolo delle variazioni (in russo). Mat. Sb. 20 (62) (1947) 3-25.

39) Wagner, V. V. (Вагнер В. В.) Teoria geometrica del problema singolare del calcolo delle variazioni nel caso semplice. (in russo) Mat. Sb. 21 (63) (1947) 321-362

40) " " La teoria dei campi di curve locali e di superficie coniche locali in X_3 e le sue applicazioni al calcolo delle variazioni e alla teoria delle equazioni differenziali parziali. (in russo) Trudy Sem. Vektor. Tenzor. Anal. (Univ. di Mosca) 6 (1948). 257-364 .

41) " " La geometria di Finsler come teoria del campo di ipersuperficie locali in $\mathfrak{X}_n$. (in russo) Trudy Sem. Vektor. Tenzor. Anal. (Mosca) 7 (1949) 65-166 .

42) " " La classificazione delle connessioni lineari rispetto ai loro gruppi d'olonomia, in una varietà composta $\mathfrak{X}_{n+(1)}$. Trudy Sem. Vektor. Tenzor. Anal. (Mosca) 7 (1949) 205-226 (in russo).

43) " " Sopra l'immersione del campo di superficie locali di $\mathfrak{X}_n$ in un campo costante di superficie nello spazio affine. (in russo) Dokl. Akad. Nauk. SSSR 66 (1949) 785-788.

44) " " Teoria delle varietà composte. (in russo) Trudy Sem. Vektor. Tenzor. Anal. (Mosca) 8 (1950) 11-72.

45) Wagner, V. V. (Вагнер, В. В.) La geometria di uno spazio dotato di metrica iperareale come teoria del campo di ipersuperficie locali in una varietà composta. (in russo)
Trudy Sem. Vektor. Tenzor. Anal. (Mosca) 8 (1950) 144-196.

46) " " La teoria di iperstrisce locali . (in russo)
Trudy Sem. Vektor. Tenzor. Anal. (Mosca) 8 (1950) 197-272.

47) " " Il calcolo delle variazioni come teoria del campo di semiconi locali. (in russo)
Annuario scientifico dell'Università di Saratov 1955 (1950) 27-34.

48) Žotikov, G. I. (Жотиков, Г. И.) Sulla teoria del campo di superficie locali nella varietà composta tangente del I ordine $E_n(\mathfrak{X}_n)$. (in russo)
Izv. Vysš. Učebn. Zaved. Matematika (Kazan) 2 (9) (1959) 69-79.

49) " " Sulla teoria del campo di superficie coniche locali nella varietà composta tangente. I II . (in russo)
Izv. Vysš. Učebn. Zaved. Matematika (Kazan).

50) " " Sullateoria del campo di superficie locali nella varietà composta tangente del I ordine. (in russo)
Trudy Sem. Vektor. Tenzor. Anal. (Mosca) 11 (1961) 189-218.

51) Kabanov, N. I. (Кабанов, Н. И.) Lo spazio di Cartan definito a meno di trasformazioni di Caratheodory. (in russo)
Note scientifiche dell'Istituto Pedagogico di Balašov 3 (1958) 47-77.

52) " " Teoria geometrica delle trasformazioni di Caratheodory nel problema di Lagrange. (in russo)
Trudy Sem. Vektor. Tenzor. Anal. (Mosca) 11 (1961) 219-240.

53) Kaganov, S. A. (Каганов, С. А) Sulla teoria geometrica del problema variazionale singolare per gli integrali (n-1)-upli. (in russo)
Dokl. Akad. Nauk SSSR, 75 (1950) 487-490.

54) " " Geometria dello spazio dotato di metrica iperareale singolare. (in russo)
Mat. Sb., 42 (84) (1957) 497-512.

55) Losik, M. V. (Лосик, М. В.) Una interpretazione geometrica di certe condizioni per il problema variazionale semplice con le derivate di ordine superiore. (in russo)
Sibirsk Mat. Ž. (in corso di stampa)

56) Ržehina, N. F. (Ржехина, Н. Ф.) Sulla teoria del campo di curve locali in $\mathfrak{X}_n$. (in russo)
Dokl. Akad. Nauk SSSR, 72 (1950) 461-464.

57) " " Teoria del campo delle ipersuperficie sviluppabili in $\mathfrak{X}_n$. (in russo)
Trudy Sem. Vektor. Tenzor. Anal. (Mosca) 9 (1952) 411-430.

58) Tokarev, P.I. (Токарев, П.И.) Teoria geometrica della variazione seconda nel problema variazionale di Lagrange. (in russo) Trudy Sem. Vektor. Tenzor. Anal. (Mosca) 9 (1952) 431-455.

59) Časečnikov, S. M. (Часечников, С.М.) - Teoria del campo di iperconi locali in $\mathfrak{X}_n$. (in russo) Dokl. Akad. Nauk SSSR, 117 (1957) 765-768.

60) " " Teoria del campo di iperstrisce locali improprie in $\mathfrak{X}_n$. (in russo) Trudy Sem. Vektor. Tenzor. Anal. (Mosca) 11 (1961) 165-188.